Principal Component Analysis

Principal Component Analysis

Edited by **Rebecca Cross**

CLANRYE
INTERNATIONAL

New Jersey

Published by Clanrye International,
55 Van Reypen Street,
Jersey City, NJ 07306, USA
www.clanryeinternational.com

Principal Component Analysis
Edited by Rebecca Cross

International Standard Book Number: 978-1-63240-415-2 (Hardback)

Printed in the United States of America.

Contents

Preface

The aim of this book is to enhance knowledge of scientists, engineers and researchers regarding the advantages of principal component analysis in data analysis. Principal component analysis involves a statistical procedure which orthogonally transforms a set of possibly correlated observations into set of values of linearly uncorrelated variables called principal components. This book elucidates the uses of PCA in distinct fields like face recognition, and image and speech processing. The book also covers core concepts and novel techniques in data analysis and feature extraction.

The researches compiled throughout the book are authentic and of high quality, combining several disciplines and from very diverse regions from around the world. Drawing on the contributions of many researchers from diverse countries, the book's objective is to provide the readers with the latest achievements in the area of research. This book will surely be a source of knowledge to all interested and researching the field.

In the end, I would like to express my deep sense of gratitude to all the authors for meeting the set deadlines in completing and submitting their research chapters. I would also like to thank the publisher for the support offered to us throughout the course of the book. Finally, I extend my sincere thanks to my family for being a constant source of inspiration and encouragement.

Editor

Two-Dimensional Principal Component Analysis and Its Extensions

Parinya Sanguansat

Faculty of Engineering and Technology, Panyapiwat Institute of Management
Thailand

1. Introduction

Normally in Principal Component Analysis (PCA) (Sirovich & Kirby, 1987; Turk & Pentland, 1991), the 2D image matrices are firstly transformed to 1D image vectors by vectorization. The vectorization of a matrix is the column vector obtain by stacking the columns of the matrix on top of one another. The covariance or scatter matrix are formulated from the these image vectors. The covariance matrix will be well estimated if and only if the number of available training samples is not far smaller than the dimension of this matrix. In fact, it is too hard to collect this the number of samples. Then, normally in 1D subspace analysis, the estimated covariance matrix is not well estimated and not full rank.

Two-Dimensional Principal Component Analysis (2DPCA) was proposed by Yang et al. (2004) to apply with face recognition and representation. Evidently, the experimental results in Kong et al. (2005); Yang & Yang (2002); Yang et al. (2004); Zhang & Zhou (2005) have shown the improvement of 2DPCA over PCA on several face databases. Unlike PCA, the image covariance matrix is computed directly on image matrices so the spatial structure information can be preserved. This yields a covariance matrix whose dimension just equals to the width of the face image. This is far smaller than the size of covariance matrix in PCA. Therefore, the image covariance matrix can be better estimated and will usually be full rank. That means the curse of dimensionality and the Small Sample Size (SSS) problem can be avoided.

In this chapter, the detail of 2DPCA's extensions will be presented as follows: The bilateral projection scheme, the kernel version, the supervised framework, the variation of image alignment and the random approaches.

For the first extension, there are many techniques were proposed in bilateral projection schemes such as $2D^2PCA$ (Zhang & Zhou, 2005), *Bilateral 2DPCA (B2DPCA)* (Kong et al., 2005), *Generalized Low-Rank Approximations of Matrices (GLRAM)* (Liu & Chen, 2006; Liu et al., 2010; Ye, 2004), *Bi-Dierectional PCA (BDPCA)* (Zuo et al., 2005) and *Coupled Subspace Analysis (CSA)* (Xu et al., 2004). The left and right projections are determined by solving two eigenvalue problems per iteration. One corresponds to the column direction and another one corresponds to the row direction of image, respectively. In this way, it is not only consider the image in both directions but also reduce the feature matrix smaller than the original 2DPCA.

As the successful of the kernel method in kernel PCA (KPCA), the kernel based 2DPCA was proposed as *Kernel 2DPCA (K2DPCA)* in Kong et al. (2005). That means the nonlinear mapping can be utilized to improve the feature extraction of 2DPCA.

Since 2DPCA is unsupervised projection method, the class information is ignored. To embed this information for feature extraction, the Linear Discriminant Analysis (LDA) is applied in Yang et al. (2004). Moreover, the 2DLDA was proposed and then applied with 2DPCA in Sanguansat et al. (2006b). Another method was proposed in Sanguansat et al. (2006a) based on class-specific subspace which each subspace is constructed from only the training samples in own class while only one subspace is considered in the conventional 2DPCA. In this way, their representation can provide the minimum reconstruction error.

Because of the image covariance matrix is the key of 2DPCA and it is corresponding to the alignment of pixels in image. Different image covariance matrix will obtain the difference information. To produce alternated version of the image covariance matrix, it can be done by rearranging the pixels. The diagonal alignment 2DPCA and the generalized alignment 2DPCA were proposed in Zhang et al. (2006) and Sanguansat et al. (2007a), respectively.

Finally, the random subspace based 2DPCA were proposed by random selecting the subset of eigenvectors of image covariance matrix as in Nguyen et al. (2007); Sanguansat et al. (2007b; n.d.) to build the new projection matrix. From the experimental results, some subset eigenvectors can perform better than others but it cannot predict by their eigenvalues. However, the mutual information can be used in filter strategy for selecting these subsets as shown in Sanguansat (2008).

2. Two-dimensional principal component analysis

Let each image is represented by a m by n matrix $\mathbf{A}$ of its pixels' gray intensity. We consider linear projection of the form

$$\mathbf{y} = \mathbf{A}\mathbf{x}, \tag{1}$$

where $\mathbf{x}$ is an n dimensional projection axis and $\mathbf{y}$ is the projected feature of this image on $\mathbf{x}$, called *principal component vector*.

In original algorithm of 2DPCA (Yang et al., 2004), like PCA, 2DPCA search for the optimal projection by maximize the total scatter of projected data. Instead of using the criterion as in PCA, the total scatter of the projected samples can be characterized by the trace of the covariance matrix of the projected feature vectors. From this point of view, the following criterion was adopt as

$$J(\mathbf{x}) = tr(\mathbf{S_x}), \tag{2}$$

where

$$\mathbf{S_x} = E[(\mathbf{y} - E\mathbf{y})(\mathbf{y} - E\mathbf{y})^T]. \tag{3}$$

The total power equals to the sum of the diagonal elements or trace of the covariance matrix, the trace of $\mathbf{S_x}$ can be rewritten as

$$\begin{aligned}
tr(\mathbf{S}_x) &= tr\{E[(\mathbf{y} - E\mathbf{y})(\mathbf{y} - E\mathbf{y})^T]\} \\
&= tr\{E[(\mathbf{A} - E\mathbf{A})\mathbf{x}\mathbf{x}^T(\mathbf{A} - E\mathbf{A})^T]\} \\
&= tr\{E[\mathbf{x}^T(\mathbf{A} - E\mathbf{A})^T(\mathbf{A} - E\mathbf{A})\mathbf{x}]\} \\
&= tr\{\mathbf{x}^T E[(\mathbf{A} - E\mathbf{A})^T(\mathbf{A} - E\mathbf{A})]\mathbf{x}\} \\
&= tr\{\mathbf{x}^T \mathbf{G}\mathbf{x}\}. \tag{4}
\end{aligned}$$

Giving that

$$\mathbf{G} = E[(\mathbf{A} - E\mathbf{A})^T(\mathbf{A} - E\mathbf{A})]. \tag{5}$$

This matrix **G** is called *image covariance matrix*. Therefore, the alternative criterion can be expressed by

$$J(\mathbf{x}) = tr(\mathbf{x}^T \mathbf{G}\mathbf{x}), \tag{6}$$

where the image inner-scatter matrix **Gx** is computed in a straightforward manner by

$$\mathbf{G} = \frac{1}{M} \sum_{k=1}^{M} (\mathbf{A}_k - \bar{\mathbf{A}})^T (\mathbf{A}_k - \bar{\mathbf{A}}), \tag{7}$$

where $\bar{\mathbf{A}}$ denotes the average image,

$$\bar{\mathbf{A}} = \frac{1}{M} \sum_{k=1}^{M} \mathbf{A}_k. \tag{8}$$

It can be shown that the vector **x** maximizing Eq. (4) correspond to the largest eigenvalue of **G** (Yang & Yang, 2002). This can be done, for example, by using the Eigenvalue decomposition or Singular Value Decomposition (SVD) algorithm. However, one projection axis is usually not enough to accurately represent the data, thus several eigenvectors of **G** are needed. The number of eigenvectors (d) can be chosen according to a predefined threshold (θ).

Let $\lambda_1 \geq \lambda_2 \geq \cdots \geq \lambda_n$ be eigenvalues of **G** which sorted in non-increasing order. We select the d first eigenvectors such that their corresponding eigenvalues satisfy

$$\theta \leq \frac{\sum_{i=1}^{d} \lambda_i}{\sum_{i=1}^{n} \lambda_i}. \tag{9}$$

For feature extraction, Let $\mathbf{x_1}, \ldots, \mathbf{x_d}$ be d selected largest eigenvectors of **G**. Each image **A** is projected onto these d dimensional subspace according to Eq. (1). The projected image $\mathbf{Y} = [\mathbf{y_1}, \ldots, \mathbf{y_d}]$ is then an m by d matrix given by:

$$\mathbf{Y} = \mathbf{A}\mathbf{X}, \tag{10}$$

where $\mathbf{X} = [\mathbf{x_1}, \ldots, \mathbf{x_d}]$ is a n by d projection matrix.

2.1 Column-based 2DPCA

The original 2DPCA can be called the row-based 2DPCA. The alternative way of 2DPCA can be using the column instead of row, column-based 2DPCA (Zhang & Zhou, 2005).

This method can be consider as same as the original 2DPCA but the input images are previously transposed. From Eq. (7), replace the image **A** with the transposed image $\mathbf{A}^T$ and call it the column-based image covariance matrix **H**, thus

$$\mathbf{H} = \frac{1}{M} \sum_{k=1}^{M} (\mathbf{A}_k^T - \bar{\mathbf{A}}^T)^T (\mathbf{A}_k^T - \bar{\mathbf{A}}^T) \tag{11}$$

$$\mathbf{H} = \frac{1}{M} \sum_{k=1}^{M} (\mathbf{A}_k - \bar{\mathbf{A}})(\mathbf{A}_k - \bar{\mathbf{A}})^T. \tag{12}$$

Similarly in Eq. (10), the column-based optimal projection matrix can be obtained by computing the eigenvectors of $\mathbf{H}(\mathbf{z})$ corresponding to the q largest eigenvalues as

$$\mathbf{Y}_{col} = \mathbf{Z}^T \mathbf{A}, \tag{13}$$

where $\mathbf{Z} = [\mathbf{z}_1, \ldots, \mathbf{z_q}]$ is a m by q column-based optimal projection matrix. The value of q can also be controlled by setting a threshold as in Eq. (9).

2.2 The relation of 2DPCA and PCA

As Kong et al. (2005) 2DPCA, performed on the 2D images, is essentially PCA performed on the rows of the images if each row is viewed as a computational unit. That means the 2DPCA of an image can be viewed as the PCA of the set of rows of an image. The relation between 2DPCA and PCA can be proven that by rewriting the image covariance matrix $\mathbf{G}$ in normal covariance matrix as

$$\begin{aligned} \mathbf{G} &= E\left[(\mathbf{A} - \bar{\mathbf{A}})^T (\mathbf{A} - \bar{\mathbf{A}})\right] \\ &= E\left[\mathbf{A}^T \mathbf{A} - \bar{\mathbf{A}}^T \mathbf{A} - \mathbf{A}^T \bar{\mathbf{A}} + \bar{\mathbf{A}}^T \bar{\mathbf{A}}\right] \\ &= E\left[\mathbf{A}^T \mathbf{A}\right] - f(\bar{\mathbf{A}}) \\ &= E\left[\mathbf{A}^T \left(\mathbf{A}^T\right)^T\right] - f(\bar{\mathbf{A}}), \end{aligned} \tag{14}$$

where $f(\bar{\mathbf{A}})$ can be neglected if the data were previously centralized. Thus,

$$\begin{aligned} \mathbf{G} &\approx E\left[\mathbf{A}^T \left(\mathbf{A}^T\right)^T\right] \\ &= \frac{1}{M} \sum_{k=1}^{M} \mathbf{A}_k^T \left(\mathbf{A}_k^T\right)^T \\ &= m \left(\frac{1}{mM} \mathbf{C}\mathbf{C}^T\right), \end{aligned} \tag{15}$$

where $\mathbf{C} = \left[\mathbf{A}_1^T \ \mathbf{A}_2^T \ \cdots \ \mathbf{A}_M^T\right]$ and the term $\frac{1}{mM}\mathbf{C}\mathbf{C}^T$ is the covariance matrix of rows of all images.

3. Bilateral projection frameworks

There are two major difference techniques in this framework, i.e. non-iterative and iterative. All these methods use two projection matrices for both row and column. The former computes these projections separately while the latter computes them simultaneously via iterative process.

3.1 Non-iterative method

The non-iterative bilateral projection scheme was applied to 2DPCA via left and right multiplying projection matrices Xu et al. (2006); Zhang & Zhou (2005); Zuo et al. (2005) as follows

$$\mathbf{B} = \mathbf{Z}^T \mathbf{A} \mathbf{X}, \tag{16}$$

where $\mathbf{B}$ is a feature matrix which extracted from image $\mathbf{A}$ and $\mathbf{Z}$ is a left multiplying projection matrix. Similar to the right multiplying projection matrix $\mathbf{X}$ in Section 2, matrix $\mathbf{Z}$ is a m by q projection matrix that obtained by choosing the eigenvectors of image covariance matrix $\mathbf{H}$ corresponding to the q largest eigenvalues. Therefore, the dimension of feature matrix is decreasing from $m \times n$ to $q \times d$ ($q < m$ and $d < n$). In this way, the computation time also be reducing. Moreover, the recognition accuracy of B2DPCA is often better than 2DPCA as the experimental results in Liu & Chen (2006); Zhang & Zhou (2005); Zuo et al. (2005).

3.2 Iterative method

The bilateral projection scheme of 2DPCA with the iterative algorithm was proposed in Kong et al. (2005); Liu et al. (2010); Xu et al. (2004); Ye (2004). Let $\mathbf{Z} \in \mathbb{R}^{m \times q}$ and $\mathbf{X} \in \mathbb{R}^{n \times d}$ be the left and right multiplying projection matrix respectively. For an $m \times n$ image $\mathbf{A}_k$ and $q \times d$ projected image $\mathbf{B}_k$, the bilateral projection is formulated as follows:

$$\mathbf{B}_k = \mathbf{Z}^T \mathbf{A}_k \mathbf{X} \tag{17}$$

where $\mathbf{B}_k$ is the extracted feature matrix for image $\mathbf{A}_k$.

The optimal projection matrices, $\mathbf{Z}$ and $\mathbf{X}$ in Eq. (17) can be computed by solving the following minimization criterion that the reconstructed image, $\mathbf{Z}\mathbf{B}_k\mathbf{X}^T$, gives the best approximation of $\mathbf{A}_k$:

$$J(\mathbf{Z}, \mathbf{X}) = \min \sum_{k=1}^{M} \left\| \mathbf{A}_k - \mathbf{Z}\mathbf{B}_k\mathbf{X}^T \right\|_{F'}^2 \tag{18}$$

where M is the number of data samples and $\| \bullet \|_F$ is the Frobenius norm of a matrix.

The detailed iterative scheme designed to compute the optimal projection matrices, $\mathbf{Z}$ and $\mathbf{X}$, is listed in Table 1. The obtained solutions are locally optimal because the solutions are dependent on the initialized $\mathbf{Z}_0$. In Kong et al. (2005), the initialized $\mathbf{Z}_0$ sets to the $m \times m$ identity matrix $\mathbf{I}_m$, while this value is set to $\begin{bmatrix} \mathbf{I}_q \\ 0 \end{bmatrix}$ in Ye (2004), where $\mathbf{I}_q$ is the $q \times q$ identity matrix.

Alternatively, The criterion in Eq. (18) is biquadratic and has no closed-form solution. Therefore, an iterative procedure to obtain the local optimal solution was proposed in Xu et al. (2004). For $\mathbf{X} \in \mathbb{R}^{n \times d}$, the criterion in Eq. (18) can be rewritten as

$$J(\mathbf{X}) = \min \sum_{k=1}^{M} \left\| \mathbf{A}_k - \mathbf{A}_k^Z \mathbf{X}\mathbf{X}^T \right\|_{F'}^2 \tag{19}$$

where $\mathbf{A}_k^Z = \mathbf{Z}\mathbf{Z}^T\mathbf{A}_k$. The solution of Eq. (19) is the eigenvectors of the eigenvalue decomposition of image covariance matrix:

$$\mathbf{G} = \frac{1}{M} \sum_{k=1}^{M} (\mathbf{A}_k^Z - \bar{\mathbf{A}}^Z)^T (\mathbf{A}_k^Z - \bar{\mathbf{A}}^Z). \tag{20}$$

Similarly, for $\mathbf{Z} \in \mathbb{R}^{m \times q}$, the criterion in Eq. (18) is changed to

$$J(\mathbf{Z}) = \min \sum_{k=1}^{M} \left\| \mathbf{A}_k - \mathbf{Z}\mathbf{Z}^T\mathbf{A}_k^X \right\|_{F'}^2 \tag{21}$$

S_1: Initialize $\mathbf{Z}$, $\mathbf{Z} = \mathbf{Z}_0$ and $i = 0$

S_2: While not convergent

S_3:   Compute $\mathbf{G} = \frac{1}{M} \sum\limits_{k=1}^{M} (\mathbf{A}_k - \bar{\mathbf{A}})^T \mathbf{Z}_{i-1} \mathbf{Z}_{i-1}^T (\mathbf{A}_k - \bar{\mathbf{A}})$

S_4:   Compute the d eigenvectors $\{\mathbf{e}_j^{\mathbf{X}}\}_{j=1}^d$ of $\mathbf{G}$

           corresponding to the largest d eigenvalues

S_5: $\mathbf{X}_i = [\mathbf{e}_1^{\mathbf{X}}, \ldots, \mathbf{e}_d^{\mathbf{X}}]$

S_6:   Compute $\mathbf{H} = \frac{1}{M} \sum\limits_{k=1}^{M} (\mathbf{A}_k - \bar{\mathbf{A}}) \mathbf{X}_i \mathbf{X}_i^T (\mathbf{A}_k - \bar{\mathbf{A}})^T$

S_7:   Compute the q eigenvectors $\{\mathbf{e}_j^{\mathbf{Z}}\}_{j=1}^q$ of $\mathbf{H}$

           corresponding to the largest l eigenvalues

S_8: $\mathbf{Z}_i = [\mathbf{e}_1^{\mathbf{Z}}, \ldots, \mathbf{e}_q^{\mathbf{Z}}]$

S_9: $i = i + 1$

S_{10}: End While

S_{11}: $\mathbf{Z} = \mathbf{Z}_{i-1}$

S_{12}: $\mathbf{X} = \mathbf{X}_{i-1}$

S_{13}: Feature extraction: $\mathbf{B}_k = \mathbf{Z}^T \mathbf{A}_k \mathbf{X}$

Table 1. The Bilateral Projection Scheme of 2DPCA with Iterative Algorithm.

where $\mathbf{A}_k^{\mathbf{X}} = \mathbf{A}_k \mathbf{X} \mathbf{X}^T$. Again, the solution of Eq. (21) is the eigenvectors of the eigenvalue decomposition of image covariance matrix:

$$\mathbf{H} = \frac{1}{M} \sum_{k=1}^{M} (\mathbf{A}_k^{\mathbf{Z}} - \bar{\mathbf{A}}^{\mathbf{Z}})(\mathbf{A}_k^{\mathbf{Z}} - \bar{\mathbf{A}}^{\mathbf{Z}})^T. \tag{22}$$

By iteratively optimizing the objective function with respect to $\mathbf{Z}$ and $\mathbf{X}$, respectively, we can obtain a local optimum of the solution. The whole procedure, namely Coupled Subspace Analysis (CSA) Xu et al. (2004), is shown in Table 2.

4. Kernel based frameworks

From Section 2.2, 2DPCA which performed on the 2D images, is basically PCA performed on the rows of the images if each row is viewed as a computational unit.

Similar to 2DPCA, the kernel-based 2DPCA (K2DPCA) can be processed by traditional kernel PCA (KPCA) in the same manner. Let $\mathbf{a}_k^i$ is the i-th row of the k-th image, thus the k-th image can be rewritten as

$$\mathbf{A} = \left[\left(\mathbf{a}_k^1\right)^T \left(\mathbf{a}_k^2\right)^T \cdots \left(\mathbf{a}_k^m\right)^T \right]^T. \tag{23}$$

From Eq. (15), the covariance matrix $\mathbf{C}$ can be constructed by concatenating all rows of all training images together. Let $\varphi : \mathbb{R}^m \to \mathbb{R}^{m'}, m < m'$ be the mapping function that map the the row vectors into a feature space of higher dimensions in which the classes can be linearly

S_1: Initialize $\mathbf{Z}$, $\mathbf{Z} = \mathbf{I}_m$
S_2: For $i = 1, 2, \ldots, T_{\max}$
S_3:   Compute $\mathbf{A}_k^{\mathbf{Z}} = \mathbf{Z}_{i-1}\mathbf{Z}_{i-1}^T\mathbf{A}_k$
S_4:   Compute $\mathbf{G} = \frac{1}{M}\sum\limits_{k=1}^{M}(\mathbf{A}_k^{\mathbf{Z}} - \bar{\mathbf{A}}^{\mathbf{Z}})^T(\mathbf{A}_k^{\mathbf{Z}} - \bar{\mathbf{A}}^{\mathbf{Z}})$
S_5:   Compute the d eigenvectors $\{\mathbf{e}_j^{\mathbf{X}}\}_{j=1}^{d}$ of $\mathbf{G}$
         corresponding to the largest d eigenvalues
S_6: $\mathbf{X}_i = [\mathbf{e}_1^{\mathbf{X}}, \ldots, \mathbf{e}_d^{\mathbf{X}}]$
S_7:   Compute $\mathbf{A}_k^{\mathbf{X}} = \mathbf{A}_k\mathbf{X}_i\mathbf{X}_i^T$
S_8:   Compute $\mathbf{H} = \frac{1}{M}\sum\limits_{k=1}^{M}(\mathbf{A}_k^{\mathbf{Z}} - \bar{\mathbf{A}}^{\mathbf{Z}})(\mathbf{A}_k^{\mathbf{Z}} - \bar{\mathbf{A}}^{\mathbf{Z}})^T$
S_9:   Compute the q eigenvectors $\{\mathbf{e}_j^{\mathbf{Z}}\}_{j=1}^{q}$ of $\mathbf{H}$
         corresponding to the largest q eigenvalues
S_{10}: $\mathbf{Z}_i = [\mathbf{e}_1^{\mathbf{Z}}, \ldots, \mathbf{e}_q^{\mathbf{Z}}]$
S_{11}:   If $t > 2$ and $\|\mathbf{Z}_i - \mathbf{Z}_{i-1}\|_F < m\varepsilon$ and $\|\mathbf{X}_i - \mathbf{X}_{i-1}\|_F < n\varepsilon$
S_{12}:   Then Go to S_3
S_{13}:   Else Go to S_{15}
S_{14}: End For
S_{15}: $\mathbf{Z} = \mathbf{Z}_i$
S_{16}: $\mathbf{X} = \mathbf{X}_i$
S_{17}: Feature extraction: $\mathbf{B}_k = \mathbf{Z}^T\mathbf{A}_k\mathbf{X}$

Table 2. Coupled Subspaces Analysis Algorithm.

separated. Therefore the element in the kernel matrix $\mathbf{K}$ can be computed by

$$
\mathbf{K} = \begin{bmatrix}
\varphi\left(\mathbf{a}_1^1\right)\varphi\left(\mathbf{a}_1^1\right)^T & \cdots & \varphi\left(\mathbf{a}_1^1\right)\varphi\left(\mathbf{a}_1^m\right)^T & \cdots & \varphi\left(\mathbf{a}_1^1\right)\varphi\left(\mathbf{a}_M^1\right)^T & \cdots & \varphi\left(\mathbf{a}_1^1\right)\varphi\left(\mathbf{a}_M^m\right)^T \\
\vdots & \ddots & \vdots & \ddots & \vdots & \ddots & \vdots \\
\varphi\left(\mathbf{a}_1^m\right)\varphi\left(\mathbf{a}_1^1\right)^T & \cdots & \varphi\left(\mathbf{a}_1^m\right)\varphi\left(\mathbf{a}_1^m\right)^T & \cdots & \varphi\left(\mathbf{a}_1^m\right)\varphi\left(\mathbf{a}_M^1\right)^T & \cdots & \varphi\left(\mathbf{a}_1^m\right)\varphi\left(\mathbf{a}_M^m\right)^T \\
\vdots & & \vdots & \ddots & \vdots & & \vdots \\
\varphi\left(\mathbf{a}_M^1\right)\varphi\left(\mathbf{a}_1^1\right)^T & \cdots & \varphi\left(\mathbf{a}_M^1\right)\varphi\left(\mathbf{a}_1^m\right)^T & \cdots & \varphi\left(\mathbf{a}_M^1\right)\varphi\left(\mathbf{a}_M^1\right)^T & \cdots & \varphi\left(\mathbf{a}_M^1\right)\varphi\left(\mathbf{a}_M^m\right)^T \\
\vdots & \ddots & \vdots & \ddots & \vdots & \ddots & \vdots \\
\varphi\left(\mathbf{a}_M^m\right)\varphi\left(\mathbf{a}_1^1\right)^T & \cdots & \varphi\left(\mathbf{a}_M^m\right)\varphi\left(\mathbf{a}_1^m\right)^T & \cdots & \varphi\left(\mathbf{a}_M^m\right)\varphi\left(\mathbf{a}_M^1\right)^T & \cdots & \varphi\left(\mathbf{a}_M^m\right)\varphi\left(\mathbf{a}_M^m\right)^T
\end{bmatrix} \tag{24}
$$

which is an mM-by-mM matrix. Unfortunately, there is a critical problem in implementation about the dimension of its kernel matrix. The kernel matrix is $M \times M$ matrix in KPCA, where M is the number of training samples, while it is $mM \times mM$ matrix in K2DPCA, where m is the number of row of each image. Thus, the K2DPCA kernel matrix is m^2 times of KPCA kernel matrix. For example, if the training set has 200 images with dimensions of 100×100 then the dimension of kernel matrix shall be 20000×20000, that is very big for fitting in memory unit. After that the projection can be formed by the eigenvectors of this kernel matrix as same as the traditional KPCA.

5. Supervised frameworks

Since the 2DPCA is the unsupervised technique, the class information is neglected. This section presents two methods which can be used to embedded class information to 2DPCA.

Firstly, Linear Discriminant Analysis (LDA) is implemented in 2D framework. Secondly, an 2DPCA is performed for each class in class-specific subspace.

5.1 Two-dimensional linear discriminant analysis of principal component vectors

The PCA's criterion chooses the subspace in the function of data distribution while Linear Discriminant Analysis (LDA) chooses the subspace which yields maximal inter-class distance, and at the same time, keeping the intra-class distance small. In general, LDA extracts features which are better suitable for classification task. However, when the available number of training samples is small compared to the feature dimension, the covariance matrix estimated by these features will be singular and then cannot be inverted. This is called singularity problem or Small Sample Size (SSS) problem Fukunaga (1990).

Various solutions have been proposed for solving the SSS problem Belhumeur et al. (1997); Chen et al. (2000); Huang et al. (2002); Lu et al. (2003); Zhao, Chellappa & Krishnaswamy (1998); Zhao, Chellappa & Nandhakumar (1998) within LDA framework. Among these LDA extensions, Fisherface Belhumeur et al. (1997) and the discriminant analysis of principal components framework Zhao, Chellappa & Krishnaswamy (1998); Zhao, Chellappa & Nandhakumar (1998) demonstrates a significant improvement when applying LDA over principal components from the PCA-based subspace. Since both PCA and LDA can overcome the drawbacks of each other. PCA is constructed around the criteria of preserving the data distribution. Hence, it is suited for representation and reconstruction from the projected feature. However, in the classification tasks, PCA only normalize the input data according to their variance. This is not efficient since the between classes relationship is neglected. In general, the discriminant power depends on both within and between classes relationship. LDA considers these relationships via the analysis of within and between-class scatter matrices. Taking this information into account, LDA allows further improvement. Especially, when there are prominent variation in lighting condition and expression. Nevertheless, all of above techniques, the spatial structure information still be not employed.

Two-Dimensional Linear Discriminant Analysis (2DLDA) was proposed in Ye et al. (2005). For overcoming the SSS problem in classical LDA by working with images in matrix representation, like in 2DPCA. In particular, bilateral projection scheme was applied there via left and right multiplying projection matrices. In this way, the eigenvalue problem was solved two times per iteration. One corresponds to the column direction and another one corresponds to the row direction of image, respectively

Because of 2DPCA is more suitable for face representation than face recognition, like PCA. For better performance in recognition task, LDA is still necessary. Unfortunately, the linear transformation of 2DPCA reduces the input image to a vector with the same dimension as the number of rows or the height of the input image. Thus, the SSS problem may still occurred when LDA is performed after 2DPCA directly. To overcome this problem, a simplified version of the 2DLDA is applied only unilateral projection scheme, based on the 2DPCA concept (Sanguansat et al., 2006b;c). Applying 2DLDA to 2DPCA not only can solve the SSS problem and the curse of dimensionality dilemma but also allows us to work directly on the image matrix in all projections. Hence, spatial structure information is maintained and the size of all scatter matrices cannot be greater than the width of face image. Furthermore, computing with this dimension, the face image do not need to be resized, since all information still be preserved.

5.2 Two-dimensional linear discriminant analysis (2DLDA)

Let $\mathbf{z}$ be a q dimensional vector. A matrix $\mathbf{A}$ is projected onto this vector via the similar transformation as Eq. (1):

$$v=Az. \tag{25}$$

This projection yields an m dimensional feature vector.

2DLDA searches for the projection axis $\mathbf{z}$ that maximizing the Fisher's discriminant criterion Belhumeur et al. (1997); Fukunaga (1990):

$$J(\mathbf{z}) = \frac{tr\,(\mathbf{S}_b)}{tr\,(\mathbf{S}_w)}, \tag{26}$$

where $\mathbf{S}_w$ is the *within-class scatter matrix* and $\mathbf{S}_b$ is the *between-class scatter matrix*. In particular, the within-class scatter matrix describes how data are scattered around the means of their respective class, and is given by

$$\mathbf{S}_w = \sum_{i=1}^{K} Pr(\omega_i) E\left[(\mathbf{Hz})(\mathbf{Hz})^T | \omega = \omega_i\right], \tag{27}$$

where K is the number of classes, $Pr(\omega_i)$ is the prior probability of each class, and $\mathbf{H} = \mathbf{A} - E\mathbf{A}$. The between-class scatter matrix describes how different classes. Which represented by their expected value, are scattered around the mixture means by

$$\mathbf{S}_b = \sum_{i=1}^{K} Pr(\omega_i) E\left[(\mathbf{Fz})(\mathbf{Fz})^T\right], \tag{28}$$

where $\mathbf{F} = E[\mathbf{A}|\omega = \omega_i] - E[\mathbf{A}]$.

With the linearity properties of both the trace function and the expectation, $J(\mathbf{z})$ may be rewritten as

$$
\begin{aligned}
J(\mathbf{z}) &= \frac{tr(\sum_{i=1}^{K} Pr(\omega_i) E\left[(\mathbf{Fz})(\mathbf{Fz})^T\right])}{tr(\sum_{i=1}^{K} Pr(\omega_i) E\left[(\mathbf{Hz})(\mathbf{Hz})^T | \omega = \omega_i\right])} \\
&= \frac{\sum_{i=1}^{K} Pr(\omega_i) E\left[tr((\mathbf{Fz})(\mathbf{Fz})^T)\right]}{\sum_{i=1}^{K} Pr(\omega_i) E\left(tr\left[(\mathbf{Hz})(\mathbf{Hz})^T | \omega = \omega_i\right]\right)} \\
&= \frac{\sum_{i=1}^{K} Pr(\omega_i) E\left[tr((\mathbf{Fz})^T(\mathbf{Fz}))\right]}{\sum_{i=1}^{K} Pr(\omega_i) E\left(tr\left[(\mathbf{Hz})^T(\mathbf{Hz}) | \omega = \omega_i\right]\right)} \\
&= \frac{tr\left(\mathbf{z}^T\left(\sum_{i=1}^{K} Pr(\omega_i) E\left[\mathbf{F}^T\mathbf{F}\right]\right)\mathbf{z}\right)}{tr\left(\mathbf{z}^T\left(\sum_{i=1}^{K} Pr(\omega_i) E\left[\mathbf{H}^T\mathbf{H} | \omega = \omega_i\right]\right)\mathbf{z}\right)} \\
&= \frac{tr(\mathbf{z}^T\tilde{\mathbf{S}}_b\mathbf{z})}{tr(\mathbf{z}^T\tilde{\mathbf{S}}_w\mathbf{z})}. \tag{29}
\end{aligned}
$$

Furthermore, $\tilde{\mathbf{S}}_b$ and $\tilde{\mathbf{S}}_w$ can be evaluated as follows:

$$\tilde{\mathbf{S}}_b = \sum_{i=1}^{K} \frac{n_i}{K} (\bar{\mathbf{A}}_i - \bar{\mathbf{A}})^T (\bar{\mathbf{A}}_i - \bar{\mathbf{A}}) \tag{30}$$

$$\tilde{\mathbf{S}}_w = \sum_{i=1}^{K} \frac{n_i}{K} \sum_{\mathbf{A}_k \in \omega_i} (\mathbf{A}_k - \bar{\mathbf{A}}_i)^T (\mathbf{A}_k - \bar{\mathbf{A}}_i), \tag{31}$$

where n_i and $\bar{\mathbf{A}}_i$ are the number of elements and the expected value of class ω_i respectively. $\bar{\mathbf{A}}$ denotes the overall mean.

Then the optimal projection vector can be found by solving the following generalized eigenvalue problem:

$$\tilde{\mathbf{S}}_b \mathbf{z} = \lambda \tilde{\mathbf{S}}_w \mathbf{z}. \tag{32}$$

Again the SVD algorithm can be applied to solve this eigenvalue problem on the matrix $\tilde{\mathbf{S}}_w^{-1} \tilde{\mathbf{S}}_b$. Note that, in this size of scatter matrices involved in eigenvalue decomposition process is also become n by n. Thus, with the limited the training set, this decomposition is more reliably than the eigenvalue decomposition based on the classical covariance matrix.

The number of projection vectors is then selected by the same procedure as in Eq. (9). Let $\mathbf{Z} = [\mathbf{z}_1, \ldots, \mathbf{z}_q]$ be the projection matrix composed of q largest eigenvectors for 2DLDA. Given a m by n matrix $\mathbf{A}$, its projection onto the principal subspace spanned by $\mathbf{z}_i$ is then given by

$$\mathbf{V} = \mathbf{AZ}. \tag{33}$$

The result of this projection $\mathbf{V}$ is another matrix of size m by q. Like 2DPCA, this procedure takes a matrix as input and outputs another matrix. These two techniques can be further combined, their combination is explained in the next section.

5.3 2DPCA+2DLDA

In this section, we apply an 2DLDA within the well-known frameworks for face recognition, the LDA of PCA-based feature (Zhao, Chellappa & Krishnaswamy, 1998). This framework consists of 2DPCA and 2DLDA steps, namely 2DPCA+2DLDA. From Section 2, we obtain a linear transformation matrix $\mathbf{X}$ on which each input face image $\mathbf{A}$ is projected. At the 2DPCA step, a feature matrix $\mathbf{Y}$ is obtained. The matrix $\mathbf{Y}$ is then used as the input for the 2DLDA step. Thus, the evaluation of within and between-class scatter matrices in this step will be slightly changed. From Eqs. (30) and (31), the image matrix $\mathbf{A}$ is substituted for the 2DPCA feature matrix $\mathbf{Y}$ as follows

$$\tilde{\mathbf{S}}_b^Y = \sum_{i=1}^{K} \frac{n_i}{K} (\bar{\mathbf{Y}}_i - \bar{\mathbf{Y}})^T (\bar{\mathbf{Y}}_i - \bar{\mathbf{Y}}) \tag{34}$$

$$\tilde{\mathbf{S}}_w^Y = \sum_{i=1}^{K} \frac{n_i}{K} \sum_{\mathbf{Y}_k \in \omega_i} (\mathbf{Y}_k - \bar{\mathbf{Y}}_i)^T (\mathbf{Y}_k - \bar{\mathbf{Y}}_i) \tag{35}$$

where $\mathbf{Y}_k$ is the feature matrix of the k-th image matrix $\mathbf{A}_k$, $\bar{\mathbf{Y}}_i$ be the average of $\mathbf{Y}_k$ which belong to class ω_i and $\bar{\mathbf{Y}}$ denotes a overall mean of $\mathbf{Y}$,

$$\bar{\mathbf{Y}} = \frac{1}{M} \sum_{k=1}^{M} \mathbf{Y}_k. \tag{36}$$

The 2DLDA optimal projection matrix $\mathbf{Z}$ can be obtained by solving the eigenvalue problem in Eq. (32). Finally, the composite linear transformation matrix, $\mathbf{L}=\mathbf{XZ}$, is used to map the face image space into the classification space by,

$$\mathbf{D} = \mathbf{AL}. \tag{37}$$

The matrix $\mathbf{D}$ is 2DPCA+2DLDA feature matrix of image $\mathbf{A}$ with dimension m by q. However, the number of 2DLDA feature vectors q cannot exceed the number of principal component vectors d. In general case ($q < d$), the dimension of $\mathbf{D}$ is less than $\mathbf{Y}$ in Section 2. Thus, 2DPCA+2DLDA can reduce the classification time compared to 2DPCA.

5.4 Class-specific subspace-based two-dimensional principal component analysis

2DPCA is a unsupervised technique that is no information of class labels are considered. Therefore, the directions that maximize the scatter of the data from all training samples might not be as adequate to discriminate between classes. In recognition task, a projection that emphasize the discrimination between classes is more important. The extension of Eigenface, PCA-based, was proposed by using alternative way to represent by projecting to Class-Specific Subspace (CSS) (Shan et al., 2003). In conventional PCA method, the images are analyzed on the features extracted in a low-dimensional space learned from all training samples from all classes. While each subspaces of CSS learned from training samples from one class. In this way, the CSS representation can provide a minimum reconstruction error. The reconstruction error is used to classify the input data via the Distance From CSS (DFCSS). Less DFCSS means more probability that the input data belongs to the corresponding class.

This extension was based on Sanguansat et al. (2006a). Let $\mathbf{G}_k$ be the image covariance matrix of the kth CSS. Then $\mathbf{G}_k$ can be evaluated by

$$\mathbf{G}_k = \frac{1}{M} \sum_{\mathbf{A}_c \in \omega_k} (\mathbf{A}_c - \bar{\mathbf{A}}_k)^T (\mathbf{A}_c - \bar{\mathbf{A}}_k), \tag{38}$$

where $\bar{\mathbf{A}}_k$ is the average image of class ω_k. The kth projection matrix $\mathbf{X}_k$ is a n by d_k projection matrix which composed by the eigenvectors of $\mathbf{G}_k$ corresponding to the d_k largest eigenvalues. The kth CSS of 2DPCA was represented as a 3-tuple:

$$\Re_k^{2DPCA} = \{\mathbf{X}_k, \bar{\mathbf{A}}_k, d_k\} \tag{39}$$

Let $\mathbf{S}$ be a input sample and $\mathbf{U}_k$ be a feature matrix which projected to the kth CSS, by

$$\mathbf{U}_k = \mathbf{W}_k \mathbf{X}_k, \tag{40}$$

where $\mathbf{W}_k = \mathbf{S} - \bar{\mathbf{A}}_k$. Then the reconstruct image $\mathbf{W}_k^r$ can be evaluates by

$$\mathbf{W}_k^r = \mathbf{U}_k \mathbf{X}_k^T. \tag{41}$$

Therefore, the DFCSS is defined by reconstruction error as follows

$$\varepsilon_k(\mathbf{W}_k^r, \mathbf{S}) = \sum_{m=1}^{n_{row}} \sum_{n=1}^{n_{col}} \left| \mathbf{w}_{(m,n)_k}^r - \mathbf{s}_{(m,n)} \right|. \tag{42}$$

If $\varepsilon_t = \min_{1 \leq k \leq K} (\varepsilon_k)$ then the input sample $\mathbf{S}$ is belong to class ω_t.

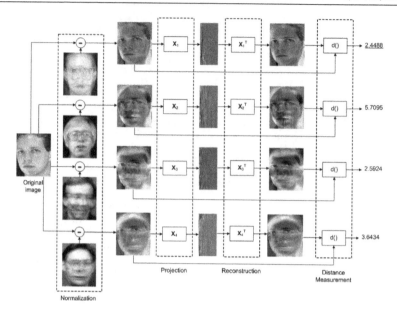

Fig. 1. CSS-based 2DPCA diagram.

For illustration, we assume that there are 4 classes, as shown in Fig. 1. The input image must be normalized with the averaging images of all 4 classes. And then project to 2DPCA subspaces of each class. After that the image is reconstructed by the projection matrices ($\mathbf{X}$) in each class. The DFCSS is used now to measure the similarity between the reconstructed image and the normalized original image on each CSS. From Fig. 1, the DFCSS of the first class is minimum, thus we decide this input image is belong to the first class.

6. Alignment based frameworks

Since 2DPCA can be viewed as the row-based PCA, that means the information contains only in row direction. Although, combining it with the column-based 2DPCA can consider the information in both row and column directions. But there still be other directions which should be considered.

6.1 Diagonal-based 2DPCA (DiaPCA)

The motivation for developing the DiaPCA method originates from an essential observation on the recently proposed 2DPCA (Yang et al., 2004). In contrast to 2DPCA, DiaPCA seeks the optimal projective vectors from diagonal face images and therefore the correlations between variations of rows and those of columns of images can be kept. Therefore, this problem can solve by transforming the original face images into corresponding diagonal face images, as shown in Fig. 2 and Fig. 3. Because the rows (columns) in the transformed diagonal face images simultaneously integrate the information of rows and columns in original images, it can reflect both information between rows and those between columns. Through the entanglement of row and column information, it is expected that DiaPCA may find some useful block or structure information for recognition in original images. The sample diagonal face images on Yale database are displayed in Fig. 4.

Experimental results on a subset of FERET database (Zhang et al., 2006) show that DiaPCA is more accurate than both PCA and 2DPCA. Furthermore, it is shown that the accuracy can be further improved by combining DiaPCA and 2DPCA together.

6.2 Image cross-covariance analysis

In PCA, the covariance matrix provides a measure of the strength of the correlation of all pixel pairs. Because of the limit of the number of training samples, thus this covariance cannot be well estimated. While the performance of 2DPCA is better than PCA, although all of the correlation information of pixel pairs are not employed for estimating the image covariance matrix. Nevertheless, the disregard information may possibly include the useful information. Sanguansat et al. (2007a) proposed a framework for investigating the information which was neglected by original 2DPCA technique, so called Image Cross-Covariance Analysis (ICCA). To achieve this point, the *image cross-covariance matrix* is defined by two variables, the first variable is the original image and the second one is the shifted version of the former. By our shifting algorithm, many image cross-covariance matrices are formulated to cover all of the information. The Singular Value Decomposition (SVD) is applied to the image cross-covariance matrix for obtaining the optimal projection matrices. And we will show that these matrices can be considered as the orthogonally rotated projection matrices of traditional 2DPCA. ICCA is different from the original 2DPCA on the fact that the transformations of our method are generalized transformation of the original 2DPCA.

First of all, the relationship between 2DPCA's image covariance matrix $\mathbf{G}$, in Eq. (5), and PCA's covariance matrix $\mathbf{C}$ can be considered as

$$\mathbf{G}(i,j) = \sum_{k=1}^{m} \mathbf{C}(m(i-1)+k, m(j-1)+k) \qquad (43)$$

where $\mathbf{G}(i,j)$ and $\mathbf{C}(i,j)$ are the ith row, j^{th} column element of matrix $\mathbf{G}$ and matrix $\mathbf{C}$, respectively. And m is the height of the image.

For illustration, let the dimension of all training images are 3 by 3. Thus, the covariance matrix of these images will be a 9 by 9 matrix and the dimension of image covariance matrix is only 3 by 3, as shown in Fig. 5.

From Eq. (43), each elements of $\mathbf{G}$ is the sum of all the same label elements in $\mathbf{C}$, for example:

$$\begin{aligned} \mathbf{G}(1,1) &= \mathbf{C}(1,1) + \mathbf{C}(2,2) + \mathbf{C}(3,3), \\ \mathbf{G}(1,2) &= \mathbf{C}(1,4) + \mathbf{C}(2,5) + \mathbf{C}(3,6), \\ \mathbf{G}(1,3) &= \mathbf{C}(1,7) + \mathbf{C}(2,8) + \mathbf{C}(3,9). \end{aligned} \qquad (44)$$

It should be note that the total power of image covariance matrix equals and traditional covariance matrix $\mathbf{C}$ are identical,

$$tr(\mathbf{G}) = tr(\mathbf{C}). \qquad (45)$$

From this point of view in Eq. (43), we can see that image covariance matrix is collecting the classification information only $1/m$ of all information collected in traditional covariance matrix. However, there are the other $(m-1)/m$ elements of the covariance matrix still be not

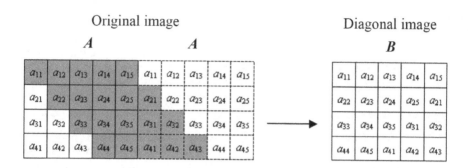

Fig. 2. Illustration of the ways for deriving the diagonal face images: If the number of columns is more than the number of rows

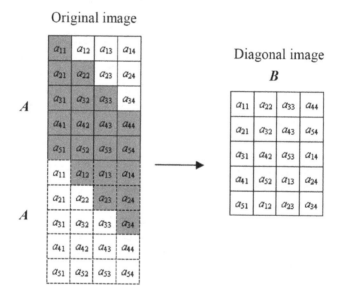

Fig. 3. Illustration of the ways for deriving the diagonal face images: If the number of columns is less than the number of rows

Fig. 4. The sample diagonal face images on Yale database.

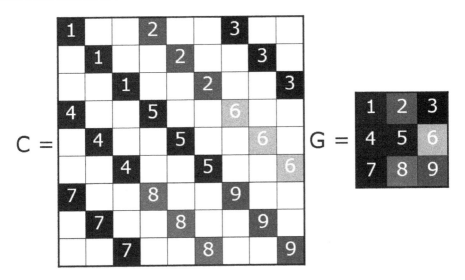

Fig. 5. The relationship of covariance and image covariance matrix.

considered. By the experimental results in Sanguansat et al. (2007a). For investigating how the retaining information in 2D subspace is rich for classification, the new **G** is derived from the PCA's covariance matrix as

$$\mathbf{G}_L(i,j) = \sum_{k=1}^{m} \mathbf{C}\left(f\left(m\left(i-1\right)+k\right), m\left(j-1\right)+k\right), \tag{46}$$

$$f(x) = \begin{cases} x+L-1, & 1 \le x \le mn-L+1 \\ x-mn+L-1, & mn-L+2 \le x \le mn \end{cases} \tag{47}$$

where $1 \le L \le mn$.

The $\mathbf{G}_L$ can also be determined by applying the shifting to each images instead of averaging certain elements of covariance matrix. Therefore, the $\mathbf{G}_L$ can alternatively be interpreted as the *image cross-covariance matrix* or

$$\mathbf{G}_L = E[(\mathbf{B}_L - E[\mathbf{B}_L])^T(\mathbf{A} - E[\mathbf{A}])] \tag{48}$$

where $\mathbf{B}_L$ is the Lth shifted version of image $\mathbf{A}$ that can be created via algorithm in Table 3. The samples of shifted images $\mathbf{B}_L$ are presented in Fig. 6.

In 2DPCA, the columns of the projection matrix, $\mathbf{X}$, are obtained by selection the eigenvectors which corresponding to the d largest eigenvalues of image covariance matrix, in Eq. (5). While in ICCA, the eigenvalues of image cross-covariance matrix, $\mathbf{G}_L$, are complex number with non-zero imaginary part. The Singular Value Decomposition (SVD) is applied to this matrix instead of Eigenvalue decomposition. Thus, the ICCA projection matrix contains a set of orthogonal basis vectors which corresponding to the d largest singular values of image cross-covariance matrix.

For understanding the relationship between the ICCA projection matrix and the 2DPCA projection matrix, we will investigate in the simplest case, i.e. there are only one training

S_1: Input $m \times n$ original image $\mathbf{A}$
     and the number of shifting L ($2 \leq L \leq mn$).
S_2: Initialize the row index, $irow = [2, \ldots, n, 1]$,
     and output image $\mathbf{B} = m \times n$ zero matrix.
S_3: For $i = 1, 2, \ldots, L - 1$
S_4:   Sort the first row of $\mathbf{A}$ by the row index, $irow$.
S_5:   Set the last row of $\mathbf{B}$ = the first row of $\mathbf{A}$.
S_6:   For $j = 1, 2, \ldots, m - 1$
S_7:   Set the jth row of $\mathbf{B}$ = the $(j+1)^{th}$ row of $\mathbf{A}$.
S_8:   End For
S_9:   Set $\mathbf{A} = \mathbf{B}$
S_{10}: End For

Table 3. The Image Shifting Algorithm for ICCA

image. Therefore, the image covariance matrix and image cross-covariance matrix are simplified to $\mathbf{A}^T\mathbf{A}$ and $\mathbf{B}_L^T\mathbf{A}$, respectively.

The image $\mathbf{A}$ and $\mathbf{B}_L$ can be decomposed by using Singular Value Decomposition (SVD) as

$$\mathbf{A} = \mathbf{U_A}\mathbf{D_A}\mathbf{V_A^T}, \tag{49}$$

$$\mathbf{B}_L = \mathbf{U_{B_L}}\mathbf{D_{B_L}}\mathbf{V_{B_L}^T}. \tag{50}$$

Where $\mathbf{V_A}$ and $\mathbf{V_{B_L}}$ contain a set of the eigenvectors of $\mathbf{A}^T\mathbf{A}$ and $\mathbf{B}_L^T\mathbf{B}_L$, respectively. And $\mathbf{U_A}$ and $\mathbf{U_{B_L}}$ contain a set of the eigenvectors of $\mathbf{A}\mathbf{A}^T$ and $\mathbf{B}_L\mathbf{B}_L^T$, respectively. And $\mathbf{D_A}$ and $\mathbf{D_{B_L}}$ contain the singular values of $\mathbf{A}$ and $\mathbf{B}_L$, respectively. If all eigenvectors of $\mathbf{A}^T\mathbf{A}$ are selected then the $\mathbf{V_A}$ is the 2DPCA projection matrix, i.e. $\mathbf{X} = \mathbf{V_A}$.

Let $\mathbf{Y} = \mathbf{A}\mathbf{V_A}$ and $\mathbf{Z} = \mathbf{B}_L\mathbf{V_{B_L}}$ are the projected matrices of $\mathbf{A}$ and $\mathbf{B}$, respectively. Thus,

$$\mathbf{B}_L^T\mathbf{A} = \mathbf{V_{B_L}}\mathbf{Z}^T\mathbf{Y}\mathbf{V_A^T}. \tag{51}$$

Denoting the SVD of $\mathbf{Z}^T\mathbf{Y}$ by

$$\mathbf{Z}^T\mathbf{Y} = \mathbf{P}\mathbf{D}\mathbf{Q}^T, \tag{52}$$

and substituting into Eq. (51) gives

$$\mathbf{B}_L^T\mathbf{A} = \mathbf{V_{B_L}}\mathbf{P}\mathbf{D}\mathbf{Q}^T\mathbf{V_A^T} \tag{53}$$
$$= \mathbf{R}\mathbf{D}\mathbf{S}^T,$$

where $\mathbf{R}\mathbf{D}\mathbf{S}^T$ is the singular value decomposition of $\mathbf{B}_L^T\mathbf{A}$ because of the unique properties of the SVD operation. It should be note that $\mathbf{B}_L^T\mathbf{A}$ and $\mathbf{Z}^T\mathbf{Y}$ have the same singular values. Therefore,

$$\mathbf{R} = \mathbf{V_{B_L}}\mathbf{P}, \tag{54}$$
$$\mathbf{S} = \mathbf{V_A}\mathbf{Q} = \mathbf{X}\mathbf{Q} \tag{55}$$

can be thought of as orthogonally rotated of projection matrices $\mathbf{V_A}$ and $\mathbf{V_{B_L}}$, respectively.

As a result in Eq. (55), the ICCA projection matrix is the orthogonally rotated of original 2DPCA projection matrix.

Fig. 6. The samples of shifted images on the ORL database.

7. Random frameworks

In feature selection, the random subspace method can improve the performance by combining many classifiers which corresponds to each random feature subset. In this section, the random method is applied to 2DPCA in various ways to improve its performance.

7.1 Two-dimensional random subspace analysis (2DRSA)

The main disadvantage of 2DPCA is that it needs many more coefficients for image representation than PCA. Many works try to solve this problem. In Yang et al. (2004), PCA is used after 2DPCA for further dimensional reduction, but it is still unclear how the dimension of 2DPCA could be reduced directly. Many methods to overcome this problem were proposed by applied the bilateral-projection scheme to 2DPCA. In Zhang & Zhou (2005); Zuo et al. (2005), the right and left multiplying projection matrices are calculated independently while the iterative algorithm is applied to obtain the optimal solution of these projection matrices in Kong et al. (2005); Ye (2004). And the non-iterative algorithm for optimization was proposed in Liu & Chen (2006). In Xu et al. (2004), they proposed the iterative procedure which the right projection is calculated by the reconstructed images of the left projection and the left projection is calculated by the reconstructed images of the right projection. Nevertheless, all of above methods obtains only the local optimal solution.

Another method for dealing with high-dimensional space was proposed in Ho (1998b), called Random Subspace Method (RSM). This method is the one of ensemble classification methods, like Bagging Breiman (1996) and Boosting Freund & Schapire (1995). However, Bagging and Boosting are not reduce the high-dimensionality. Bagging randomly select a number of samples from the original training set to learn an individual classifier while Boosting specifically weight each training sample. The RSM can effectively exploit the high-dimensionality of the data. It constructs an ensemble of classifiers on independently selected feature subsets, and combines them using a heuristic such as majority voting, sum rule, etc.

There are many reasons the Random Subspace Method is suitable for face recognition task. Firstly, this method can take advantage of high dimensionality and far away from the curse of dimensionality (Ho, 1998b). Secondly, the random subspace method is useful for critical training sample sizes (Skurichina & Duin, 2002). Normally in face recognition, the dimension of the feature is extremely large compared to the available number of training samples. Thus applying RSM can avoid both of the curse of dimensionality and the SSS problem. Thirdly, The nearest neighbor classifier, a popular choice in the 2D face-recognition domain (Kong et al., 2005; Liu & Chen, 2006; Yang et al., 2004; Ye, 2004; Zhang & Zhou, 2005; Zuo et al., 2005), can be very sensitive to the sparsity in the high-dimensional space. Their accuracy is often far from optimal because of the lack of enough samples in the high-dimensional space. The RSM brings significant performance improvements compared to a single classifier Ho (1998a); Skurichina & Duin (2002). Finally, since there is no hill climbing in RSM, there is no danger of being trapped in local optima Ho (1998b).

The RSM was applied to PCA for face recognition in Chawla & Bowyer (2005). They apply the random selection directly to the feature vector of PCA for constructing the multiple subspaces. Nevertheless, the information which contained in each element of PCA feature vector is not equivalent. Normally, the element which corresponds to the larger eigenvalue, contains more useful information. Therefore, applying RSM to PCA feature vector is seldom appropriate.

S_1: Project image, **A**, by Eq. (10).
S_2: For $i = 1$ to the number of classifiers
S_3:   Randomly select a r dimensional random subspace, $\mathbf{Z}_i^r$,
        from **Y** $(r < m)$.
S_4:   Construct the nearest neighbor classifier, $\mathbf{C}_i^r$.
S_5: End For
S_6: Combine the output of each classifiers by using majority voting.

Table 4. Two-Dimensional Random Subspace Analysis Algorithm

Different from PCA, the 2DPCA feature is a matrix form. Thus, RSM is more suitable for 2DPCA, because the column direction does not depend on the eigenvalue.

A framework of Two-Dimensional Random Subspace Analysis (2DRSA) (Sanguansat et al., n.d.) is proposed to extend the original 2DPCA. The RSM is applied to feature space of 2DPCA for generating the vast number of feature subspaces, which be constructed by an autonomous, pseudorandom procedure to select a small number of dimensions from a original feature space. For a m by n feature matrix, there are 2^m such selections that can be made, and with each selection a feature subspace can be constructed. And then individual classifiers are created only based on those attributes in the chosen feature subspace. The outputs from different individual classifiers are combined by the uniform majority voting to give the final prediction.

The Two-Dimensional Random Subspace Analysis consists of two parts, 2DPCA and RSM. After data samples was projected to 2D feature space via 2DPCA, the RSM are applied here by taking advantage of high dimensionality in these space to obtain the lower dimensional multiple subspaces. A classifier is then constructed on each of those subspaces, and a combination rule is applied in the end for prediction on the test sample. The 2DRSA algorithm is listed in Table 4, the image matrix, **A**, is projected to feature space by 2DPCA projection in Eq. (10). In this feature space, it contains the data samples in matrix form, the $m \times d$ feature matrix, **Y** in Eq. (10). The dimensions of feature matrix **Y** depend on the height of image (m) and the number of selected eigenvectors of the image covariance matrix **G** (d). Therefore, only the information which embedded in each element on the row direction was sorted by the eigenvalue but not on the column direction. It means this method should randomly pick up some rows of feature matrix **Y** to construct the new feature matrix **Z**. The dimension of **Z** is $r \times d$, normally r should be less than m. The results in Ho (1998b) have shown that for a variety of data sets adopting half of the feature components usually yields good performance.

7.2 Two-dimensional diagonal random subspace analysis (2D^{2}RSA)

The extension of 2DRSA was proposed in Sanguansat et al. (2007b), namely the Two-Dimensional Diagonal Random Subspace Analysis. It consists of two parts i.e. DiaPCA and RSM. Firstly, all images are transformed into the diagonal face images as in Section 6.1. After that the transformed image samples was projected to 2D feature space via DiaPCA, the RSM are applied here by taking advantage of high dimensionality in these space to obtain the lower dimensional multiple subspaces. A classifier is then constructed on each of those subspaces, and a combination rule is applied in the end for prediction on the test sample. Similar to 2DRSA, the 2D^{2}RSA algorithm is listed in Table 5.

S_1: Transforming images into diagonal images.

S_2: Project image, $\mathbf{A}$, by Eq. (10).

S_3: For $i = 1$ to the number of classifiers

S_4:   Randomly select a r dimensional random subspace, $\mathbf{Z}_i^r$, from $\mathbf{Y}$ $(r < m)$.

S_5:   Construct the nearest neighbor classifier, $\mathbf{C}_i^r$.

S_6: End For

S_7: Combine the output of each classifiers by using majority voting.

Table 5. Two-Dimensional Diagonal Random Subspace Analysis Algorithm.

7.3 Random subspace method-based image cross-covariance analysis

As discussed in Section 6.2, not all elements of the covariance matrix is used in 2DPCA. Although, the image cross-covariance matrix can be switching these elements to formulate many versions of image cross-covariance matrix, the $(m - 1)/m$ elements of the covariance matrix are still not advertent in the same time. For integrating this information, the Random Subspace Method (RSM) can be using here via randomly select the number of shifting L to construct a set of multiple subspaces. That means each subspace is formulated from difference versions of image cross-covariance matrix. And then individual classifiers are created only based on those attributes in the chosen feature subspace. The outputs from different individual classifiers are combined by the uniform majority voting to give the final prediction. Moreover, the RSM can be used again for constructing the subspaces which are corresponding to the difference number of basis vectors d. Consequently, the number of all random subspaces of ICCA reaches to $d \times L$. That means applying the RSM to ICCA can be constructed more subspaces than 2DRSA. As a result, the RSM-based ICCA can alternatively be apprehended as the generalized 2DRSA.

8. Conclusions

This chapter presents the extensions of 2DPCA in several frameworks, i.e. bilateral projection, kernel method, supervised based, alignment based and random approaches. All of these methods can improve the performance of traditional 2DPCA for image recognition task. The bilateral projection can obtain the smallest feature matrix compared to the others. The class information can be embedded in the projection matrix by supervised frameworks that means the discriminant power should be increased. The alternate alignment of pixels in image can reveal the latent information which is useful for the classifier. The kernel based 2DPCA can achieve to the highest performance but the appropriated kernel's parameters and a huge of memory are required to manipulate the kernel matrix while the random subspace method is good for robustness.

9. References

Belhumeur, P. N., Hespanha, J. P. & Kriegman, D. J. (1997).   Eigenfaces vs. Fisherfaces: Recognition using class specific linear projection, *IEEE Trans. Pattern Anal. and Mach. Intell.* 19: 711–720.

Breiman, L. (1996). Bagging predictors, *Machine Learning* 24(2): 123–140.

Chawla, N. V. & Bowyer, K. (2005). Random subspaces and subsampling for 2D face recognition, *Computer Vision and Pattern Recognition*, Vol. 2, pp. 582–589.

Chen, L., Liao, H., Ko, M., Lin, J. & Yu, G. (2000). A new LDA based face recognition system which can solve the small sample size problem, *Pattern Recognition* 33(10): 1713–1726.

Freund, Y. & Schapire, R. E. (1995). A decision-theoretic generalization of on-line learning and an application to boosting, *European Conference on Computational Learning Theory*, pp. 23–37.

Fukunaga, K. (1990). *Introduction to Statistical Pattern Recognition*, second edn, Academic Press.

Ho, T. K. (1998a). Nearest neighbors in random subspaces, *Proceedings of the 2nd Int'l Workshop on Statistical Techniques in Pattern Recognition*, Sydney, Australia, pp. 640–648.

Ho, T. K. (1998b). The random subspace method for constructing decision forests, *IEEE Trans. Pattern Anal. and Mach. Intell.* 20(8): 832–844.

Huang, R., Liu, Q., Lu, H. & Ma, S. (2002). Solving the small sample size problem of LDA, *Pattern Recognition* 3: 29–32.

Kong, H., Li, X., Wang, L., Teoh, E. K., Wang, J.-G. & Venkateswarlu, R. (2005). Generalized 2D principal component analysis, *IEEE International Joint Conference on Neural Networks (IJCNN)* 1: 108–113.

Liu, J. & Chen, S. (2006). Non-iterative generalized low rank approximation of matrices, *Pattern Recognition Letters* 27: 1002–1008.

Liu, J., Chen, S., Zhou, Z.-H. & Tan, X. (2010). Generalized low-rank approximations of matrices revisited, *Neural Networks, IEEE Transactions on* 21(4): 621 –632.

Lu, J., Plataniotis, K. N. & Venetsanopoulos, A. N. (2003). Regularized discriminant analysis for the small sample size problem in face recognition, *Pattern Recogn. Lett.* 24(16): 3079–3087.

Nguyen, N., Liu, W. & Venkatesh, S. (2007). Random subspace two-dimensional pca for face recognition, *Proceedings of the multimedia 8th Pacific Rim conference on Advances in multimedia information processing*, PCM'07, Springer-Verlag, Berlin, Heidelberg, pp. 655–664.
URL: *http://portal.acm.org/citation.cfm?id=1779459.1779555*

Sanguansat, P. (2008). 2dpca feature selection using mutual information, *Computer and Electrical Engineering, 2008. ICCEE 2008. International Conference on*, pp. 578 –581.

Sanguansat, P., Asdornwised, W., Jitapunkul, S. & Marukatat, S. (2006a). Class-specific subspace-based two-dimensional principal component analysis for face recognition, *International Conference on Pattern Recognition*, Vol. 2, Hong Kong, China, pp. 1246–1249.

Sanguansat, P., Asdornwised, W., Jitapunkul, S. & Marukatat, S. (2006b). Two-dimensional linear discriminant analysis of principle component vectors for face recognition, *IEICE Trans. Inf. & Syst. Special Section on Machine Vision Applications* E89-D(7): 2164–2170.

Sanguansat, P., Asdornwised, W., Jitapunkul, S. & Marukatat, S. (2006c). Two-dimensional linear discriminant analysis of principle component vectors for face recognition, *IEEE International Conference on Acoustics, Speech, and Signal Processing*, Vol. 2, Toulouse, France, pp. 345–348.

Sanguansat, P., Asdornwised, W., Jitapunkul, S. & Marukatat, S. (2007a). Image cross-covariance analysis for face recognition, *IEEE Region 10 Conference on Convergent Technologies for the Asia-Pacific*, Taipei, Taiwan.

Sanguansat, P., Asdornwised, W., Jitapunkul, S. & Marukatat, S. (2007b). Two-dimensional diagonal random subspace analysis for face recognition, *International Conference on Telecommunications, Industry and Regulatory Development*, Vol. 1, pp. 66–69.

Sanguansat, P., Asdornwised, W., Jitapunkul, S. & Marukatat, S. (n.d.). Two-dimensional random subspace analysis for face recognition, *7th International Symposium on Communications and Information Technologies*.

Shan, S., Gao, W. & Zhao, D. (2003). Face recognition based on face-specific subspace, *International Journal of Imaging Systems and Technology* 13(1): 23–32.

Sirovich, L. & Kirby, M. (1987). Low-dimensional procedure for characterization of human faces, *J. Optical Soc. Am.* 4: 519–524.

Skurichina, M. & Duin, R. P. W. (2002). Bagging, boosting and the random subspace method for linear classifiers, *Pattern Anal. Appl.* 5(2): 121–135.

Turk, M. & Pentland, A. (1991). Eigenfaces for recognition, *J. of Cognitive Neuroscience* 3(1): 71–86.

Xu, A., Jin, X., Jiang, Y. & Guo, P. (2006). Complete two-dimensional PCA for face recognition, *International Conference on Pattern Recognition*, Vol. 3, pp. 481–484.

Xu, D., Yan, S., Zhang, L., Liu, Z. & Zhang, H. (2004). Coupled subspaces analysis, *Technical report*, Microsoft Research.

Yang, J. & Yang, J. Y. (2002). From image vector to matrix: A straightforward image projection technique IMPCA vs. PCA, *Pattern Recognition* 35(9): 1997–1999.

Yang, J., Zhang, D., Frangi, A. F. & yu Yang, J. (2004). Two-dimensional PCA: A new approach to appearance-based face representation and recognition, *IEEE Trans. Pattern Anal. and Mach. Intell.* 26: 131–137.

Ye, J. (2004). Generalized low rank approximations of matrices, *International Conference on Machine Learning*, pp. 887–894.

Ye, J., Janardan, R. & Li, Q. (2005). Two-dimensional linear discriminant analysis, *in* L. K. Saul, Y. Weiss & L. Bottou (eds), *Advances in Neural Information Processing Systems 17*, MIT Press, Cambridge, MA, pp. 1569–1576.

Zhang, D. & Zhou, Z. H. (2005). $(2D)^2$PCA: 2-directional 2-dimensional PCA for efficient face representation and recognition, *Neurocomputing* 69: 224–231.

Zhang, D., Zhou, Z.-H. & Chen, S. (2006). Diagonal principal component analysis for face recognition, *Pattern Recognition* 39(1): 133–135.

Zhao, W., Chellappa, R. & Krishnaswamy, A. (1998). Discriminant analysis of principle components for face recognition, *IEEE 3rd Inter. Conf. on Automatic Face and Gesture Recognition*, Japan.

Zhao, W., Chellappa, R. & Nandhakumar, N. (1998). Empirical performance analysis of linear discriminant classifiers, *Computer Vision and Pattern Recognition*, IEEE Computer Society, pp. 164–171.

Zuo, W., Wang, K. & Zhang, D. (2005). Bi-dierectional PCA with assembled matrix distance metric, *International Conference on Image Processing*, Vol. 2, pp. 958–961.

2

Principal Component Analysis: A Powerful Interpretative Tool at the Service of Analytical Methodology

Maria Monfreda

Italian Customs Agency
Central Directorate for Chemical Analysis and Development of Laboratories, Rome,
Italy

1. Introduction

PCA is one of the most widely employed and useful tools in the field of exploratory analysis. It offers a general overview of the subject in question, showing the relationship that exists among objects as well as between objects and variables.

An important application of PCA consists of the characterization and subsequent differentiation of products in relation to their origin (known as traceability). PCA is often applied in order to characterize some products obtained via a manufacturing process and the transformation of some raw materials. In this case, there are two kinds of elements linkable to the differentiation of products in relation to their origin: the variability associated to the raw material and the differences in various production techniques used around the world. In this study, two examples of PCA application to some products obtained via a manufacturing process are presented. These products, belonging to completely different fields (foodstuffs and petroleum based fuel) show one element in common: their traceability is correlated to the raw material and the production process.

The strength of PCA is that it provides the opportunity to visualize data in reference to objects described by more than 3 variables. Indeed, PCA allows us to study and understand such systems, helping the human eye to see in two or three dimension systems that otherwise would necessarily have to be seen in more than three dimensions in order to be studied. PCA allows data to maintain their original structure, making only an orthogonal rotation of variables, which helps to simplify the visualization of all the information already contained in the data. Consequently, PCA can be considered the best technique to begin to approach any qualitative multivariate problem, be it unsupervised or supervised. Needless to say, supervised problems - following a primary study by PCA - require the application of either a classification or a class modeling method. In this study, three cases regarding supervised problems which involved the preliminary application of PCA are put forward. Results from PCA have been compared to those obtained from classification or class modeling tools.

2. PCA and traceability

PCA is widely used to characterize foodstuffs according to their geographical origin (Alonso-Salces et al., 2010; Diaz et al., 2005; Gonzalvez et al. 2009; Marini et al., 2006). Such a requirement is becoming prominent in the control field, especially in the marketing of products with PDO (Protected Denomination or Origin) or PGI (Protected Geographical Indication) markings. The PDO marking is awarded to products linked strictly to a typical area. Both the production of raw materials and their transformation into the final product must be carried out in the region that lends its name to the product. As a consequence, some analytical methods, whose results could be directly linked to the sample origin, would be extremely useful in the legal battle against the fraudulent use of PDO or PGI marking.

The local nature of a food product, strongly associated with its geographical location, can be correlated to the quality of the raw material used and its production techniques. Environmental conditions in a specific geographical area also provide the raw material with set characteristics, becoming a factor of primary importance in determining the final product "typicality". The production technique is of primary importance for both agricultural products and so-called transformed products, where culture, the instruments used, the ability and experience of the operator and the addition of particular ingredients create a unique product. Brescia et al. (2005) characterized buffalo milk mozzarella samples with reference to their geographical origin (two provinces, namely Foggia, in Apulia and Caserta, in Campania, were considered), by comparing several analytical and spectroscopic techniques. Some analyses were also performed on the raw milk (from which mozzarella had been obtained) with the purpose of evaluating how the differences among milk samples had transferred to the final product. In this study, a further PCA was applied only to those analytical variables measured on both milk and mozzarella samples: fat, ash, Li, Na, K, Mg, Ca, $\delta^{15}N/^{14}N$ e $\delta^{13}C/^{12}C$, disregarding all the analyses carried out only on mozzarella samples for which any comparison with milk samples could not be performed and vice versa. The biplots relative to PCA carried out on milk and mozzarella samples are reported in figures 1 and 2 respectively. It is easy to see that the milk samples are completely separated, according to their origin, on the PC1 (figure 1), whilst mozzarella samples lose such a strong separation, even though they maintain a good trend in their differentiation.

As already stated by Brescia et al., milk samples from Campania have a higher ^{13}C content, whilst samples from Apulia have a greater Li, Na and K content. If PCA results relative to mozzarella samples are compared to those from milk samples, it can be deduced that geographical differences, very clearly defined in the raw material, tend to drop slightly in the final product. There is a factor (K content) whose distribution is inverted between the raw material and the final product (positive loading on PC1 for milk samples and negative loading on PC1 for mozzarella samples). Another factor (Na content) was a discriminator for the raw milk (high positive loading on PC1) but its loading in mozzarella samples rises on the PC2 (the direction perpendicular to the geographical separation) and becomes negative on PC1. As Na content is known to be linked to the salting process of a cheese, the production technique is thought to reduce some differences originating from the raw materials. In other words, the differences that exist between buffalo mozzarella from Campania and Apulia are mainly determined by the differences between the two types of raw milk, rather than between manufacturing processes.

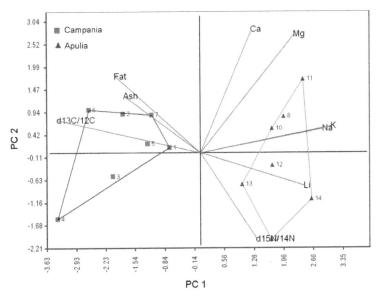

Fig. 1. Score plot of PC2 versus PC1 for milk samples.

Tables 1 and 2 show variances and cumulative variances associated to the principal components with eigenvalues greater than 1 for milk and mozzarella samples respectively. 4 PCs were extracted for both data set, which explain 86% of the variance for milk samples and 83% of variance for mozzarella samples.

PC	Variance %	Cumulative %
1	40.23	40.23
2	20.93	61.16
3	13.57	74.73
4	11.33	86.06

Table 1. PCs with eigenvalues greater than 1, extracted applying PCA to milk samples.

PC	Variance %	Cumulative %
1	35.95	35.95
2	20.45	56.40
3	15.10	71.50
4	11.83	83.33

Table 2. PCs with eigenvalues greater than 1, extracted applying PCA to mozzarella samples.

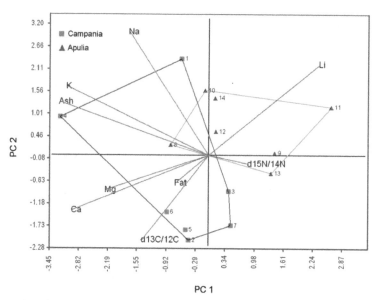

Fig. 2. Score plot of PC2 versus PC1 for mozzarella samples.

From this example, it can be deduced that the application of PCA to results obtained from chemical analyses of the raw material from which a transformed product has been obtained allows a characterization of the raw material in relation to its geographical origin. Secondly, the transformed product characterization allows to see how geographical differences among the raw materials have been spread out in the final product. In particular, it can be seen whether production techniques amplified or, indeed, reduced the pre-existing differences among the varying classes of the raw material. In other words, the application of PCA to the chemical analyses of a food product – as well as the raw material from which it has been made - allows to understand what the main elements are that provide a product characterization in relation to its origin: i.e. the quality of the raw material, the production techniques, or in fact a combination of both.

The characterization of products in relation to their origin is, however, not only important for food products. In forensic investigations, for example, it is becoming increasingly essential to identify associations among accelerants according to their source. Petroleum-based fuels (such as gasoline, kerosene, and diesel), which are often used as accelerants as they increase the rate and spread of fire, are also in fact transformed products from raw material (petroleum). Differentiation of such products in relation to their source (brand or refinery) depends both on the origin of the petroleum and the specific production techniques used during the refining process. Monfreda and Gregori (2011) differentiated 50 gasoline samples belonging to 5 brands (indicated respectively with the letters A, B, C, D and E) according to their refinery. Samples were analyzed by solid-phase microextraction (SPME) and gas chromatography-mass spectrometry (GC-MS). Some information on the origin of the crude oil was available but only for two of the brands: A samples were obtained from crude oil coming from only one country, whilst D samples were produced from crude oil coming from several countries. In addition A samples were tightly clustered in the score plots while D samples were fairly well spread out in the same score plots. This evidence was explained by considering that crude oil coming from only one place might have consistent chemical properties, compared to crude oils coming from several countries. Therefore differences existing between the raw materials had been transferred to the final products, determining very clustered samples with consistent chemical properties (for A brand) and samples with a greater variability within the class (for D brand). The score plot of PC2 versus PC1, shown in figure 3, was obtained by Monfreda and Gregori (2011).

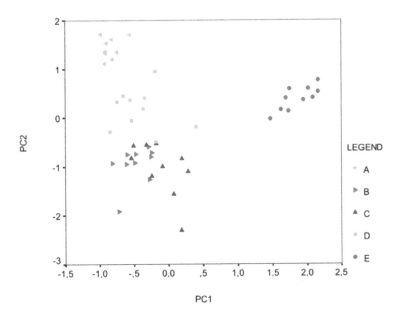

Fig. 3. Score plot of PC2 versus PC1 for gasoline samples (obtained by Monfreda & Gregori, 2011).

In the study presented here, 25 diesel samples belonging to the same 5 brands studied by Monfreda and Gregori were analysed using the same analytical procedure, SPME-GC-MS. As in the previous work, chromatograms were examined using the TCC approach (Keto & Wineman, 1991, 1994; Lennard at al., 1995). Peak areas were normalized to the area of the base peak (set to 10000), which was either tridecane, tetradecane or pentadecane, depending on the sample. Three independent portions for each sample of diesel were analyzed and peak areas were averaged. Analysis of variance was carried out before the multivariate statistical analysis, in order to eliminate same variables whose variance between classes was not significantly higher than the variance within class. Tetradecane, heptadecane, octadecane and hexadecane tetramethyl were then excluded from multivariate statistical analysis. PCA was finally applied to a data set of 25 samples and 33 variables, as listed in table 3.

Variable	COMPOUND
1	Nonane
2	Octane, 2,6-dimethyl
3	Benzene, 1-ethyl, 2-methyl
4	Decane
5	Benzene, 1,2,3-trimethyl
6	Benzene, 1-methyilpropyl
7	Nonane, 2,6-dimethyl
8	Benzene, 1-methyl-2-(1-methylethyl)
9	Benzene, 1,2,3-trimethyl
10	Cyclohexane, butyl
11	Benzene, 1-methyl-3-propyl
12	benzene, 4-ethyl-1,2-dimethyl
13	benzene, 1-methyl-2-propyl
14	Benzene, 1-methyl-4-(1-methylethyl)
15	Benzene, 4-ethyl-1,2-dimethyl
16	Undecane
17	Benzene, 1-ethyl-2,3-dimethyl
18	Benzene, 1,2,3,5-tetramethyl
19	Benzene, 1,2,3,4-tetramethyl
20	Cyclohexane, pentyl
21	Dodecane
22	Undecane 3,6-dimethyl
23	Cycloexane, hexyl
24	Tridecane
25	Naphthalene, 2-methyl
26	Naphthalene, 1-methyl
27	Pentadecane
28	Hexadecane
29	Pentadecane tetramethyl
30	Nonadecane
31	Eicosane
32	Heneicosane
33	Docosane

Table 3. Target compounds used as variables in multivariate statistical analysis of diesel samples.

Three PCs were extracted, with eigenvalues greater than 1, accounting for 92.16% of the total variance, as shown in table 4. From the score plot of PC2 versus PC1 (figure 4), it can be seen that a separation of samples according to the refinery was achieved, because each group stands in a definite area in the plane of PC1 and PC2. A samples are more clustered than D samples, according to the results obtained for gasoline samples.

PC	Variance %	Cumulative %
1	59.48	59.48
2	20.70	80.18
3	11.98	92.16

Table 4. PCs with eigenvalues greater than 1, extracted applying PCA to diesel samples.

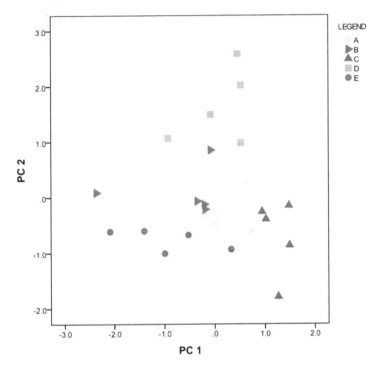

Fig. 4. Score plot of PC2 versus PC1 for diesel samples.

Results of both studies, carried out respectively on gasoline and diesel samples coming from the same five refineries, allow to achieve a traceability of these products according to their brands, that is to say that production techniques give well-defined features to these products. Properties of crude oil, otherwise, show a strong influence on the homogeneity of samples distribution within their class, based on information availability (only for two of five refineries).

3. The PCA role in classification studies

3.1 Case 1

The gasoline data matrix has been used in real cases of arson to link a sample of unevaporated gasoline, found at a fire scene in an unburned can, to its brand or refinery. This helped to answer, for example, questions posed by a military body about the origin of an unevaporated gasoline sample taken from a suspected arsonist. The gasoline sample under investigation was analyzed with the same procedure adopted by Monfreda and Gregory (2011) and using the same devices. Analyses were carried out almost in the same period in which the 50 samples of the previous work had been analyzed. Three independent portions of the sample were analyzed and from the Total Ion Chromatogram (TIC) of each analysis, a semi-quantitative report of peak areas of the same target compounds (TCs) used by Monfreda and Gregori was obtained. Areas were normalized to the area of the base peak (benzene, 1,2,3-trimethyl), set to 10000, as in the previous study. The average areas (of the three portions analyzed) corresponding to the aromatic compounds were appended to the data matrix of 50 gasoline samples analyzed by Monfreda and Gregori. A PCA was then applied to a data set of 51 samples and 16 variables. Results are shown in the scatter plots of figures 5, 6 and 7. From these scatter plots it can be seen that the sample under investigation is significantly different from those of the A and E brands. As a consequence, these two refineries could be excluded from further investigations by the relevant authorities because the membership of the unknown sample to A or E brands was less likely than it belonging to other classes. The score plot of PC2 versus PC1 (figure 5) shows the unknown sample among the classes B, C and D. From the score plot of PC3 versus PC1 (figure 6), it can be seen that the unknown sample is very close to those of class B, and quite distant from class C. The unknown sample, however, falls into an area where some samples of D brand are also present. Finally, from the scatter plot of PC3 versus PC2 and PC1 (figure 7), the sample under investigation would appear to fall between the B and D classes.

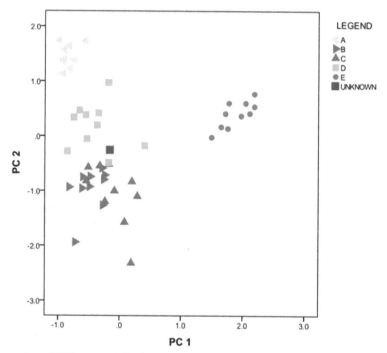

Fig. 5. Score plot of PC2 versus PC1 for 51 gasoline samples.

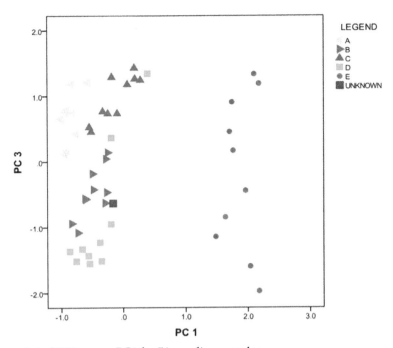

Fig. 6. Score plot of PC3 versus PC1 for 51 gasoline samples.

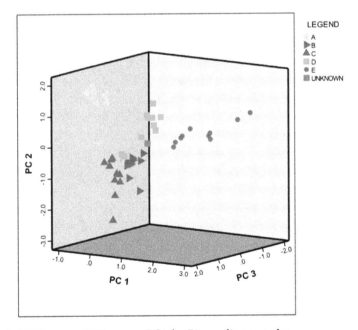

Fig. 7. Score plot of PC3 versus PC2 versus PC1 for 51 gasoline samples.

The application of PCA was especially useful for an initial visualization of data, however the question posed by the military body also needed to be handled with some supervised methods; in other words, discriminant analysis or class modeling tools. In such a way, the system is forced to create a boundary between classes and eventually the unknown sample is processed. For this kind of problem, class modeling tools are clearly preferable to discriminant analysis, in that they first create a model for each category as opposed to creating a simple delimiter between classes. The modeling rule discriminates between the studied category and the rest of the universe. As a consequence, each sample can be assigned to a single category, or to more than one category (if more than one class is modeled) or, alternatively, considered as an outlier if it falls outside the model. Discriminant analysis tends, however, to classify in any case the unknown sample in one of the studied categories even though it may not actually belong to any of them. In this case, the class modeling technique known as SIMCA (Soft Independent Models of Class Analogy) was applied to the data set under investigation. SIMCA builds a mathematical model of the category with its principal components and a sample is accepted by the specific category if its distance to the model is not significantly different from the class residual standard deviation. This chemometric tool was applied considering a 95% confidence level to define the class space and the unweighted augmented distance (Wold & Sjostrom, 1977). A cross validation with 10 cancellation groups was then carried out and 8 components were used to build the mathematical model of each class. The boundaries were forced to include all the objects of the training set in each class, which provided a sensitivity (the percentage of objects belonging to the category which are correctly identified by the mathematical model) of 100%. Results are shown in the Cooman's plots (figures 8, 9 and 10), where classes are labeled with the numbers 1 to 5 instead of the letters A to E respectively. The specificity (the percentage of objects from other categories which are classified as foreign) was also 100%.

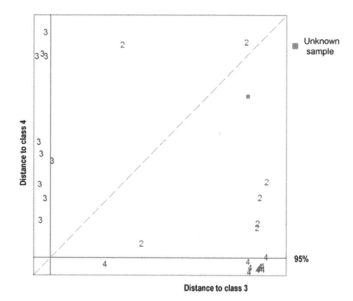

Fig. 8. Cooman's plot for the classes 3 (C) and 4 (D).

From the Cooman's plot of classes 3 (C) and 4 (D) (figure 8), the unknown sample (red square) results in an outlier but is closer to class 4 than to class 3. In figure 9, the distances from classes 2 (B) and 4 (D) are displayed and the sample under investigation remains an outlier, but its distance from class 2 is shorter than the equivalent from class 4. In figure 10, where the distances from classes 1 (A) and 5 (E) are plotted, the unknown sample is missing as it is too far from both classes.

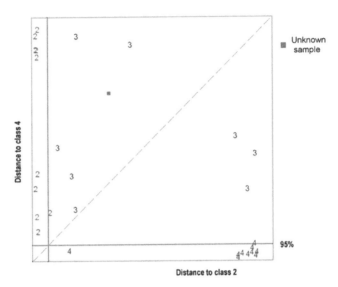

Fig. 9. Cooman's plot for the classes 2 (B) and 4 (D).

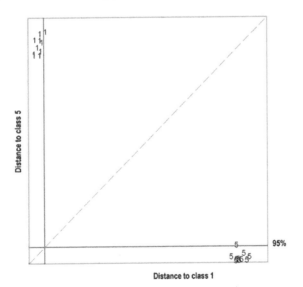

Fig. 10. Cooman's plot for the classes 1 (A) and 5 (E).

SIMCA confirms, therefore, the results obtained with PCA in so far as the unknown sample was significantly different from the A and E samples. Regarding classes B, C and D, SIMCA allows to conclude that the unknown sample, outlier for all classes, is nevertheless closer to class B (figure 9) than to the others. Finally, it can be concluded that the sample under investigation does not belong to any of the classes studied (for example, it comes from another refinery, not included in the data matrix); otherwise the sample could belong to one of the classes studied (the most probable class is number 2, followed by class 4) but the variability within each class might not have been sufficiently represented in the data matrix used.

3.2 Case 2

There are other examples that show the importance of PCA for data visualization. In forensic investigations, there is often the need to compare very similar samples. These comparisons invariably require the use of specific devices. One example was a specific request to compare three paint samples (A, B & C) in order to discover whether sample C was more similar to A or B. At first glance, all three samples appeared to be very alike. The analytical method that might have allowed to answer this question is pyrolysis, followed by GC, but the laboratory in question wasn't equipped with the necessary devices. Therefore, FT-IR (Fourier Transform Infrared Spectroscopy) analyses were carried out in transmission on 10 portions for each sample (these are both quick and relatively cheap) in order to characterize each sample variability: in other words, each sample was treated as if it were a class with 10 samples. PCA was applied to a data set relative to 30 samples and variables obtained from a data spacing of 64 cm^{-1} (with a smooth of 11, corresponding to 21.213 cm^{-1}) of FT-IR transmittances, in order to obtain a first data visualization. From the score plot of PC1 vs PC2 vs PC3, shown in figure 11, a trend can be seen in the separation between samples of classes A and B, while C samples are more frequently close to A samples than to B samples. Therefore, the similarity between C and A classes is assumed to be bigger than the one between C and B classes.

As the analytical problem required the classification of C sample to one of two classes, A or B, a discriminant analysis tool was then applied, with discriminant functions calculated only for A and B classes, while C samples were considered as unknown. The aim of this analysis was to verify in which of the two classes (A or B) samples C were more frequently classified. Discriminant analysis always classifies an unknown sample in one class (even if it is an outlier or it belongs to a different class from those implemented), because it calculates only a delimiter between the known classes. For the purpose of this case study, this tool was therefore preferable to a class modeling tool, which builds, on the other hand, a defined mathematical model for each class. Discriminant analysis was performed calculating canonical discriminant functions and using the leave-one-out method; this method is an extension of Linear Discriminant Analysis (LDA), which finds a number of variables that reflect as much as possible the difference between the groups.

The results of discriminant analysis, apart from indicating a classification ability of 100% for both classes A and B and a prediction ability of 70% and 80% respectively, show that seven C samples were classified in class A against three samples classified in class B. To conclude, the results obtained perfectly reflected those achieved in a laboratory equipped with pyrolysis devices.

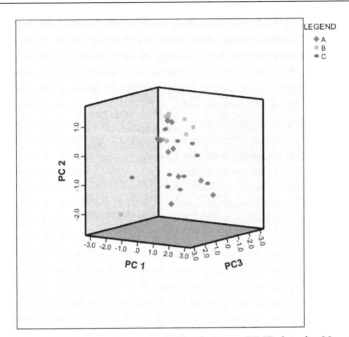

Fig. 11. Score plot of PC3 versus PC2 versus PC1 relative to FT-IR data for 30 paint samples.

A further observation about how discriminant analysis was applied in this case needs to be made. Indeed, this chemometric tool was applied to 5 principal components which account for 97,48% of the total variance, instead of the original variables. Such a procedure was adopted because the original variables were more than the samples used for building the classification rule between A and B classes (20). PCA is, therefore, imperative in classification problems where the number of variables is greater than that of the samples. In these cases, the application of discriminant analysis to the original variables would cause some overfitting problems; in other words, a sound and specific model would be obtained only for the training set used for its construction. The application of this model in real cases (like this one) would not prove very reliable. In reality, with overfitting, classification ability tends to increase while prediction ability tends to decrease. The best approach to take in these cases is to apply discriminant analysis to the PCs, by using a number of PCs (obviously less than the number of original variables) that explain a fair quantity of the variance contained in the original data. DA provides reliable results if the ratio between the number of samples and variables is more than 3.

3.3 Case 3

Another (forensic) case which involved a comparison between very similar samples was the comparison between a piece of a packing tape used on a case containing drugs with a roll of packing tape found during a house search, in order to establish whether the packing tape could have been ripped from the roll. Finding such evidence would have been of utmost importance in building a strong case against the suspect. Both exhibits, analyzed by FTIR in transmission, revealed an adhesive part of polybutylacrylate and a support of polypropylene. Both supports and adhesive parts showed significant similarity in IR absorptions. This similarity, though necessary, was not sufficient in itself to establish whether the packing tape had been ripped from the exact same roll seized at the suspect's home. The compatibility between the two exhibits was studied through a multivariate approach, analyzing, via FTIR, 10 independent portions of the adhesive part for each exhibit. 10 portions of the adhesive part (in polybutylacrylate) of two other rolls of packing tape (not linked to the case) were also analyzed. PCA was then applied to a data set relative to 40 samples and variables obtained from a data spacing of 16 cm⁻¹ (with a smooth of 11, corresponding to 21.213 cm⁻¹) of FT-IR transmittances. Six PCs were extracted, with eigenvalues greater than 1, explaining 98,15% of the total variance.

The score plot of the first three principal components is shown in figure 12, where samples taken from the seized roll are indicated as class 1, the other two rolls are indicated respectively as classes 2 and 3, while the piece of packing tape is indicated as class 4. From the score plot it can be seen that points of class 4 are fairly close to those of class 1, indicating a decent similarity between the two classes of interest. However, points of class 4 are also rather close to points of class 3, suggesting a similarity also between classes 4 and 3, while points of class 2 appear more distant, showing a lower similarity between classes 2 and 4. In this case, PCA gave a first display of data, but could not be used as definitive proof to establish the compatibility between classes 1 and 4 because class 4 appears also to be consistent with class 3.

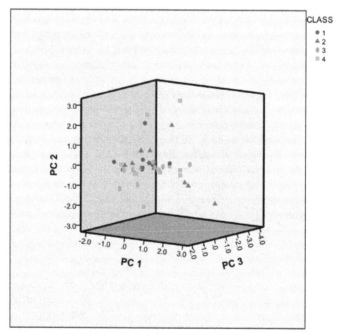

Fig. 12. Score plot of PC3 versus PC2 versus PC1 relative to FT-IR data for 40 packing tape samples.

SIMCA was then applied, considering a 95% confidence level to define class space and the unweighted augmented distance (Wold & Sjostrom, 1977). A cross validation with 10 cancellation groups was carried out and 8 components were used to build the mathematical model of each class. The boundaries were forced to include all the objects of the training set in each class, which provided a sensitivity of 100% for each class. With regard to the specificity, class 4 showed a specificity of 90% towards class 2, 80% for class 3 and 10% towards class 1. Such results can be visualized in the Cooman's plots.

For classes 1 and 4, the Cooman's plot is shown in figure 13. It can be seen that 9 samples of class 1 fall in the common area between classes 1 and 4 (the specificity of class 4 towards class 1 was in fact 10%). This kind of result indicates a significant similarity between classes 1 and 4, that is between the roll of packing tape found in the suspect's house and the piece of packing tape stuck on the case containing drugs.

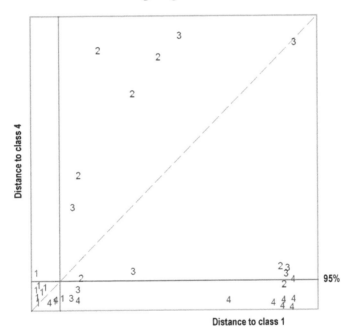

Fig. 13. Cooman's plot for the classes 1 and 4.

From the Cooman's plot relative to classes 2 and 4 (figure 14), it can be deduced that only one sample from class 2 is classified in the common area between classes 2 and 4 (specificity of class 4 towards class 2 equal to 90%), while no samples from class 4 are classified also in class 2. The similarity between classes 2 and 4 can therefore be considered insignificant.

Finally, from the Cooman's plot relative to classes 3 and 4 (figure 15), it is clearly visible that only 2 samples of class 3 fall in the overlapping area with class 4 (the specificity of class 4 towards class 3 was in fact 80%), whilst there are no samples from class 4 that fall in the overlapping area with class 3. From this last figure it can be deduced that the similarity between classes 1 and 4 is significantly higher than the similarity between classes 3 and 4.

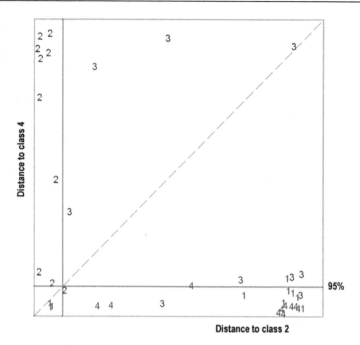

Fig. 14. Cooman's plot for the classes 2 and 4.

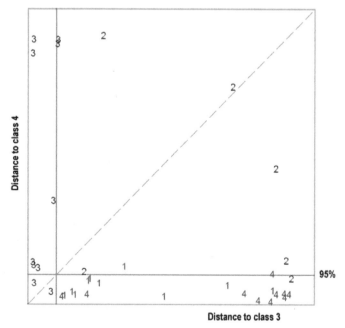

Fig. 15. Cooman's plot for the classes 3 and 4.

In conclusion, SIMCA analysis allowed the comparison between a piece of a packing tape and three rolls of packing tape that had the same chemical composition, finding the most significant similarity with the seized roll. Such a degree of similarity was measured in terms of specificity of the tape class (4) with the roll classes (1, 2 and 3): the lower the specificity is, the higher the similarity between the two classes under study is. SIMCA results are fairly consistent with PCA results, which gave a simple visualization of data. Both techniques found that class 4 had the lowest similarity with class 2. In addition, SIMCA, as a class modeling tool, gave better results than PCA with regards classes 1, 3 and 4.

4. Conclusions

This study shows the importance of PCA in traceability studies which can be carried out on different kind of matrices. As the majority of products come about from a transformation of some raw material, traceability has components deriving from both the fingerprint geographical characteristics transfer to the raw material and the production techniques developed in a specific context.

Moreover, PCA is a very useful tool for dealing with some supervised problems, due to its capability of describe objects without altering their native structure. However, it must be noted that, especially in forensics, results originating from a multivariate statistical analysis need to be presented and considered in a court of law with great care. For these kinds of results, the probabilistic approach is different from the one generally adopted for analytical results. In fact, in univariate analytical chemistry, the result of a measurement is an estimate of its true value, with its uncertainty set at a stated level of confidence. On the other hand, the use of multivariate statistical analysis in a court of law would imply a comparison between an unknown sample and a data set of known samples belonging to a certain number of classes. However, there remains the real possibility that the unknown sample might belong to yet another class, different from those of the known samples. In case 1, for example, the unknown sample might have been produced in a refinery that had not been included in the data matrix used for the comparison, or in case 3, the piece of packing tape, might not have belonged to any of the rolls analyzed. (Case 2 appears to be different, because sample C was specifically required to be classified in class A or B).

In these cases, an initial approach to the analytical problem by using PCA is fundamental because it allows the characteristics of the unknown sample to be compared with those of samples of which the origin is known. Depending on the results obtained at this step, a potential similarity between the unknown sample and samples from some specific classes may be excluded, or the class presenting the best similarity with the unknown sample might be found.

Results derived from PCA present a real picture of the situation - without any data manipulation or system forcing - and as such can form the basis for further deduction and the application of any other multivariate statistical analysis. A second step might be the application of some discriminant analysis or class modeling tool and an attempt to classify the sample in one of the classes included in the data matrix. A good result is achieved when PCA results fit those of supervised analysis. However, in a court of law these results would only become compelling alongside other strong evidence from the investigation, because, as already stated, the sample would have been compared with samples belonging to some distinct classes (and not all existing ones) and the data matrix might not adequately show the variability within each class.

5. References

Alonso-Salces, R.M. Héberger, K. Holland M.V., Moreno-Rojas J.M., C. Mariani, G. Bellan, F. Reniero, C. Guillou. (2010). Multivariate analysis of NMR fingerprint of the unsaponifiable fraction of virgin olive oils for authentication purposes. *Food Chemistry*, Vol. 118 pp. 956-965.

Brescia, M.A. Monfreda, M. Buccolieri, A. & Carrino, C. (2005). Characterisation of the geographical origin of buffalo milk and mozzarella cheese by means of analytical and spectroscopic determinations. *Food Chemistry*, Vol. 89, pp. 139-147.

Diaz, T.G. Merás, I.D. Casas, J.S. & Franco, M.F.A. (2005). Characterization of virgin olive oils according to its triglycerides and sterols composition by chemometric methods. *Food Control*, Vol. 16 pp. 339-347.

Gonzalvez, A. Armenta, S. De la Guardia, M. (2009). Trace-element composition and stable - isotope ratio for discrimination of foods with Protected Designation of Origin. *Trends in Analytical Chemistry*, Vol. 28 No.11, 2009.

Keto, R.O. & Wineman, PL. (1991). Detection of petroleum-based accelerants in fire debris by target compound gas chromatography/mass spectrometry. *Analitycal Chemistry*, Vol. 63 pp. 1964-71.

Keto, R.O. & Wineman, PL. (1994). Target-compound method for the analysis of accelerant residues in fire debris. *Analytica Chimica Acta*, Vol.288 pp.97-110.

Lennard, C.J. Tristan Rochaix, V. Margot, P. & Huber, K. (1995). A GC–MS Database of target compound chromatograms for the identification of arson accelerants. *Science & Justice*; Vol. 35 No.1 pp.19–30.

Marini, F. Magrì, A.L. Bucci, R. Balestrieri, F. & Marini, D. (2006). Class –modeling techniques in the authentication of Italian oils from Sicily with a Protected Denomination of Origin (PDO). *Chemometrics and Intelligent Laboratory Systems*, Vol. 80, pp. 140-149.

Monfreda, M. & Gregori, A. (2011). Differentiation of Unevaporated Gasoline Samples According to Their Brands, by SPME-GC-MS and Multivariate Statistical Analysis. *Journal of Forensic Sciences*, Vol. 56 (No. 2), pp. 372-380, March 2011.

Wold, S. & Sjostrom, M. (1977). SIMCA: A Method for Analyzing Chemical Data in Terms of Similarity and Analogy In: *Chemometrics, Theory and Application*, Kowalsky B.R. pp. 243-282, American Chemical Society Symposium Series No. 52 Washington.

3

Application of Principal Component Analysis to Elucidate Experimental and Theoretical Information

Cuauhtémoc Araujo-Andrade et al.*
Unidad Académica de Física, Universidad Autónoma de Zacatecas
México

1. Introduction

Principal Component Analysis has been widely used in different scientific areas and for different purposes. The versatility and potentialities of this unsupervised method for data analysis, allowed the scientific community to explore its applications in different fields. Even when the principles of PCA are the same in what algorithms and fundamentals concerns, the strategies employed to elucidate information from a specific data set (experimental and/or theoretical), mainly depend on the expertise and needs of each researcher.

In this chapter, we will describe how PCA has been used in three different theoretical and experimental applications, to explain the relevant information of the data sets. These applications provide a broad overview about the versatility of PCA in data analysis and interpretation. Our main goal is to give an outline about the capabilities and strengths of PCA to elucidate specific information. The examples reported include the analysis of matured distilled beverages, the determination of heavy metals attached to bacterial surfaces and interpretation of quantum chemical calculations. They were chosen as representative examples of the application of three different approaches for data analysis: the influence of data pre-treatments in the scores and loadings values, the use of specific optical, chemical and/or physical properties to qualitatively discriminate samples, and the use of spatial orientations to group conformers correlating structures and relative energies. This reason fully justifies their selection as case studies. This chapter also pretends to be a reference for those researchers that, not being in the field, may use these methodologies to take the maximum advantage from their experimental results.

*Claudio Frausto-Reyes[2], Esteban Gerbino[3], Pablo Mobili[3], Elizabeth Tymczyszyn[3], Edgar L. Esparza-Ibarra[1], Rumen Ivanov-Tsonchev[1] and Andrea Gómez-Zavaglia[3]
[1]Unidad Académica de Física, Universidad Autónoma de Zacatecas*
[2]Centro de Investigaciones en Óptica, A.C. Unidad Aguascalientes*
[3]Centro de Investigación y Desarrollo en Criotecnología de Alimentos (CIDCA)*
[1,2]México*
[3]Argentina*

2. Principal component analysis of spectral data applied in the evaluation of the authenticity of matured distilled beverages

The production of distilled alcoholic beverages can be summarised into at least three steps: *i)* obtaining and processing the raw materials, *ii)* fermentation and distillation processes, and *iii)* maturation of the distillate to produce the final aged product (Reazin, 1981). During the obtaining and fermentation steps, no major changes in the chemical composition are observed. However, throughout the maturation process, distillate undergoes definite and intended changes in aromatic and taste characteristics.

These changes are caused by three major types of reactions continually occurring in the barrel: *1)* extraction of complex wood substances by liquid (*i.e.*: acids, phenols, aldehydes, furfural, among others), *2)* oxidation of the original organic substances and of the extracted wood material, and *3)* reaction between various organic substances present in the liquid to form new products (Baldwin et al., 1967; Cramptom & Tolman,1908; Liebman & Bernice, 1949; Rodriguez-Madera et al., 2003; Valaer & Frazier,1936). Because of these reactions occurring during the maturation process, the stimulation and odour of ethanol in the distillate are reduced, and consequently, its taste becomes suitable for alcoholic beverages (Nishimura & Matsuyama, 1989). It is known that the concentration of extracts from wood casks in matured beverages seriously depend on the casks conditions (Nose et al., 2004). Even if their aging periods are the same, the use of different casks for the maturation process, strongly conditions the concentration of these extracts. (Philip, 1989; Puech, 1981; Reazin, 1981). Diverse studies on the maturation of distillates like whiskey, have demonstrated that colour, acids, esters, furfural, solids and tannins increase during the aging process. Except for esters, the greatest rate of change in the concentration of these compounds occurs during the first year (Reazin, 1981). For this reason, the extracts of wood and the chemically produced compounds during the aging process confer some optical properties that can be used to evaluate the authenticity and quality of the distillate in terms of its maturation process (Gaigalas et al., 2001; Walker, 1987).

The detection of economic fraud due to product substitution and adulteration, as well as health risk, requires an accurate quality control. This control includes the determination of changes in the process parameters, adulterations in any ingredient or in the whole product, and assessment that flavours attain well defined standards. Many of these quality control issues have traditionally been assessed by experts, who were able to determine the quality by observing their colour, texture, taste, aroma, etc. However, the acquisition of these skills requires years of experience, and besides that, the analysis may be subjective. Therefore, the use of more objective tools to evaluate maturation becomes essential. Nevertheless, it is difficult to find direct sensors for quality parameters. For this reason, it is necessary to determine indirect parameters that, taken individually, may weakly correlate to the properties of interest, but as a whole give a more representative picture of these properties. In this regard, different chromatographic techniques provide reliable and precise information about the presence of volatile compounds and the concentration of others (*i.e.*: ethanol, methanol, superior alcohols or heavy metals, etc.), thus proving the quality and authenticity of distilled alcoholic beverages (Aguilar-Cisneros, et al., 2002; Bauer-Christoph et al., 2003; Ragazzo et al.,2001; Savchuk et al., 2001; Pekka et al., 1999; Vallejo-Cordoba et al., 2004). In spite of that, chromatographic techniques, generally destroy the sample under study and also require equipment installed under specific protocols and installations

(Abbott & Andrews, 1970). On the other hand, the use of spectroscopic techniques such as infrared (NIR and FTIR), Raman, ultraviolet/visible together with multivariate methods, has already been used for the quantification of the different components of distilled beverages (*i.e.*: ethanol, methanol, sugar, among others). This approach allows the evaluation of quality and authenticity of these alcoholic products in a non-invasive, easy, fast, portable and reliable way (Dobrinas et al., 2009; Nagarajan et al., 2006). However, up to our knowledge, none of these reports has been focused on the evaluation of the quality and authenticity of distilled beverages in terms of their maturation process.

Mezcal is a Mexican distilled alcoholic beverage produced from agave plants from certain regions in Mexico (NOM-070-SCFI-1994), holding origin denomination. As many other similar matured distilled beverages, mezcal can be adulterated in the flavour and appearance (colour), these adulterations aiming to imitate the sensorial and visual characteristics of the authentic matured beverage (Wiley, 1919). Considering that the maturation process in distillate beverages has a strong impact on their taste and price, adulteration of mezcal beverage pursuit obtaining the product in less time. However, the product is of lower quality. In our group, a methodology based in the use of UV-absorption and fluorescence spectroscopy has been proposed for the evaluation of the authenticity of matured distilled beverages, and focused in mezcal. We took advantage of the absorbance/emission properties of woods extracts and molecules added to the distilled during maturation in the wood casks. In this context, principal component analysis method appears as a suitable option to analyse spectral data aiming to elucidate chemical information, thus allowing discrimination of authentic matured beverages from those non-matured or artificially matured.

In this section, we present the PCA results obtained from the investigation of two sets of spectroscopic data (UV absorption and fluorescence spectra), collected from authentic mezcal samples at different stages of the maturation: *white or young* (non-maturated), *rested* (matured ≥ 2 months in wood casks), and *aged* (≥1 year in wood casks). Samples belonging to false matured mezcals (artificially matured) are labelled as: *abocado* (white or young mezcal artificially coloured and flavoured) and *distilled* (coloured white mezcal). These samples were included with the aim of discriminating authentic matured mezcals from those artificially matured. The discussion is focused on the influence of the pre-treatments of spectra on the scores and loadings values. The criteria used for the scores and loadings interpretation are also discussed.

2.1 Spectra pre-treatment

Prior to PCA, spectra were smoothed. Additionally, both spectra data sets were mean centred (MC) prior the analysis as a default procedure. In order to evaluate the effect of the standardization pre-treatment (1/Std) over the scores and loadings values, PCA was also conducted over the standardized spectra. Multivariate spectra analysis and data pre-treatment were carried out using The Unscrambler ® software version 9.8 from CAMO company.

2.2 Collection of UV absorption spectra

Spectra were collected in the 285-450 nm spectral range, using an UV/Vis spectrometer model USB4000 from the Ocean Optics company, coupled to the Deuterium tungsten halogen light source and cuvette holder by means of optical fibers, and with a spectral resolution of ~1.5 nm. The mezcal samples were deposited in disposable 3.0 mL cuvettes, specially designed for UV/Vis spectroscopy under a transmission configuration, which remained constant for all measurements.

2.2.1 PCA-scores

Fig. 1 (a) and (b), depict the distribution of objects (samples/spectra) corresponding to the two pre-treatment options (MC and 1/Std) in the PC-space. In both graphs, a similar distribution of objects along PC1-axis was observed. The groupings along PC1 indicate a good discrimination between matured and non-matured mezcals. Additionally, samples corresponding to mezcals *a priori* known as artificially matured (*i.e.* abocado and distilled samples) and other few, labeled as rested but presumably artificially matured, cluster together with the non-maturated ones. This indicates that the UV absorbance properties of compounds and molecules naturally generated in the wood cask, are significantly different from those from other compounds used with counterfeit purposes (Boscolo et al, 2002).

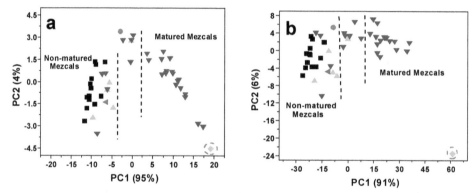

Fig. 1. PCA-Scores plots obtained from raw UV absorption spectra (a) mean centred, (b) standardized. (■) White/young, (●) White w/worm, (▲) abocado or artificially matured, (◄) distilled (white/young coloured), (▼) rested and (●) aged.

A central region, delimited with dashed lines and mainly including samples corresponding to rested mezcals and a few artificially mature samples (abocado and white mezcal w/worm) can be considered as an "indecisive zone". However, taking into account that some samples analysed in this study were directly purchased from liquor stores, it may be possible that few of them, claimed as authentic rested, have been artificially matured. In addition, the sample corresponding to aged mezcal is separated from all the other samples in both graphs, but always clustering together with the rested samples. This indicates that the cluster of objects/samples is related not only with their maturation stage, but also with their maturation time. This behaviour points out that the standardization pre-treatment does not affect significantly the distribution of objects in the scores plots. However, there are some issues that must be considered: in Fig. 1 (a), the aged sample is located close to the rested group, but non as part of it. This can be explained in terms of their different times of maturation. On the other hand, in Fig. 1 (b), the aged sample seems to be an outlier or a sample non-related with the other objects. This unexpected observation can be explained considering the similarity between the spectra of the rested and aged mezcals [see Fig. 2 (a)]. For this reason, the PCA-scores plot corresponding to standardized spectra, must be considered cautiously since they can lead to incorrect interpretations.

2.2.2 PCA-loadings

Once the distribution of objects in the PC-space has been interpreted, the analysis of the one-dimensional loadings plots has been carried out in order to find the relationship between the original variables (wavelength) and the scores plots (Esbensen, 2005; Geladi & Kowalski, 1986; Martens & Naes, 1989). In this case, PC1 is the component discriminating mezcal samples according to their maturation stage and time. Consequently, the PC1-loadings provide information about the spectral variables contributing to that discrimination. Fig.2 (a) shows four representative absorption spectra in the 290-450 nm ultraviolet range for white, abocado, rested and aged mezcals. According to the Figure, the absorption spectra of white and abocado samples look similar, and different from those corresponding to the rested and aged mezcals.

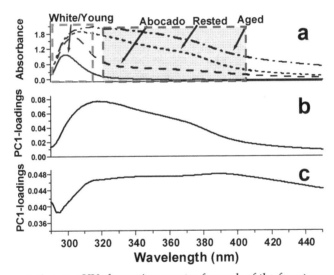

Fig. 2. (a) Representative raw UV absorption spectra for each of the four types of mezcals, (b) PC1-loadings plot for the centred spectra, and (c) PC1-loadings plot for the centred and standardized spectra.

The loading plots indicate that the 320-400 nm region [blue dashed rectangle, Fig. 2 (a)], is the best region to evaluate the authenticity of matured mezcals because the wood compounds extracted, produced and added to mezcals during the aging process absorb in this region. The 290-320 nm range [red dashed rectangle, Fig. 2 (a)], provides the signature for non-maturated mezcals. Fig. 2 (b) and (c) depict one-dimensional PC1 loadings plots corresponding to mean centred and standardized spectra, respectively. From Fig. 2 (b), it is feasible to observe the great similarity between the one-dimensional PC1 loadings plot and the representative spectrum of rested and aged mezcals, suggesting that PC1 mainly models the spectral features belonging to authentic matured mezcals. On the other hand, one-dimensional loadings plot obtained from standardized spectra [Fig. 2 (c)], lacks in the spectral information provided, thus limiting its uses for interpretation purposes. In spite of that, standardization may be useful for certain applications (*i.e.* calibration of prediction/classification by PLS or PLS-DA) (Esbensen, 2005).

2.3 Collection of fluorescence spectra

Taking into account that the emission spectra of organic compounds can provide information about them and about their concentration in mixed liquids, this spectroscopic technique appears as a complementary tool allowing the evaluation of the authenticity of matured alcoholic beverages (Gaigalas et al., 2001; Martínez et al, 2007; Navas & Jimenez, 1999; Walker, 1987). Fluorescence spectra were collected in the 540-800 nm spectral range, using a spectrofluorometer model USB4000-FL from the Ocean Optics company, coupled to a laser of 514 nm wavelength and cuvette holder by optical fibers. The spectral resolution was ~10 nm. The mezcal samples were put into 3.0 mL quartz cuvettes, in a 90 degrees configuration between the excitation source and the detector. This orientation remained constant during the collection of all the spectra. The laser power on the samples was 45 mW.

2.3.1 PCA-scores

Fig. 3 (a) and (b) depict the scores plots obtained from the mean centred spectra. According to Fig. 3 (a), PC1 explains 90 % of the variance. Two groups can be observed along PC1-axis, one of them including white mezcals and ethanol, and the other one, including rested, abocado and distilled mezcals. This indicates that data structure is mainly influenced by the presence or absence of certain organic molecules (not necessarily extracted from wood), all of them having similar emission features.

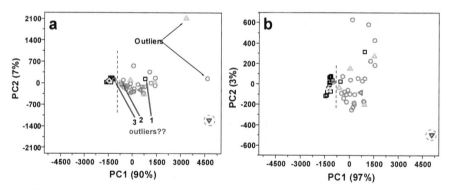

Fig. 3. PCA-scores plots obtained for the mean centred fluorescence spectra. Samples correspond to different stages of maturation. (a) Scores plot before the removal of outliers, (b) scores plot after the removal of outliers. (□) White or young, (△) abocado, (○) rested, (▽) aged, (◁) distilled and (◇) ethanol.

Three isolated objects, corresponding to rested, abocado and aged, can also be observed along PC1-axis. Among them, the first two can be considered as outliers. On the contrary, in the case of the aged sample, the higher concentration of wood extracts in comparison to the rested samples originates a noticeably different spectrum, thus explaining the observation of this aged sample as an isolated object. In order to improve the distribution of objects, samples detected as outliers were removed. Fig. 3 (b) shows the scores plot after outliers removal. Similarly to Fig. 3(a), two main groups can be observed. However, the percentage of explained variance for PC1 increases to 97%. This fact indicates that, even when the object distribution does not depict significant variations, the removal of these outliers, allows the model to describe the data structure in a more efficient way.

There are other samples that can be considered as outliers [indicated with numbers 1-3 in Fig. 3(a)]. Among them, number 1 corresponds to white mezcal with worm (some mezcal producers add a worm to their product as a distinctive), the worm probably providing certain organic molecules. These organic molecules would have similar emission properties than those of the rested ones. Hence, number 1 can be taken as outlier or not. We decide to take it, as correctly classified. On the contrary, objects 2 and 3, were considered as outliers, because they do not have any particular characteristics like object 1. Furthermore, when they were removed from the PCA (data not shown), distribution of objects and explained variance percentages remained similar. For this reason, we decided not to remove them.

In conclusion, only the joint analysis of the scores plot, the raw spectra, the loading plots and all other available information of the samples can give a correct interpretation of our results.

Figure 4 (a) and (b) shows the PCA-scores plots obtained from standardized spectra before and after the removal of outliers. As it was described before, the similar object distribution observed in Fig. 3 and 4 indicates that standardization does not provide any additional benefit for the spectral data analysis.

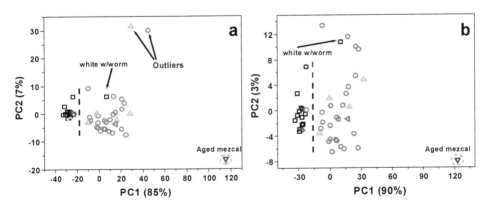

Fig. 4. PCA-scores plots obtained for the mean centred and standardized fluorescence spectra. Samples correspond to different stages of maturation. (a) Scores plot before the removal of outliers, (b) scores plot after the removal of outliers. (□) White or young, (△) abocado, (○) rested, (▽) aged, (◁) distilled and (◇) ethanol.

2.3.2 PCA-loadings

Fig. 5 (a) depicts six representative fluorescence spectra corresponding to each type of mezcal analysed. A high similarity between the fluorescence spectrum of ethanol and white/young mezcal is observed. On the other hand, rested, abocado and distilled mezcals, have similar spectra. Finally, the huge differences between the intensity of the emission spectra corresponding to aged mezcal, and that of the other types of mezcal, can be attributed to the higher concentration of organic molecules coming from the wood cask during the maturation process.

This also explains the grouping along PC1-axis. From these results, it can be concluded that PC1 can discriminate between naturally/artificially matured samples and white mezcal samples. On the contrary, the one-dimensional PC1-loading plot obtained from standardized spectra; does not provide clear information about the objects distribution.

In this sense, and according with the results described above, the standardization pre-treatment does not improve the discrimination between samples in the scores plots. On the contrary, it leads to a misinterpretation of the loading plots.

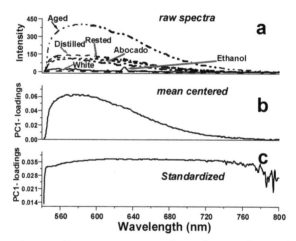

Fig. 5. (a) Representative raw fluorescence spectra for each type of mezcals, (b) PC1-loadings plot for the centred spectra, and (c) PC1-loadings plot for standardized spectra.

2.4 Final remarks

The results described above, showed that PCA conducted over a set of UV absorption spectra from different types of mezcals, allows an efficient and reliable discrimination between artificially and naturally matured mezcals in wood casks, the data pre-treatments playing an important role for the correct interpretation. This discrimination power is based on the differential absorbance spectra of the compounds naturally produced from wood during maturation stage and those corresponding to the compounds used for adulteration (*i.e.:* colorants to confer a matured appearance to the beverage).

On the other hand, PCA conducted over fluorescence spectra allowed the identification of two main groups correlated with the presence or absence of certain organic molecules, not necessarily correlated with maturation in the wood casks. Thus, fluorescence spectroscopy did not demonstrate to have enough sensibility to discriminate between authentic and artificially matured mezcals.

3. PCA of Raman spectra for the determination of heavy metals attached to bacterial surfaces

Toxic metals are not degradable and tend to accumulate in the exposed organisms causing serious health effects. The use of biological agents to remove or neutralize contaminants (biorremediation) is a very important tool for the removal of such toxics. In particular, the use of inactivated microorganisms as adsorbents (biosorption) has been suggested as an effective and economical way to remove heavy metals from water and food intended to human or animal consumption (Davis et al., 2003; Haltunen et al., 2003, 2007, 2008; Ibrahim et al. 2006; Mehta & Gaur, 2005; Mrvčić et al., 2009; Shut et al., 2011; Volesky & Holan, 1995).

Metal biosorption is usually evaluated by means of analytical methods or atomic absorption spectrometry. These methods allow the quantification of free metal ions in the supernatants of bacterial/metal samples (Ernst et al., 2000; Haltunen et al., 2003, 2007, 2008; Velazquez et al., 2009; Zolotov et al., 1987).

In this sense, a method involving vibrational spectroscopic techniques (*i.e.*: Raman spectroscopy) and multivariate methods (both unsupervised and/or supervised), would represent an advantage over the standard procedures, due to the possibility of quantifying the metal ions directly from the bacterial sample, and at the same time, obtaining structural information (Araujo-Andrade et al., 2004, 2005, 2009; Ferraro et al., 2003; Gerbino et al. 2011; Jimenez Sandoval, 2000).

PCA carried out on the Raman spectra represents the first step in the construction of a calibration model allowing the quantification of metal ions attached to bacterial surfaces. This analysis allows obtaining a correlation between the spectral variations and the property of interest (*i.e.* the metal ion concentration), identifying the optimal spectral region/s for the calibration of quantification models, and also detecting erroneous measurements leading to reduce the predictive ability of the model.

In this section, we present the PCA results obtained from the Raman spectra corresponding to bacterial samples (*Lactobacillus kefir*) before and after the interaction with four heavy metals (Cd^{2+}, Pb^{2+}, Zn^{2+}, Ni^{2+}) in three different concentrations each.

Even when the main objective of this study was to calibrate models for the quantification of metal ions attached to bacterial surfaces using supervised methods (*i.e.*: PLS), the calibration of prediction models goes beyond of the intention of this chapter. For this reason, we focused this section just on the discussion of the PCA results.

3.1 Collection of spectral data set

The Raman spectra of the bacterial samples before and after the interaction with metal ions were measured by placing them onto an aluminum substrate and then under a Leica microscope (DMLM) integrated to a Renishaw micro-Raman system model 1000B. In order to retain the most important spectral information from each sample, multiple scans were conducted in different points of the bacterial sample moving the substrate on an X-Y stage.

The Raman system was calibrated with a silicon semiconductor using the Raman peak at 520 cm^{-1}, and further improved using samples of chloroform ($CHCl_3$) and cyclohexane (C_6H_{12}). The wavelength of excitation was 830 nm and the laser beam was focused on the surface of the sample with a 50X objective.

The laser power irradiation over the samples was 45 mW. Each spectrum was registered with an exposure of 30 seconds, two accumulations, and collected in the 1800-200 cm^{-1} region with a spectral resolution of 2 cm^{-1}.

3.2 Spectral data pre-treatment

Raman spectra analyzed were collected over dry solid bacterial samples before and after the interaction with different concentrations of metal ions. Therefore, it is highly probable that our measurements include some light scattering effects (background scattering). These effects are in general composed of multiplicative and additive effects (Martens & Naes, 1989).

Spectra collected and analyzed in this section were baseline corrected in order to subtract the fluorescence contribution. To perform this correction, a polynomial function was approximated to the spectrum baseline, and after that, subtracted from the spectrum. Also, the spectra were smoothed using Savitzky-Golay method. Light scattering effects were corrected using the multiplicative scatter correction (MSC) algorithm and then, the spectra were mean centred. Data pre-treatment and multivariate spectra analysis were carried out with Origin version 6.0 from Microcal Company, and The Unscrambler® software version 9.8 from CAMO company.

3.3 Analysis and discussion of PCA results

PCA was performed on the pre-treated Raman spectra of each bacteria/metal sample in order to correlate metal concentrations with the spectral information.

The criteria used for PCA-scores and loadings interpretation are depicted in the next subsection. Even when the presented data set corresponds to the bacteria/Cd^{+2} interaction, the same methodology was employed for the analysis of the other bacteria/metal samples.

3.3.1 Scores interpretation

Fig. 6 depicts the PCA-scores plots obtained before and after outliers exclusion (panels a and b, respectively). Three main groups, labeled as I, II and III, can be observed in Fig. 6 (a). These groups can be represented by their PC-coordinates as follows: $(+i, +j)$, $(-i, -j)$ and $(-i, +j)$, where i and j represent the i-esime and j-esime score value for PC1 and PC2, respectively.

These three groups or clusters are highly related with the three different concentrations of cadmium trapped on the bacterial surfaces. However, there are several potential outliers that should be removed to get a better description of the data structure. In this case, the outliers were selected on the basis of the dispersion existent among objects of the same group along PC1, the component explaining the major percentage of variance of the data set.

After the removal of outliers, the three groups identified in Fig. 6 (a), became much better defined in the PC-space. However, a different cluster distribution in the PC-space was observed [Fig. 6 (b)]. Table 1 describes these changes in terms of their PC- coordinates before and after the removal of outliers.

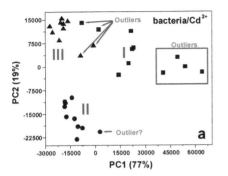 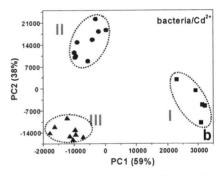

Fig. 6. PCA-Scores plots obtained from pre-treated Raman spectra corresponding to three concentrations of bacteria/Cd^{+2} samples: (■) 0.059 mM, (●) 0.133 mM, (▲) 0.172 mM. (a) With outliers, (b) without outliers.

Additionally, a different distribution of the individual percentage of explained variances was observed in both PCs (PC1, 77% to 59%, and PC2 19% to 38%). However, the total percentage of explained variances before and after the removal of outliers was similar (96%before and 97% after the removal of outliers). This indicates that the removal of outliers did not reduce the information about the data structure provided by both PCs.

	$(PC1_i, PC2_j)$ coordinates before outlier removal	$(PC1_i, PC2_j)$ coordinates after outlier removal
Cluster I	$(+i, +j)$	$(+i, -j)$
Cluster II	$(-i, -j)$	$(-i, +j)$
Cluster III	$(-i, +j)$	$(-i, -j)$

Table 1. Cluster coordinates in the PC-space before and after the removal of outliers.

According to Fig 6 (b), a good discrimination between the lowest (group I) and the medium/highest cadmium concentrations (groups II and III) was observed along PC1-axis.

In summary, it can be concluded that PC1 allows a gross discrimination (due to the huge difference in the concentration of samples clustered in I and samples clusters in II and III). Lower differences in Cd^{+2} concentrations are modelled by PC2 (clusters II and III are well separated along this PC).

3.3.2 Loadings interpretation

Once that distribution of objects in the scores plot was interpreted and correlated with the cadmium concentration attached to the bacterial biomass, the one-dimensional loadings before and after the removal of outliers were analysed to correlate Cd^{+2} concentrations with changes in the original spectra (*i.e.* Raman shift, wavenumber, wavelength, etc.).

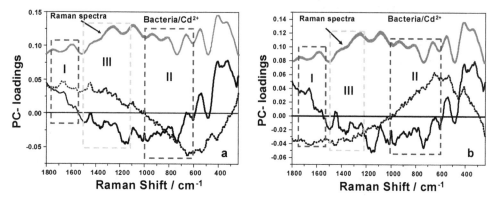

Fig. 7. One-dimensional loadings plots obtained from the PCA corresponding to different bacteria/Cd^{+2} concentrations; (a) before and (b) after the removal of outliers. Solid line corresponds to PC1-loadings and dashed line to PC2-loadings.

Considering that PCs can be represented as a linear combination of the original unit vectors, where the loadings are the coefficients in these linear combinations, distribution and/or localization of each object in the PC-space has a direct relation with their respective PC-loadings values (Esbensen, 2005).

Fig. 7 (a) depicts the one-dimensional loadings plots corresponding to PC1 and PC2 before the removal of outliers. The influent spectral regions for the distribution of objects in the PC-space were underlined using dashed frames. The main spectral differences between objects of cluster I and objects of clusters II and III were found in the 1800-1500 cm⁻¹ spectral region. This region was selected taking into account the loadings values and the PC-coordinates for each cluster.

For instance, cluster I has PC-coordinates (+i, +j), then we selected the region where both loadings, PC1 and PC2 have positive values (in this case, the 1800-1500 cm⁻¹ region). For cluster II, whose coordinates are -i, -j, we selected the region with negative loadings values for both PCs (1000-600 cm⁻¹). Finally, for cluster III, we selected the region with negative and positive loadings values for PC1 and the PC2, respectively (1500-1100 cm⁻¹) whose coordinates are -i, +j. The same strategy was adopted for the loading analysis after the removal of outliers [Figure 7 (b)]. Even when the loading values are different, the spectral regions representing each cluster are the same.

In summary, it can be concluded that the main spectral differences between objects of cluster I and objects of clusters II and III, can be observed in the 1800-1500 cm⁻¹ region. Very interestingly, the carboxylate (COO⁻) groups absorb in this region. In our previous work we have reported that metal ions can be attached to the bacterial surface through the COO⁻ groups (Gerbino et al., 2011). Spectral differences between clusters II and III were found in the 1000-600 cm⁻¹ and 1500-1100 cm⁻¹ regions, respectively.

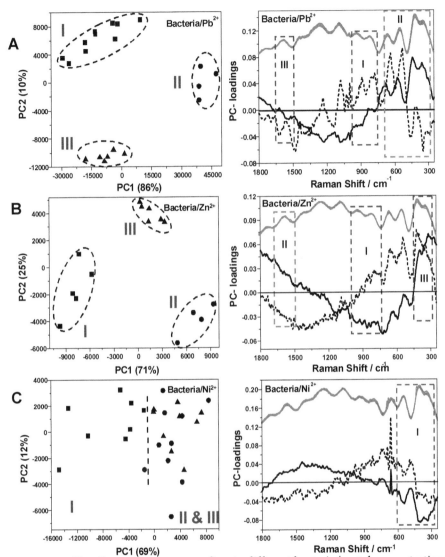

Fig. 8. Scores and loadings plots corresponding to different bacteria/metal concentrations. a) lead: (■) 0.028 mM, (●) 0.181 mM, (▲) 0.217 mM, b) zinc: (■) 0.114 mM, (●) 0.307 mM, (▲) 0.350 mM and c) nickel: (■) 0.022 mM, (●) 0.109 mM, (▲) 0.181 mM. Solid line corresponds to PC1-loadings and dashed line to PC2-loadings.

These two regions provide vibrational information about the phosphate groups and superficial polysaccharides related with certain bacteria superficial structures that have been previously reported as responsible for the bacteria/metal interaction (Mobili et al., 2010; Sara & Sleytr, 2000).

3.3.3 PCA results for other metal ions

The same approach performed for bacteria/Cd^{+2} interaction was employed to analyse the interaction of the other metal ions with the bacterial surface. The scores and loadings plots obtained after the removal of outliers from PCA, carried out on the spectral data sets corresponding to samples of different concentrations of the three metal ions attached to the bacterial surfaces are shown in Fig. 8. A clear discrimination between objects corresponding to different concentrations of lead and zinc trapped on the bacterial surface was observed along the PC-space [Figure 8(a) & (b)]. The information provided by the loadings explains the distribution of samples in clusters surrounded by ellipses and labeled as I, II and III.

The dashed rectangles depicted in the loadings plots indicate the spectral regions that are statistically influent in the distribution of objects. For lead, region labeled as I corresponds to the superficial polysaccharides (1000-800 cm^{-1} region), region labeled as II, to the fingerprint region (700-250 cm^{-1}), and region labeled as III, to the amide I region (1650-1500 cm^{-1}). For zinc, region I corresponds to superficial polysaccharides (1000-750 cm^{-1}), region II, to amide I (1700-1500 cm^{-1}), and region III, to the fingerprint region (450-250 cm^{-1}).

It is important to point out that in both bacteria/Pb^{+2} and bacteria/Zn^{+2} interactions, the same spectral regions were identified as influent for the objects distribution observed in PCA. This indicates that similar molecular structures may be involved in the bacteria/metal interaction. The scores plot corresponding to the bacteria/Ni^{+2} samples indicate a poor discrimination among objects [Fig. 8 (c)]. According to the loadings plots of the bacteria/Ni^{+2} samples, the spectral similarities and differences between samples are mainly in the fingerprint region (region surrounded by a blue dashed line).

3.4 Final remarks

From the results discussed and described above, it can be concluded that Raman spectra allow obtaining chemical information related with bacteria/metal interactions, and also with metal ion concentrations. PCA allowed an efficient elucidation of the information obtained from the spectra. Furthermore, the one-dimensional loadings analysis allowed identifying the influent spectral regions, as well as the molecular structures involved in the objects/samples distribution in the scores plots.

4. PCA applied to the interpretation of quantum chemical calculations

Quantum chemical calculations in large flexible molecules represent a challenge, due to the high number of combinations of conformationally relevant parameters and computational resources/capabilities required. This is an extremely complicated task, unless a systematic approach is carried out. In this section, a PCA-based methodology allowing the correlation between molecular structures and properties of a conformationally flexible molecule (arbutin) is described. This procedure is simple and requires relatively modest computational facilities (Araujo-Andrade et al. 2010).

Arbutin is an abundant solute in the leaves of many freezing- or desiccation-tolerant plants. It has been used pharmaceutically in humans for centuries, either as plant extracts or, in more recent decades, in the purified form. Arbutin acts as an antiseptic or antibacterial agent on the urinary mucous membranes while converting into hydroquinone in the kidney

(Witting et al., 2001). It is also used as a depigmenting agent (skin whitening agent) as it inhibits melanin synthesis by inhibition of tyrosinase activity.

From a chemical point of view, arbutin is a flexible molecule composed by a glucopyranoside moiety bound to a phenol ring (Fig. 9). It has eight conformationally relevant dihedral angles, five of them related with the orientation of the hydroxyl groups and the remaining three taking part in the skeletal of the molecule.

Up to our knowledge, no attempts to use of a PCA based methodology for the structural analysis of quantum chemical information were reported.

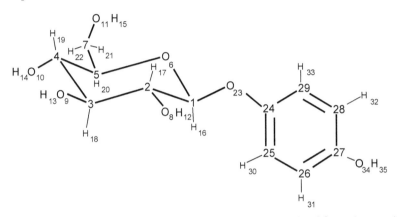

Fig. 9. Arbutin molecule, with atom numbering scheme. (copyrighted from Araujo-Andrade et al., 2010)

4.1 Quantum chemical calculations

The semi-empirical PM3 method (Stewart, 1989) was used to perform a systematic preliminary conformational search on the arbutin potential energies surface (PES), which were later on taken into account in the subsequent, more reliable analysis performed at higher level of theory. This preliminary conformational search was carried out using the HyperChem Conformational Search module (Howard & Kollman, 1988; HyperChem, Inc. © 2002; Saunders, 1987, 1990).

The eight dihedral angles defining the conformational isomers of arbutin (Fig. 9) were considered in the random search: $C_2C_1O_{23}C_{24}$, $C_1O_{23}C_{24}C_{25}$, $O_6C_5C_7O_{11}$, $C_5C_7O_{11}H_{15}$, $C_3C_4O_{10}H_{14}$, $C_2C_3O_9H_{13}$, $C_1C_2O_8H_{12}$ and $C_{26}C_{27}O_{34}H_{35}$. Conformations with energies lower than 50 kJ mol[-1] were stored while higher-energy conformations or duplicate structures were discarded. The structures obtained from this conformational search were used as start points for the construction of the input files later used in the higher level quantum chemical calculations. These latter were performed with Gaussian 03 (Gaussian, 2003) at the DFT level of theory, using the 6-311++G(d,p) basis set (Frisch et al, 1990) and the three-parameter density hybrid functional abbreviated as B3LYP, which includes Becke's gradient exchange correction (Becke, 1988) and the Lee, Yang and Parr (Lee et al, 1988) and Vosko, Wilk and Nusair correlation functionals (Vosko et al., 1980). Conformations were optimized using the Geometry Direct Inversion of the Invariant Subspace (GDIIS) method (Csaszar & Pulay,

1984). The optimized structures of all conformers were confirmed to correspond to true minimum energy conformations on the PES by inspection of the corresponding Hessian matrix. Vibrational frequencies were calculated at the same level of theory. PCA were performed using The Unscrambler® software (v9.8).

4.2 Theoretical data set and pre-treatment

The group of Cartesian coordinates corresponding to the 35 atoms of arbutin (see Fig. 9), for each of the 130 conformers found after the conformational analysis was used as data set in this study. In other words, our data set consisted of a matrix of 130 x 105 elements, corresponding to the arbutin conformers and the x, y, z coordinates of each atom of the molecule, respectively. Data were mean centred prior PCA.

4.3 Data analysis

In order to provide a general and fast procedure to perform the PCA on the conformational data sets, the next strategy was followed: 1) In the Cartesian referential, all conformers were oriented, in such a way that the structurally rigid fragment of arbutin (the glucopyranoside ring) was placed as close as possible to the axes origin; 2) All Cartesian coordinates of the 130 conformers of arbutin were then used to perform the PCA. The table of data (data matrix) was built as follows: each row corresponds to a conformer and the columns to the Cartesian coordinates: the first 35 columns, to the x- coordinates of the 35 atoms of arbutin, the second 35 columns, to the y coordinates, and the last 35 columns, to the z coordinates.

4.3.1 Scores and loadings analysis

Fig. 10 depicts the distribution of the arbutin conformers, in the PC-space. Three well separated groups were observed. Group A (squares), appears well separated from groups B (circles) and C (triangles) along the PC1-axis, which explain 77% of total variance. Group B is separated from Group C along the PC2-axis. In order to elucidate the main structural parameters for this separation, the one-dimensional loadings values were analyzed. The loading values of PC1 and PC2, indicate the atoms' coordinates contributing the most to structurally distinguish the conformers of arbutin in the chosen reference system [Fig. 10 (b) and (c), respectively].

According to these values, the orientation of the atoms 25 to 35, related with the spatial orientation of the phenol ring relatively to the reference glucopyranoside fragment, is the main contributing factor allowing for the differentiation among conformers. Consequently, the relative spatial orientation of the phenol ring is determined by the dihedrals interconnecting the glucopyranoside and phenol rings, $C_2C_1O_{23}C_{24}$ and $C_1O_{23}C_{24}C_{25}$, which are then shown to be of first importance in structural terms. Fig. 10 (d) shows the means and standard errors of the means (standard deviation of the sample divided by the square root of the sample size) of the 8 conformationally relevant dihedral angles of arbutin (the angles were first converted to the 0-360° range). From this graph, it can be clearly observed that $C_2C_1O_{23}C_{24}$ and $C_1O_{23}C_{24}C_{25}$ dihedral angles, describe the distribution of the three groups along the PC1 and PC2 axis, respectively. In other words, PC1 is related with the $C_2C_1O_{23}C_{24}$ dihedral angle, and allows the discrimination of conformers belonging to group A. In these conformers, the phenol ring is placed above and nearly perpendicular to the glucopyranoside ring. On the contrary, in all conformers belonging to groups B and C, the phenol ring is pointing out of the glucopyranoside moiety, and oriented to the side of the oxygen atom from the glucopyranoside ring. PC2 allows a specific discrimination among the three groups of conformers. This specificity factor is given by the values of $C_1O_{23}C_{24}C_{25}$.

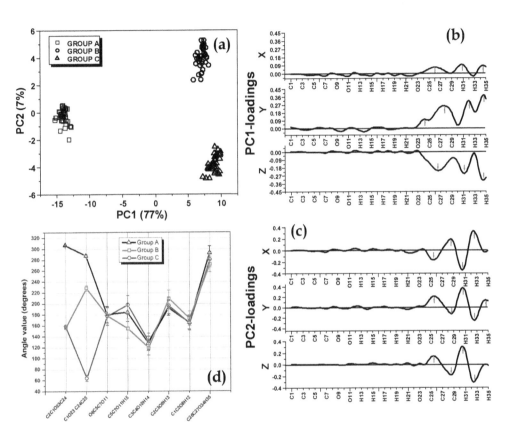

Fig. 10. (a) PCA-scores and, (b, c) the corresponding loadings grouping arbutin conformers in terms of structural similarity. (d) total average values and standard deviations of the 8 conformationally relevant dihedral angles of arbutin in the 3 groups of conformers. (copyrighted from Araujo-Andrade et al., 2010)

The relationship between the energetic and conformational parameters related with each of the three groups identified in the scores plot, was also investigated. Fig. 11 depicts the relative energy values (taken as reference the energy of the conformational ground state) for each conformer according to the group they belong. From an energetic point of view, groups B and C are equivalent. However, no conformer with relative energy below 15 kJ mol^{-1} belonging to Group A. This trend can be correlated with the orientations adopted by the phenol ring relatively to the glucopyranoside ring, as was described before.

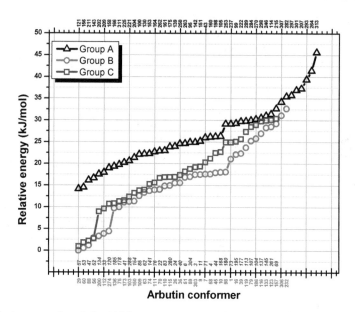

Fig. 11. Relative energies of the 130 lowest energy conformers of arbutin (the energy of the conformational ground state was taken as reference). (copyrighted from Araujo-Andrade et al., 2010)

Once the influence of the relative position of the two rings in arbutin on the relative energy of the conformers was evaluated, the preferred conformations assumed by the substituents of the glucopyranoside ring and their influence on energies were investigated in deeper detail. To this aim, PCA was conducted on each of the previously determined groups of conformers (A, B, C), excluding the x, y, z, coordinates corresponding to the phenol ring (atoms 23-35). This strategy allowed for the elimination of information that is not relevant for a conformational analysis within the glucopyranoside ring. PCA-scores/loadings analysis and interpretation was realized by using the methodology described above for the whole arbutin molecule. The results of this analysis are shown in Fig. 12-15. The PCA scores plot for Group A [Fig. 12 (a)] shows four well defined groups, labeled as subgroups A1, A2.a, A2.b and A3. If all elements of these four subgroups are projected over the PC1 axis, three groups can be distinguished, with one of them constituted by subgroups A1 and A3, other formed by subgroup A2.a and the third one by subgroup A2.b. On the other hand, projecting the elements over the PC2-axis allows also to distinguish three groups, but this time corresponding to A1, (A2.a, A2.b) and A3.

The observation of the one-dimensional loadings plots for PC1 and PC2 [Fig. 12 (b) & (c)], allows us to conclude that the positions of atoms C_7, H_{21}, H_{22}, O_{11} and H_{15} are highly related with the conformers distribution in the PCA scores plot, $i.e.$, the conformation exhibited by the CH_2OH substituent at C_5 is the main discriminating factor among subgroups. In consonance with this observation, when the mean values of the conformationally relevant dihedral angles associated with the substituted glucopyranoside ring in each subgroup are plotted [Fig. 12 (d)], it is possible to observe that the dihedral angles associated with the

CH$_2$OH substituent (O$_6$-C$_5$-C$_7$-O$_{11}$ and C$_5$-C$_7$-O$_{11}$-H$_{15}$) are those allowing the discrimination among the four subgroups. In the case of the C$_5$-C$_7$-O$_{11}$-H$_{15}$ dihedral, one can promptly correlate the three groups corresponding to the projection of the PCA subgroups over the PC1-axis with the three dihedral mean values shown in the plot: ca. 60, 160 and 275° (-85°), respectively for A2.b, A2.a and (A1, A3).

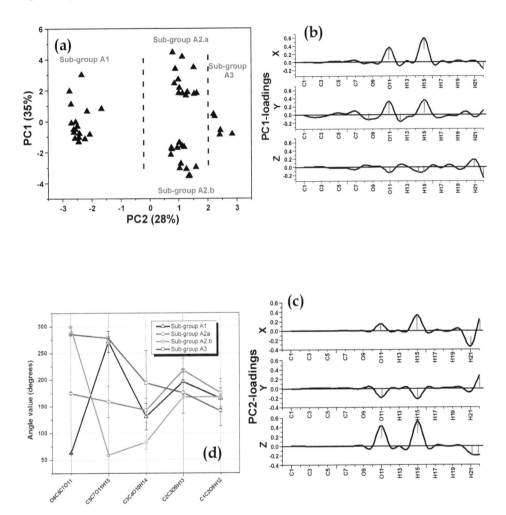

Fig. 12. (a) PCA-scores and, (b,c) the corresponding loadings and belonging to Group A in terms of structural similarity in the conformations of the substituents of the glucopyranoside ring. (d) total average values and standard errors of the means of the 5 conformationally relevant dihedral angles of the glucopyranoside ring arbutin in the 4 subgroups. (copyrighted from Araujo-Andrade et al., 2010)

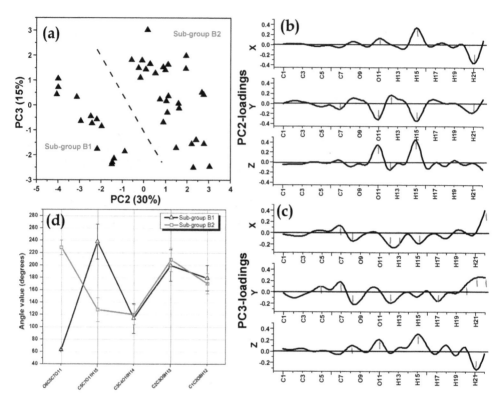

Fig. 13. (a) PCA-scores and, (b, c) the corresponding loadings belonging to Group B in terms of structural similarity in the conformations of the substituents of the glucopyranoside ring. (d) total average values and standard errors of the means of the 5 conformationally relevant dihedral angles of the glucopyranoside ring arbutin in the 2 subgroups of conformers. (copyrighted from Araujo-Andrade et al., 2010)

The PCA-scores plot for the conformers belonging to the Group B [Fig. 13 (a)] shows only two clear groupings of conformers, where PC2 is the component separating these two groups the best. The loadings plot of PC2 [Fig. 13 (b)] shows that the clusters are also determined by the positions of atoms C_7, H_{21}, H_{22}, O_{11} and H_{15}, $i.e.$, by the conformation of the CH_2OH fragment. As expected, these observations are in agreement with the dihedral angles' mean values plot [Fig. 13 (d)], which clearly reveals that there, the values of the O_6-C_5-C_7-O_{11} and C_5-C_7-O_{11}-H_{15} dihedral angles are the ones that mainly discriminate internal coordinates among the conformers belonging to subgroups B1or B2.

A similar analysis made for conformers belonging to Group C allows concluding that three subgroups (C1, C2 and C3) can be defined [Fig. 14 (a)], once again resulting mainly from different conformations assumed by the CH_2OH substituent [Fig. 14 (b) & (d)]. Regarding the energies of the conformers, subgroups are not strongly discriminative. However, subgroups A3, C2 and, in less extent B1, include conformers gradually less stable than the remaining subgroups of each main group (data not shown).

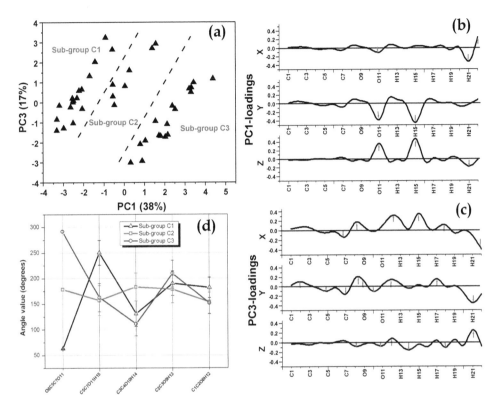

Fig. 14. (a) PCA-scores, and (b, c) the corresponding loadings, belonging to Group C. (d) total average values and standard errors of the means of the 5 relevant dihedral angles of the glucopyranoside ring arbutin in the 3 subgroups. (copyrighted from Araujo-Andrade et al., 2010)

4.4 Final remarks

PCA analyses based on atomic Cartesian coordinates of the properly oriented in the Cartesian system conformers of arbutin allowed the grouping of these conformers by structural analogies, which could be related with the conformationally relevant dihedral angles. Among them, the dihedrals interconnecting the glucopyranoside and phenol rings and those associated with the CH_2OH fragment were found to be the most relevant ones.

In summary, this work represents a new simple approach for the structural analysis of complex molecules and its aim was also to show another application of PCA.

5. Conclusion

The results reported in this chapter for each experimental and theoretical application of PCA, demonstrate the versatility and capabilities of this unsupervised method to analyse samples from different origins. Three different examples were selected to show the relevance of PCA to elucidate specific information from a data collection in several fields. Among these issues, the following aspects must be underlined: a) the influence of the data pre-treatment on the scores and loadings values; b) the *a-priori* knowledge of the data source to select the appropriate data pre-processing; c) the strategies and criteria used for the scores and loadings plots interpretation and, d) criteria used for outliers detection, and their influence in the PCA model.

The amalgamation of the different sections included in this chapter can be used as a starting point for those researchers who are not specialists in the field, but that are interested in using these methodologies, to take the maximum advantage from their results.

6. Acknowledgments

This work was supported by CONACyT, Mexico [Projects No. 119491 (2009) and No. 153066 (2010)] and PROMEP, Mexico (Project UAZ-PTC-092), Agencia Nacional de Promoción Científica y Tecnológica, Argentina (Projects PICT/2008/145 and PICT/2010/2145), CYTED Program (Ciencia y Tecnología para el Desarrollo) Network P108RT0362 and CONACyT-CONICET (México, Argentina) (bilateral project res. N° 962/07-05-2009), PIFI, México (project P/PIFI 2010-32MSU0017H-06). AGZ, PM and EET are members of the research career CONICET (National Research Council, Argentina). EG is doctoral fellow from CONICET.

7. References

Abbott, D. & Andrews, R. S. (1970). *Introduction to Chromatography*, Longman group LTD, ISBN 978-0582321946, London.

Aguilar-Cisneros, B. O.; López, M. G.; Richling, E.; Heckel, F.; Schreier, P. (2002). Tequila authenticity assessment by headspace SPME-HRGC-IRMS analysis of 13C/12C and 18O/16O ratios of ethanol. *J. Agric. Food Chem.*, Vol. 50, No.6, pp.7520-7523.

Araujo-Andrade, C., Ruiz, F., Martinez-Mendoza, J.R., and Terrones, H. (2004). Non-invasive in-vivo blood glucose levels prediction using near infrared spectroscopy. *AIP Conf. Proc.* Vol.724, pp. 234-239.

Araujo-Andrade, C., Campos-Cantón, I., Martínez, J.R., Ortega, G., and Ruiz, F. (2005). Prediction model based on multivariate analysis to determine concentration of sugar in solution. *Rev. Mex. Fis. E,* Vol.51, pp. 67-73.

Araujo-Andrade, C., Ruiz, F., Martínez-Mendoza, J.R., Padrón, F., and Hernández-Sierra, J. (2009). Feasibility for non invasive estimation of glucose concentration in newborns using NIR spectroscopy and PLS. *Trends Appl. Spectrosc.* Vol.7, pp. 27-37.

Araujo Andrade, C.; Lopes, S.; Fausto, R. and Gómez-Zavaglia, A. (2010). Conformational study of arbutin by quantum chemical calculations and multivariate analysis. *Journal of Molecular Structure.* Vol.975, pp. 100-109.

Baldwin, S. R., Black, A., Andreasen, A. A. & Adams, S. L. (1967). Aromatic congener formation in maturation of alcoholic distillates, *J. Agr. Food Chem.*, Vol.15, No.3, pp. 381-385.

Bauer-Christoph, C.; Christoph, N.; Aguilar-Cisneros, B. O.; López, M. G.; Richling, E.; Rossmann, A.; Schreier, P. (2003). Authentication of tequila by gas chromatography and stable isotope ratio analyses. *Eur. Food Res. Technol.*, Vol.217, No.5, pp. 438-443.

Becke, A. (1988). Density-functional exchange-energy approximation with correct asymptotic behavior, *Phys. Rev. A*, Vol.38, pp. 3098-3100.

Boscolo, M.; Andrade-Sobrinho, L. G.; Lima-Neto B. S.; Franco D. W. & Castro Ferreira M. M. (2002). Spectrophotometric Determination of Caramel Content in Spirits Aged in Oak Casks, *Journal of AOAC International* Vol.85, No.3, pp. 744-750.

Crampton, C. A. & Tolman, L. M. (1908). A study of the changes taking place in Whyskey stored in wood, *J. Am. Chem. Soc.* Vol.30, No.1, pp. 98–136.

Csaszar, P. & Pulay, P. (1984). Geometry optimization by direct inversion in the iterative subspace, *J. Mol. Struct. (Theochem)*, Vol.114, pp. 31-34.

Davis, T.A., Volesky, B., & Mucci, A. (2003). A review of the biochemistry of heavy metal biosorption by brown algae. *Water Research* Vol.37, pp. 4311-4330.

Dobrinas, S.; Stanciu, G. & Soceanu, A. (2009). Analytical characterization of three distilled drinks, *Ovidius University Annals Chemistry*, Vol.20, No.1, pp. 48-52.

Ernst, T.; Popp, R. & Van Eldik, R. (2000). Quantification of heavy metals for the recycling of waste plastics from electrotechnical applications, *Talanta* Vol.53, No.2, pp. 347–357.

Esbensen, K. H. (2005). *Multivariate Data Analysis - In Practice*, CAMO Software AS, ISBN 82-993330-3-2, Esbjerg, Denmark.

Ferraro, J. R.; Nakamoto, K. & Brown, C. W. (2003). Introductory Raman Spectroscopy, Academic Press, ISBN 978-0122541056, London.

Frisch,M.; Head-Gordon, M. & Pople, J. (1990). Semidirect Algorithms for the Mp2 Energy and Gradient, *Chem. Phys. Lett.* Vol.166, pp. 281-289.

Gaigalas, A.; K., Li, L.; Henderson, O.; Vogt, R.; Barr, J.; Marti, G.; Weaver, J.; Schwartz, A. (2001). The development of fluorescence intensity standards, *J. Res. Natl. Inst. Stand. Technol.* Vol.106, No.2, pp. 381-389.

Gaussian 03, Revision C.02, M. J. Frisch, G. W. Trucks, H. B. Schlegel, G. E. Scuseria, M. A. Robb, J. R. Cheeseman, J. A. Montgomery, Jr., T. Vreven, K. N. Kudin, J. C. Burant, J. M. Millam, S. S. Iyengar, J. Tomasi, V. Barone, B. Mennucci, M. Cossi, G. Scalmani, N. Rega, G. A. Petersson, H. Nakatsuji, M. Hada, M. Ehara, K. Toyota, R. Fukuda, J. Hasegawa, M. Ishida, T. Nakajima, Y. Honda, O. Kitao, H. Nakai, M. Klene, X. Li, J. E. Knox, H. P. Hratchian, J. B. Cross, V. Bakken, C. Adamo, J. Jaramillo, R. Gomperts, R. E. Stratmann, O. Yazyev, A. J. Austin, R. Cammi, C. Pomelli, J. W. Ochterski, P. Y. Ayala, K. Morokuma, G. A. Voth, P. Salvador, J. J. Dannenberg, V. G. Zakrzewski, S. Dapprich, A. D. Daniels, M. C. Strain, O. Farkas, D. K. Malick, A. D. Rabuck, K. Raghavachari, J. B. Foresman, J. V. Ortiz, Q. Cui, A. G. Baboul, S. Clifford, J. Cioslowski, B. B. Stefanov, G. Liu, A. Liashenko, P. Piskorz, I. Komaromi, R. L. Martin, D. J. Fox, T. Keith, M. A. Al-Laham, C. Y. Peng, A. Nanayakkara, M. Challacombe, P. M. W. Gill, B. Johnson, W. Chen, M. W. Wong, C. Gonzalez, and J. A. Pople, Gaussian, Inc., Wallingford CT, 2004. Copyright C 1994-2003, Gaussian, Inc.

Geladi, P. & Kowalski, B. R. (1986). Partial Least Square Regression: A tutorial, *Anal. Chim. Acta*, Vol.185, pp. 1-17.

Gerbino, E., Mobili, P., Tymczyszyn, E.E., Fausto, R., and Gómez-Zavaglia, A. (2011). FTIR spectroscopy structural analysis of the interaction between Lactobacillus kefir S-layers and metal ions. *Journal of Molecular Structure*. Vol.987, pp. 186-192.

Halttunen, T., Kankaanpää, P., Tahvonen, R., Salminen, S. & Ouwehand, A.C. (2003). Cadmium removal by lactic acid bacteria. *Bioscience Microflora* Vol. 22, pp.93-97.

Halttunen, T., Salminen, S. & Tahvonen, R. (2007). Rapid removal of lead and cadmium from water by specific lactic acid bacteria. *International Journal of Food Microbiology* Vol.114, pp. 30-35.

Halttunen, T., Salminen, S., Meriluoto, J., Tahvonen, R., & Lertola, K. (2008). Reversible surface binding of cadmium and lead by lactic acid & bifidobacteria. *International Journal of Food Microbiology* Vol.125, pp.170-175.

Hincha, D.; Oliver, A. E. & Crowe, J. H. (1999). Lipid composition determines the effects of arbutin on the stability of membranes, *Biophys. J.* Vol.77, No.4. pp.2024-2034.

Howard, A. E.; & Kollman, P. A. (1988). An analysis of current methodologies for conformational searching of complex molecules, *J. Med. Chem.* Vol.31, No.9, pp.1669-1675.

HyperChem Conformational Search module (2002). Tools for Molecular Modeling. Hypercube, Inc., 1115 NW 4th St., Gainesville, FL 32608 (USA)

Ibrahim, F., Halttunen, T., Tahvonen, R., & Salminen, S. (2006). Probiotic bacteria as potential detoxification tools: assessing their heavymetal binding isotherms. *Canadian Journal of Microbiology* Vol.52, pp. 877-885.

Jimenez Sandoval, S. (2000). Micro-Raman spectroscopy: a powerful technique for materials research, *Microelectronic Journal*, Vol.31, pp. 419-427.

Lee, C.; Yang, W. & Parr, R. (1988). Development of the Colle-Salvetti correlation-energy formula into a functional of the electron density, *Phys. Rev. B.* Vol.37, pp. 785-789.

Liebmann, A. J. & Scherl, B. (1949). Changes in whisky while maturing, *Industrial and engineering chemistry*, Vol.41, No.3, pp. 534-543.

Martens, H. & Næs, T. (1989). *Multivariate Calibration*, Wiley & Sons, ISBN 0-471-90979-3, Chichester, England.

Martínez, J. R.; Campos-Cantón, I.; Martínez-Castañón, G.; Araujo-Andrade, C. & Ruiz, F. (2007). Feasibility of laser induced fluorescence as a rapid method for determination of the time stored of aged alcoholic beverages, *Trends in applied spectroscopy*, Vol.6, pp. 28-33.

Martínez, J. R.; Campos-Cantón, I.; Araujo-Andrade, C.; Martínez-Castañón, G. & Ruiz, F. (2007). Analysis of Mexican spirit drinks mezcal using near infrared spectroscopy, *Trends in Applied Spectroscopy*, Vol.6, pp. 35-41.

Mehta, S.K. & Gaur, J.P. (2005). Use of algae for removing heavy metal ions from wastewater: progress and prospects. *Critical Reviews in Biotechnology* Vol.25, pp.113-152.

Mobili, P.; Londero, A.; De Antoni, G.; Gomez-Zavaglia, A.; Araujo-Andrade, C.; Ávila-Donoso, H.; Ivanov-Tzonchev, R.; Moreno-Hernandez, I. & Frausto Reyes, C. (2010). Multivariate analysis of Raman spectra applied to microbiology: discrimination of microorganisms at the species level, *Revista Mexicana de Física*, Vol.56, No.5, pp. 378–385.

Mobili, P.; Araujo-Andrade, C.; Londero, A.; Frausto-Reyes, C.; Ivanov-Tzonchev, R.; De Antoni, G.L.; & Gómez-Zavaglia, A. (2011). Development of a method based on chemometric analysis of Raman spectra for the discrimination of heterofermentative lactobacilli. *Journal of Dairy Research* Vol.78, pp. 233-241.

Mrvčić, J., Prebeg, T., Barišić, L., Stanzer, D., Bačun-Družina, V. & Stehlik-Tomas, V. (2009). Zinc binding by lactic acid bacteria. *Food Technology Biotechnology* Vol.47, pp. 381-388.

Nagarajan, R.; Gupta, A.; Mehrotra, R.; & Bajaj, M. M. (2006). Quantitative Analysis of Alcohol, Sugar, and Tartaric Acid in Alcoholic Beverages Using Attenuated Total Reflectance Spectroscopy, Journal of Automated Methods and Management in Chemistry, *J Autom Methods Manag Chem.* Vol.2006, No. pp. 1–5.

Navas, M. J. & Jiménez, A. M. (1999). Chemoluminescent methods in alcoholic beverage analysis, *J. Agric. Food Chem.* Vol.47, No.1, pp. 183-189.

Nishimura, K. & Matsuyama R. (1989). Maturation and maturation chemistry, In: *the science and technology of Whiskies*, J.R. Piggott, R. Sharp, E. E. B. Duncan, (Ed.), 244-253, Longman Scientific & Technical, ISBN 978-0582044289, Essex, U.K.

Norma Oficial Mexicana NOM-070-SCFI-1994, Bebidas alcohólicas-Mezcal-Especificaciones.

Nose, A., Hojo, M., Suzuki, M. & Ueda, T. J. (2004). Solute effects on the interaction between water and etanol in aged whiskey, Agric. Food Chem. Vol.(52), pp. 5359-5365.

Pekka, J. L.; LaDena A. K. & Eero T. A-M. (1999). Multi-method analysis of matured distilled alcoholic beverages for brand identification, *Z Lebensm Unters Forsch*, Vol. 208, pp. 413-417.

Philp, J. M. (1989). Cask quality and warehouse condition, In: *The Science and Technology of Whiskies*; Piggott, J.R., Sharp, R., Duncan, R.E.B., (Ed.), 270-275, Longman Scientific & Technical, ISBN 978-0582044289, Essex, U.K.

Puech, J.L. (1981). Extraction and evolution of lignin products in Armagnac matured in oak, *Am. J. Enol. Vitic.* Vol.32, No.2, pp. 111-114.

Ragazzo, J. A.; Chalier, P.; Crouzet, J.; Ghommidh, C. (2001). Identification of alcoholic beverages by coupling gas chromatography and electronic nose. *Spec. Publ.-R. Soc. Chem.: Food FlaVors Chem.*, Vol. 274, pp. 404-411.

Reazin, G. H. (1981). Chemical mechanisms of whiskey maturation, *Am. J. Enol. Vitic.* Vol.32, No.4, pp. 283-289.

Rodriguez Madera, R., Blanco Gomis, D. & Mangas Alonso, J. J. (2003). Influence of distillation system, oak wood type, and aging time on composition of cider brandy in phenolic and furanic compounds, *J. Agr. Food. Chem.* Vol.51, No.27, pp. 7969-7973.

Sara, M. & Sleytr, U.B. (2000). S-layer proteins. *Journal of Bacteriology* Vol.182, pp.859-868.

Saunders, M. (1987). Stochastic Exploration of Molecular Mechanics Energy Surfaces. Hunting for the Global Minimum, *J. Am. Chem. Soc.* Vol.109, No.10, pp.3150-3152.

Saunders, M.; Houk, K. N.; Wu, Y.-D.; Still, W. C.; Lipton, J. M.; Chang ,G. & Guidal, W. C. (1990). Conformations of cycloheptadecane. A comparison of methods for conformational searching, *J. Am. Chem. Soc.* Vol.112, No.4, pp. 1419-1427.

Savchuk, S. A.; Vlasov, V. N.; Appolonova, S. A.; Arbuzov, V. N.; Vedenin, A. N.; Mezinov, A. B.; Grigor'yan, B. R. (2001). Application of chromatography and spectrometry to the authentication of alcoholic beverages. *J. Anal. Chem.* Vol.56, No.3, pp. 214- 231.

Schut, S., Zauner, S., Hampel, G., König, H., & Claus, H. (2011). Biosorption of copper by wine-relevant lactobacilli. *International Journal of Food Microbiology* Vol.145, pp. 126-131.

Stewart, J. J. P. (1989). Optimization of Parameters for Semi-Empirical Methods I-Method, *J. Comput. Chem.* Vol.10, pp. 209-220.

Turner, S. R.; Senaratna, T.; Touchell, D. H.; Bunn, E.; Dixon, K. W. & Tan, B. (2001b). Stereochemical arrangement of hydroxyl groups in sugar and polyalcohol molecules as an important factor in effective cyropreservation. *Plant Science* Vol.160, pp.489-497.

Valaer, P. & Frazier, W. H. (1936). Changes in whisky stored for four years, *Industrial and Engineering chemistry*, Vol.28, No.1, pp. 92-105.

Vallejo-Cordoba, B.; González-Córdova, A. F.; Estrada- Montoya, M. del C., (2004). Tequila volatile characterization and ethyl ester determination by solid-phase microextraction gas chromatography/ mass spectrometry analysis. *J. Agric. Food Chem.*, Vol.52, No.18, pp. 5567-5571.

Velazquez, L. & Dussan, J. (2009). Biosorption and bioaccumulation of heavy metals on dead and living biomass of Bacillus sphaericus. *Journal of Hazardous Material.* Vol.167, pp.713-716.

Volesky, B. & Holan, Z.R. (1995). Biosorption of heavy metals. *Biotechnology Progress* Vol.11, pp.235-250.

Vosko, S.; Wilk, L. & Nusair, M. (1980). Accurate spin-dependent electron liquid correlation energies for local spin density calculations: a critical analysis, *Can. J. Phys.* Vol.58, No.8, pp. 1200-1211.

Walker, D. A. (1987). A fluorescence technique for measurements of concentration in mixing liquids, *J. Phys. E: Sci. Instrum.* Vol.20, No.2, pp. 217-24.

Wiley, H. W. (1919). *Beverages and their adulteration*, Campbell Press. ISBN 978-1443755740.

Wittig, J.; Wittemer, S. & Veit, M. (2001). Validated method for the determination of hydroquinone in human urine by high-performance liquid chromatography-coulometric-array detection, *J. Chromatogra. B*, Vol.761, No.1, pp. 125-132.

Zolotov, Y. A.; Malofeeva, G. I.; Petrukhin, O. M. & Timerbaev, A. R. (1987). New methods for preconcentration and determination of heavy metals in natural, *waterPure & Appl. Chem.*, Vol.59, No.4, pp. 497-504.

Application of Linear and Nonlinear Dimensionality Reduction Methods

Ramana Vinjamuri[1,4], Wei Wang[1,4], Mingui Sun[2] and Zhi-Hong Mao[3]
[1]Department of Physical Medicine and Rehabilitation
[2]Department of Neurological Surgery
[3]Department of Electrical and Computer Engineering
[4]Center for Neural Basis of Cognition
University of Pittsburgh, Pittsburgh, PA
USA

1. Introduction

Dimensionality reduction methods have proved to be important tools in exploratory analysis as well as confirmatory analysis for data mining in various fields of science and technology. Where ever applications involve reducing to fewer dimensions, feature selection, pattern recognition, clustering, dimensionality reduction methods have been used to overcome the curse of dimensionality. In particular, Principal Component Analysis (PCA) is widely used and accepted linear dimensionality reduction method which has achieved successful results in various biological and industrial applications, while demanding less computational power. On the other hand, several nonlinear dimensionality reduction methods such as kernel PCA (kPCA), Isomap and local linear embedding (LLE) have been developed. It has been observed that nonlinear methods proved to be effective only for specific datasets and failed to generalize over real world data, even at the cost of heavy computational burden to accommodate nonlinearity.

We have systematically investigated the use of linear dimensionality reduction methods in extracting movement primitives or synergies in hand movements in Vinjamuri et al. (2010a;b; 2011). In this chapter, we applied linear (PCA and Multidimensional Scaling (MDS)) and nonlinear (kPCA, Isomap, LLE) dimensionality reduction methods in extracting kinematic synergies in grasping tasks of the human hand. At first, we used PCA and MDS on joint angular velocities of the human hand, to derive synergies. The results obtained indicated ease and effectiveness of using PCA. Then we used nonlinear dimensionality reduction methods for deriving synergies. The synergies extracted from both linear and nonlinear methods were used to reconstruct the joint angular velocities of natural movements and ASL postural movements by using an l_1-minimization algorithm. The results suggest that PCA outperformed all three nonlinear methods in reconstructing the movements.

2. Synergies

The concept of synergies (in Greek *synergos* means working together) was first represented numerically by Bernstein Bernstein (1967). Although synergies were originally defined by Bernstein as high-level control of kinematic parameters, different definitions of synergies exist and the term has been generalized to indicate the shared patterns observed in the behaviors of muscles, joints, forces, actions, etc. Synergies in hand movements especially present a complex optimization problem as to how the central nervous system (CNS) controls the hand with over 25 degrees of freedom(DoF) (Mackenzie & Iberall (1994)). Yet, the CNS handles all the movements effortlessly and at the same time dexterously. Endeavoring to solve the DoF problem, many researchers have proposed several concepts of synergies such as the following:

(i) Postural synergies: In Jerde et al. (2003); Mason et al. (2001); Santello et al. (1998; 2002); Thakur et al. (2008); Todorov & Ghahramani (2004), researchers found that the entire act of grasp can be described by a small number of dominant postures, which were defined as postural synergies.

(ii) Kinematic synergies: Studies in Grinyagin et al. (2005); Vinjamuri et al. (2007) expressed the angular velocities of finger joints as linear combinations of a small number of kinematic synergies, which were also angular velocities of finger joints but were extracted from a large set of natural movements. Kinematic synergies are not limited to hand movements. In (d'Avella et al. (2006)), d'Avella et al. reported that kinematic synergies were found in tracking 7 DoF arm movements.

(iii) Dynamic synergies: Dynamic synergies were defined as stable correlations between joint torques that were found during precision grip movements in Grinyagin et al. (2005).

The above classification was already presented in Vinjamuri et al. (2010b). In addition to synergies proposed in postures, and kinematics which are of relevance to the current study, synergies were also proposed in muscle activities d'Avella et al. (2006).

2.1 What are temporal postural synergies?

The synergies presented in this chapter are purely kinematic synergies. These are the synergies derived from angular velocities of the finger joints of the human hand collected during grasping tasks. For example, in the Fig. 1 two synergies (s^1, s^2) combine using a weighted linear combination ($w_1 s^1 + w_2 s^2$) to achieve a grasping hand movement. w_1 and w_2 represent weights of control signals. Each row of a synergy corresponds to the angular velocity profile of a finger joint; For example, the first synergy represents the synchronous large movement of first joint and medium movement of the second joint followed by a small movement of the third joint. In this example, s^1 (blue) and s^2 (brown) form a weighted ($w_1 = w_2 = 0.5$) combination to result in the aggregate movement(black) on the right hand side. For illustration purposes, only 3 of 10 joints of the hand are shown in the synergies and the reconstructed movement. Also shown in the figure are the hand postures of the reconstructed movement across time. As these synergies preserve both the temporal structure and the postural information these are termed as *temporal postural synergies* (Vinjamuri et al. (2010a;b)).

2.2 Applications of synergies

In our attempt to apply linear and nonlinear dimensionality reduction methods to solve the problem of extraction of synergies, let us first know how these synergies are being used in the real world applications in the areas of prosthesis and rehabilitation.

(i) Prosthetics: Apart from neuro-physiological significance, synergies are viewed to be crucial design elements in future generation prosthetic hands. Biologically inspired synergies have already taken prime place in artificial hands (Popovic & Popovic (2001)). Synergies based

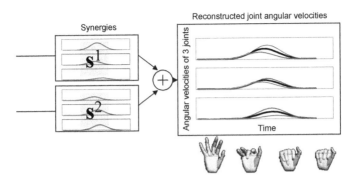

Fig. 1. Two distinct synergies (s^1, s^2) use a weighted linear combination ($w_1 s^1 + w_2 s^2$) to achieve a grasping hand movement. w_1 and w_2 represent weights of control signals. Each row of a synergy corresponds to the angular velocity profile of a finger joint; For example, the first synergy represents the synchronous large movement of first joint and medium movement of the second joint followed by a small movement of the third joint. In this example, s^1 (blue) and s^2 (brown) form a weighted ($w_1 = w_2 = 0.5$) combination to result in the aggregate movement(black) on the right hand side. For illustration purposes, only 3 of 10 joints of the hand are shown in the synergies and the reconstructed movement. Also shown in the figure are the hand postures of the reconstructed movement across time. Adapted from Vinjamuri et al. (2011)

on the principles of data reduction and dimensionality reduction, are soon to find place in tele-surgery and tele-robotics (Vinjamuri et al. (2007)). Synergies are projected to be miniature windows to provide immense help in next generation rehabilitation. Recently our group has demonstrated a synergy based brain machine interface where two control signals calculated from the spectral powers of the brain signals controlled two synergies, that commanded a 10 DoF virtual hand (Vinjamuri et al. (2011)). This showed promising results for controlling a synergy-based neural prosthesis.

(ii) Diagnostics: Applying similar concepts of synergies on the hand movements of the individuals with movement disorders, the sources that contain the tremor were isolated. Using blind source separation and dimensionality reduction methods, the possible neural sources that contained tremor were extracted from the hand movements of individuals with Essential Tremor (Vinjamuri et al. (2009)). This led to an efficient quantification of tremor.

(iii) Robotics: Biologically inspired synergies are being used in balance control of humanoid robots (Hauser et al. (2007)). Based on the principle that biological organisms recruit kinematic synergies that manage several joints, a control strategy for balance of humanoid robots was developed. This control strategy reduced computational complexity following a biological framework that central nervous system reduces the computational complexity of managing numerous degrees of freedom by effectively utilizing the synergies. Biologically inspired neural network controller models (Bernabucci et al. (2007)) that can manage ballistic arm movements have been developed. The models simulated the kinematic aspects, with bell-shaped wrist velocity profiles, and generated movement specific muscular synergies for the execution of movements.

(iv) Rehabilitation: Bimanual coordination is damaged in brain lesions and brain disorders Vinjamuri et al. (2008). Using a small set of modifiable and adjustable synergies tremendously simplifies the task of learning new skills or adapting to new environments. Constructing internal neural representations from a linear combination of a reduced set of basis functions might be crucial for generalizing to novel tasks and new environmental conditions (Flash & Hochner (2005); Poggio & Bizzi (2004)).

2.3 Extraction of synergies

Synergies or movement primitives are viewed as small building blocks of movement that are present inherently within the movements and are shared across several movements. In other words, for example, in a set of hundred grasping movements, there might be a five or six synergies that are shared and common across all the movements. So it is to say that these hundred hand movements are composed of synergies. How do we decompose these hundred hand movements to a few building blocks of movement? This is the problem we are trying to solve.

In order to extract these primitives, several methods have been used. Several researchers view this as a problem of extracting basis functions. In fact, PCA can be viewed as extracting basis functions that are orthogonal to each other. Radial basis functions were also used as synergy approximations. Gradient descend method and non-negative matrix factorization methods (d'Avella et al. (2003)), multivariate statistical techniques (Santello et al. (2002)) were used in extracting the synergies. Different from the above interpretations of synergies, Todorov & Ghahramani (2004) suggested that synergistic control may not mean dimensionality reduction or simplification, but might imply task optimization using optimal feedback control.

In the coming sections we will use linear and nonlinear dimensionality reduction methods in extracting the synergies.

2.4 Dimensionality reduction methods for extracting synergies

In the previous section, we listed different methods used to extract the synergies. In this section these methods were limited to dimensionality reduction methods as these are of relevance to this chapter.

Based on the principal component analysis, Jerde et al. (Jerde et al. (2003)) found support for the existence of postural synergies of angular configuration. The shape of human hand can be predicted using a reduced set of variables and postural synergies. Similarly, Santello et al. (1998) showed that a small number of postural synergies were sufficient to describe how human subjects grasped a large set of different objects. Moreover, Mason et al. (2001) used singular value decomposition (SVD) to demonstrate that a large number of hand postures during reach-to-grasp can be constructed by a small number of principal components or eigen postures.

With PCA, Braido & Zhang (2004) examined the temporal co-variation between finger-joint angles. Their results supported the view that the multi-joint acts of the hand are subject to stereotypical motion patterns controlled via simple kinematic synergies. In the above mentioned study of eigen postures, Mason et al. (2001) also investigated the temporal evolutions of the eigen postures and observed similar kinematic synergies across subjects and grasps. In addition, kinematic synergies have been observed in the spatiotemporal coordination between thumb and index finger movements and co-ordination of tip-to-tip finger movements (Cole & Abbs (1986)).

Another concept of synergies was proposed by d'Avella et al. (2003). Although their work was not directly related to the hand movements, they investigated the muscle synergies of frogs during a variety of motor behaviors such as kicking. Using a gradient descent method, they decomposed the muscle activities into linear combinations of three task-independent time-varying synergies. They also observed that these synergies were very much related to movement kinematics and that similarities existed between synergies in different tasks.

3. Preparing the hand kinematics for dimensionality reduction

In this section, we first recorded the joint angles when ten subjects participated in an experiment of reaching and grasping tasks while wearing a dataglove. Then we transformed the recorded joint angles into joint angular velocities and further preprocessed it to prepare datasets to be used as inputs to the dimensionality reduction methods.

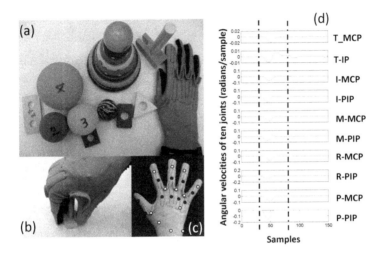

Fig. 2. (a) Objects grasped by the subjects. (b) An example of a rapid grasp of a wooden toy nut. (c) Sensors of the CyberGlove used for analysis (dark circles.) (d) A sample of rapid movement profile (finger-joint-angular-velocity profile). Onset and end of movements are marked in the figure. Abbreviations: T, thumb; I, index finger; M, middle finger; R, ring finger; P, pinky finger; CMC, carpometacarpal joint; MCP, metacarpophalangeal joint; IP, interphalangeal joint; PIP, proximal interphalangeal joint; DIP, distal interphalangeal joint. The dotted lines showed the onset and end of movement.

3.1 Experiment

The experimental setup consisted of a right-handed CyberGlove (CyberGlove Systems LLC, San Jose, CA, USA) equipped with 22 sensors which can measure angles at all the finger joints. For the purpose of reducing computational burden, in this study we only considered 10 of the sensors which correspond to the metacarpophalangeal (MCP) and interphalangeal (IP) joints of the thumb and the MCP and proximal interphalangeal (PIP) joints of the other four fingers as shown in Fig. 2(c). These ten joints can capture most characteristics of the hand in grasping tasks.

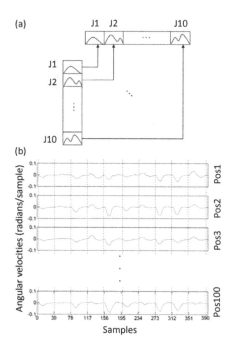

Fig. 3. Angular-velocity matrix. (a) Cascading angular-velocity profiles of 10 joints to form a row of the angular-velocity matrix. (b) Each row of the angular-velocity matrix represents a grasping task. In each row the angular-velocity profiles of 10 joints are separated by dotted red lines. Hundred such tasks put together is an angular-velocity matrix.

A typical task consisted of grasping the objects of various shapes and sizes as shown in Fig. 2(a). Objects (wooden and plastic) of different shapes (spheres, circular discs, rectangles, pentagons, nuts, and bolts) and different dimensions were used in the grasping tasks and were selected based on two strategies. One was gradually increasing sizes of similar shaped objects, and the other was using different shapes. Start and stop times of each task were signaled by computer-generated beeps. In each task, the subject was in a seated position, resting his/her right hand at a corner of a table and upon hearing the beep, grasped the object placed on the table. At the time of the start beep hand was in rest posture, and then the subject grasped the object and held it until the stop beep. Between the grasps, there was enough time for the subjects to avoid the affects due to fatigue on succeeding tasks. The experiment was split into two phases, training phase and testing phase, the difference in these two being the velocity of grasps and types of grasps.

3.2 Training

In the training phase, subjects were instructed to rapidly grasp 50 objects, one at a time. This was repeated for the same 50 objects, and thus the whole training phase obtained 100 rapid grasps. Only these 100 rapid grasps were used in extracting synergies.

3.3 Testing

In the testing phase, subjects were instructed to grasp the above 50 objects naturally (slower than the rapid grasps) then repeat the same again. So far the tasks involved only grasping action. To widen the scope of applicability of the synergies, subjects were also asked to pose 36 American Sign Language (ASL) postures. Here subjects started from an initial posture and stopped at one ASL posture. These postures consisted of 10 numbers (0-9) and 26 alphabets (A-Z). Note that these movements are different from grasping tasks. This is the testing phase which consisted of 100 natural grasps and 36 ASL postural movements. The synergies were derived from the hand movements collected in the training phase using linear and nonlinear dimensionality reduction methods. Then they were used in the reconstruction of movements collected during the testing phase.

3.4 Preprocessing

After obtaining the joint angles at various times from the rapid grasps, angular velocities were calculated. These angular velocities were filtered from noise. Only the relevant projectile movement (about 0.45 second or 39 samples at a sampling rate of 86 Hz) of the entire angular-velocity profile was preserved and the rest was truncated (Fig. 2(d)).

Next an angular-velocity matrix, denoted V, was constructed for each subject. Angular-velocity profiles of the 10 joints corresponding to one rapid grasp were cascaded such that each row of the angular-velocity matrix represented one movement in time. The matrix consisted of 100 rows and $39 \times 10 = 390$ columns:

$$V = \begin{bmatrix} v_1^1(1) & \cdots & v_1^1(39) & \cdots & v_{10}^1(1) & \cdots & v_{10}^1(39) \\ \vdots & \vdots & \vdots & \vdots & \vdots & \vdots & \vdots \\ v_1^g(1) & \cdots & v_1^g(39) & \cdots & v_{10}^g(1) & \cdots & v_{10}^g(39) \\ \vdots & \vdots & \vdots & \vdots & \vdots & \vdots & \vdots \\ v_1^{100}(1) & \cdots & v_1^{100}(39) & \cdots & v_{10}^{100}(1) & \cdots & v_{10}^{100}(39) \end{bmatrix} \tag{1}$$

where $v_i^g(t)$ represents the angular velocity of joint i ($i = 1,...,10$) at time t ($t = 1,...,39$) in the g-th rapid-grasping task ($g = 1,...,100$). An illustration of this transformation was shown in the Fig. 3.

4. Linear dimensionality reduction methods

In this section we derived synergies using two unique linear dimensionality reduction methods, namely, PCA and MDS. The angular-velocity matrix computed in preprocessing was used as input to these methods. Linear methods are easy to use and demand less computational power when compared to nonlinear methods, hence this first exercise.

4.1 Principal component analysis

The winning advantage of PCA is less time for computation and equally effective results when compared to gradient descent methods(Vinjamuri et al. (2007)). PCs are essentially the most commonly used patterns across the data. In this case, PCs are the synergies which are most commonly used across different movements. Moreover these PCs when graphically visualized revealed anatomical implications of physiological properties of human hand prehension(Vinjamuri et al. (2010a;b)).

There are several ways to implement PCA. Two most widely used methods were shown below. First method has three steps: (1) Subtract mean from the data (2) Calculate covariance matrix (3) Compute eigen values and eigen vectors of covariance matrix. Principal components are eigen vectors. Second method uses singular value decomposition (SVD). Third method is a function readily available in Statistics Tool Box of MATLAB which essentially implements first method.

```
load hald; %Load sample dataset in MATLAB
Data = ingredients';
%Fetch dimensions of the data (M Dimensions x N observations)
[M,N] = size(Data);
MeanofData = mean(Data,2); %Calculate Mean of the Data
Data = Data - repmat(MeanofData,1,N); %Subtract Mean from the Data
Covariance = 1 / (N-1) * Data * Data'; %Calculate the covariance of the Data
[pc, latent] = eig(Covariance); %Find the eigenvectors and eigenvalues
latent = diag(latent); %Take only diagonal elements that are eigen values
[rd, rindices] = sort(-1*latent); %Sort them in decreasing order
latent = latent(rindices); %Extract eigen values corresponding to indices
pc = pc(:,rindices); %Extract principal components corresponding to indices
score = (pc' * Data)'; %Project the Data into PC space

load hald; %Load sample dataset in MATLAB
Data = ingredients';
%Fetch dimensions of the data (M Dimensions x N observations)
[M,N] = size(Data);
MeanofData = mean(Data,2); %Calculate Mean of the Data
Data = Data - repmat(MeanofData,1,N); %Subtract Mean from the Data
Y = Data' / sqrt(N-1); %Normalized Data
[u,S,pc] = svd(Y); %Peform SVD
S = diag(S); %Extract diagonal elements corresponding to eigen values
latent = S .* S; %Calculate the eigen values
score = (pc' * Data)'; %Project the Data into PC space

load hald; %Load sample dataset in MATLAB
[pc,score,latent,tsquare] = princomp(ingredients); %Peform PCA
```

Here PCA using SVD Jolliffe (2002) was performed on the angular-velocity matrix V of each subject:

$$V = U \Sigma S \tag{2}$$

where U is a 100-by-100 matrix, which has orthonormal columns so that $U'U = I_{100 \times 100}$ (100-by-100 identity matrix); S is a 100-by-390 matrix, which has orthonormal rows so that $SS' = I_{100 \times 100}$; and Σ is a 100-by-100 diagonal matrix: $\text{diag}\{\lambda_1, \lambda_2, ..., \lambda_{100}\}$ with $\lambda_1 \geq \lambda_2 \geq \cdots \geq \lambda_{100} \geq 0$. Matrix V can be approximated by another matrix $\tilde{V}$ with reduced rank m by replacing Σ with Σ_m, which contains only the m largest singular values, i.e., $\lambda_1, ..., \lambda_m$ (the other singular values are replaced by zeros). The approximation matrix $\tilde{V}$ can be written in a more compact form:

$$\tilde{V} = U_m \, \text{diag}\{\lambda_1, ..., \lambda_m\} \, S_m \tag{3}$$

where U_m is a 100-by-m matrix containing the first m columns of U and S_m is a m-by-390 matrix containing the first m rows of S. Denoting $W = U_m \operatorname{diag}\{\lambda_1, ..., \lambda_m\}$, we have

$$V \approx \hat{V} = W S_m. \tag{4}$$

Then each row of S_m is called a *principal component* (PC), and W is called the weight matrix.

For easy comparison, let us name the elements of S_m in a way similar to (1):

$$S_m \equiv \begin{bmatrix} s_1^1(1) & \cdots & s_1^1(39) & \cdots & s_{10}^1(1) & \cdots & s_{10}^1(39) \\ \vdots & \vdots & \vdots & \vdots & \vdots & \vdots & \vdots \\ s_1^m(1) & \cdots & s_1^m(39) & \cdots & s_{10}^m(1) & \cdots & s_{10}^m(39) \end{bmatrix} \tag{5}$$

and name the elements of W in the following way:

$$W = \begin{bmatrix} w_1^1 & \cdots & w_m^1 \\ \vdots & \vdots & \vdots \\ w_1^g & \cdots & w_m^g \\ \vdots & \vdots & \vdots \\ w_1^{100} & \cdots & w_m^{100} \end{bmatrix}. \tag{6}$$

According to (4), each row of V can be approximated by a linear combination of m PCs, and according to (4), (1), (5), and (6), we have

$$v_i^g(t) \approx \sum_{j=1}^m w_j^g s_i^j(t) \tag{7}$$

for $i = 1, ..., 10$, $g = 1, ..., 100$, and $t = 1, ..., 39$.

Thus the above SVD procedure has found a solution to the synergy-extraction problem: The angular-velocity profiles (obtained by rearranging all joints row-wise for the PCs)

$$\begin{bmatrix} s_1^j(1) & \cdots & s_1^j(39) \\ s_2^j(1) & \cdots & s_2^j(39) \\ \vdots & \vdots & \vdots \\ s_{10}^j(1) & \cdots & s_{10}^j(39) \end{bmatrix}, \quad j = 1, ..., m$$

can be viewed as synergies. According to (4) or (7), these synergies can serve as "building blocks" to reconstruct joint-angular-velocity profiles of hand movements.

To decide m, the number of PCs or synergies that we want to use in reconstruction of the testing movements, we consider the accuracy of approximation in (4) or (7). The approximation accuracy can be measured by an index defined as

$$\frac{\lambda_1^2 + \lambda_2^2 + \cdots + \lambda_m^2}{\lambda_1^2 + \lambda_2^2 + \cdots + \lambda_{100}^2}.$$

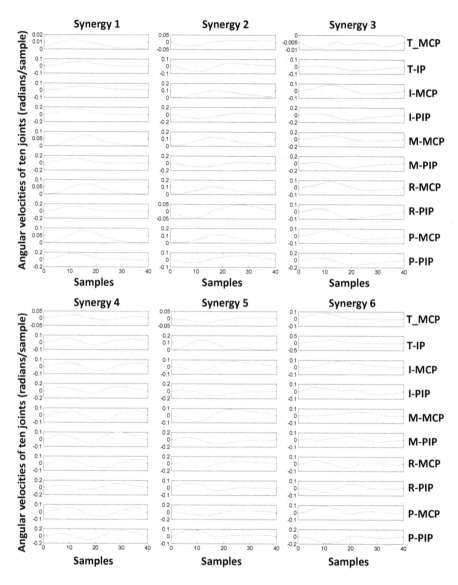

Fig. 4. Six kinematic synergies obtained for subject 1 using PCA. Each synergy is about 0.45 s in duration (39 samples at 86 Hz). Abbreviations: T, thumb; I, index finger; M, middle finger; R, ring finger; P, pinky finger; MCP, metacarpophalangeal joint; IP, interphalangeal joint; PIP, proximal IP joint.

The larger this index is, the closer the approximation is. This index also provides indication of the fraction of total variance of the data matrix accounted by the PCs. To ensure satisfactory approximation, the index should be greater than some threshold. In this study, we used 95% as the threshold (a commonly used threshold Jolliffe (2002)) to determine the number of PCs or synergies (i.e. m). With this threshold we found the six synergies can account for 95% of variance in the postures. Fig. 4 shows six kinematic synergies obtained for subject 1 using PCA.

4.2 Multidimensional scaling

Classical Multidimensional Scaling (MDS) can still be grouped under linear methods. This was introduced here to the reader to give a different perspective of dimensionality reduction in a slightly different analytical approach when compared to PCA discussed previously. The two methods PCA and MDS are unique as they perform dimensionality reduction in different ways. PCA operates on covariance matrix where as MDS operates on distance matrix. In MDS, a Euclidean distance matrix is calculated from the original matrix. This is nothing but a pairwise distance matrix between the variables in the input matrix. This method tries to preserve these pairwise distances in a low dimensional space, thus allowing for dimensionality reduction and preserving the inherent structure of the data simultaneously. PCA and MDS were compared using a simple example in MATLAB below.

```
load hald; %Load sample dataset in MATLAB
[pc,score,latent,tsquare] = princomp(ingredients); %Peform PCA

D = pdist(ingredients); %Calculate pairwise distances between ingredients
[Y, e] = cmdscale(D); %Perform Classical MDS
```

score in PCA represented the data that was projected in the PC space. Compare this to Y calculated in MDS. These are same. Similarly in place of sample dataset when the posture matrix V was used as input to MDS, it yielded the same synergies as PCA. This was introduced here because we build upon this method for the nonlinear methods coming up in the next section.

5. Nonlinear dimensionality reduction methods

So far, we have investigated the use of linear dimensionality reduction methods (PCA and MDS) in extracting synergies. In this section we used nonlinear dimensionality reduction methods for the same purpose. The motivation to explore nonlinear methods was that physiologists who studied motor control have propounded that there were inherent nonlinearities in the human motor system. By using nonlinear methods we could probably achieve improved precision in reconstruction of natural movements. The nonlinear methods applied in this chapter are Isomap, local linear embedding (LLE), and kernel PCA (kPCA). The first two methods Isomap and LLE, are built on the framework of classical multidimensional scaling discussed in the previous section. kPCA is built on the framework of PCA.

5.1 Isomap

Isomap is similar to PCA and MDS. Although Isomap does linear estimations in the data point neighborhoods, the synergies extracted are nonlinear because these small neighborhoods are stitched together without trying to maintain linearity.

The following were the steps involved in estimating nonlinear synergies using Isomap:

1. Define neighbors for each data point

2. Find D, a matrix of inter-point distances

3. Find eigenvectors of $\tau(D)$, where $\tau(D) = -HSH/2$, $S_{ij} = (D_{ij})^2$ and $H_{ij} = d_{ij} - 1/N$, where N is number of data points and d is Kronecker delta function.

In PCA we estimated the eigen values and eigen vectors of covariance of the data. Similarly, here, we took a nonlinear approach to preserve inter-point distances on the manifold. The matrix D is similar to covariance matrix in PCA. D can actually be thought of as the covariance matrix in higher dimensions. Since in an N-dimensional space, the dimensions are the data points, the covariance for a particular pair of dimensions is the distance between the data points that define those dimensions.

Although this method looks linear like PCA, the source of nonlinearity is the method in which inter-point distances are calculated. For Isomap, we do not use the Euclidean distances between the points. If we use, it becomes classical MDS discussed in previous section. Rather, we use those distances only for points considered neighbors. The rest of inter-point distances are calculated by finding the shortest path through the graph on the manifold using Floyd's algorithm (Tenenbaum et al. (2000)). The goal of the Isomap is to preserve the geodesic distances rather than the euclidian distances. Geodesic distances are calculated by moving along the approximate nonlinear manifold with given data point and interpolation between them.

We used drtoolbox in MATLAB by van der Maaten et al. (2009) to perform Isomap on the angular-velocity matrix to extract synergies. Fig. 5showed the top six synergies extracted using this method. Similar to PCA, all the nonlinear methods also yield the nonlinear synergies in descending order of their significance. The synergies extracted using this method had more submovements when compared to those in PCA.

5.2 Local Linear Embedding

Locally Linear Embedding (LLE) as the name suggests, tries to find a nonlinear manifold by stitching together small linear neighborhoods (Roweis & Saul (2000)). This is very similar to Isomap. The difference between the two algorithms is in how they do the stitching. Isomap does this by doing a graph traversal by preserving geodesic distances while LLE does it by finding a set of weights that perform local linear interpolations that closely approximate the data.

The following were the steps involved in estimating nonlinear synergies using LLE:

1. Define neighbors for each data point

2. Find weights that allow neighbors to interpolate original data accurately

3. Given those weights, find new data points that minimize interpolation error in lower dimensional space

We used drtoolbox in MATLAB by van der Maaten et al. (2009) to perform LLE on the angular-velocity matrix to extract synergies. Fig. 6showed the top six synergies extracted using this method. Similar to Isomap, in this method also we found more submovements in synergies than those from PCA.

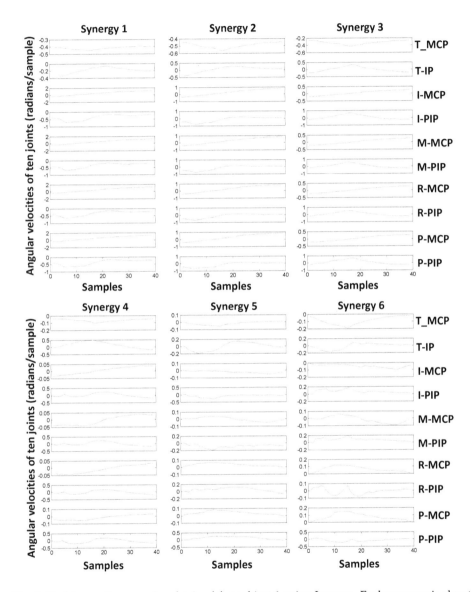

Fig. 5. Six kinematic synergies obtained for subject 1 using Isomap. Each synergy is about 0.45 s in duration (39 samples at 86 Hz). Abbreviations: T, thumb; I, index finger; M, middle finger; R, ring finger; P, pinky finger; MCP, metacarpophalangeal joint; IP, interphalangeal joint; PIP, proximal IP joint.

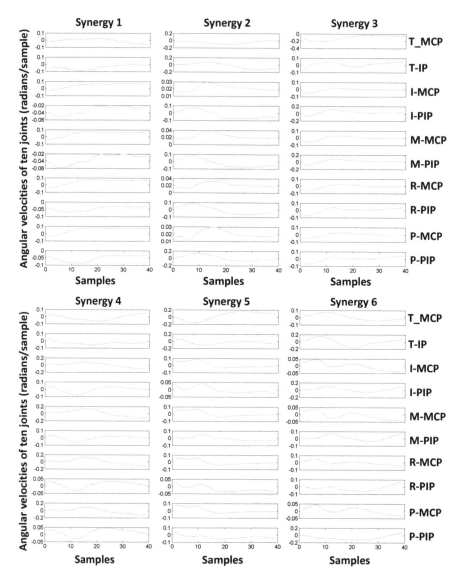

Fig. 6. Six kinematic synergies obtained for subject 1 using LLE. Each synergy is about 0.45 s in duration (39 samples at 86 Hz). Abbreviations: T, thumb; I, index finger; M, middle finger; R, ring finger; P, pinky finger; MCP, metacarpophalangeal joint; IP, interphalangeal joint; PIP, proximal IP joint.

5.3 Kernel PCA

Kernel PCA (kPCA) is an extension of PCA in a high-dimensional space (Scholkopf et al. (1998)). A high-dimensional space is first constructed by using a kernel function. Instead of directly doing a PCA on the data, the kernel based high dimensional feature space is used as input. In this chapter, we have used a gaussian kernel function. Kernel PCA computes the principal eigenvectors of the kernel matrix, rather than those of the covariance matrix. A kernel matrix is similar to the inner product of the data points in the high dimensional space that is constructed using the kernel function. The application of PCA in the kernel space provides Kernel PCA the property of constructing nonlinear mappings. We used drtoolbox in MATLAB by van der Maaten et al. (2009) to perform kPCA on the angular-velocity matrix to extract synergies. Fig. 7showed the top six synergies extracted using this method. These synergies were similar to those obtained from PCA.

6. Reconstruction of natural and ASL movements

The synergies extracted from linear and nonlinear dimensionality reduction methods were used in reconstruction of natural movements. l_1-norm minimization was used to reconstruct natural and ASL movements from the extracted synergies. This method with illustrations was already presented in Vinjamuri et al. (2010a). We have included a brief explanation here for readability and for the sake of completeness. We ask the readers to refer Vinjamuri et al. (2010a) for further details.

Briefly, these were the steps involved in l_1-norm minimization algorithm that was used for reconstruction of natural and ASL movements. Let us assume for a subject m synergies were obtained. The duration of the synergies is t_s samples ($t_s = 39$ in this study). Consider an angular-velocity profile of the subject, $\{\mathbf{v}(t), t = 1, ..., T\}$, where T ($T = 82$ in this study) represents the movement duration (in samples). This profile can be rewritten as a row vector, denoted $\mathbf{v}_{\text{row}}$:

$$\mathbf{v}_{\text{row}} = [v_1(1), ..., v_1(T), ..., v_{10}(1), ..., v_{10}(T)].$$

Similarly, a synergy $\mathbf{s}^j(\cdot)$ can be rewritten as the following row vector:

$$[s_1^j(1), ..., s_1^j(t_s), 0, ..., 0, ..., s_{10}^j(1), ..., s_{10}^j(t_s), 0, ..., 0].$$

We add $T - t_s$ zeros after each $s_i^j(t_s)$ ($i = 1, ..., 10$) in the above vector in order to make the length of the vector the same as that of $\mathbf{v}_{\text{row}}$. If the synergy is shifted in time by t_{jk} ($t_{jk} \le T - t_s$) samples, then we obtain the following row vector:

$$[0, ..., 0, s_1^j(1), ..., s_1^j(t_s), 0, ..., 0, ...,$$
$$0, ..., 0, s_{10}^j(1), ..., s_{10}^j(t_s), 0, ..., 0]$$

with t_{jk} zeros added before each $s_i^j(1)$ and $T - t_s - t_{jk}$ zeros added after each $s_i^j(t_s)$.

Then we construct a matrix as shown in Fig. 8 consisting of the row vectors of the synergies and all their possible shifts with $1 \le t_{jk} \le T - t_s$.

With the above notation, we are trying to achieve a linear combination of synergies that can reconstruct the velocity profiles as in the following equation.

$$\mathbf{v}_{\text{row}} = \mathbf{c}B \tag{8}$$

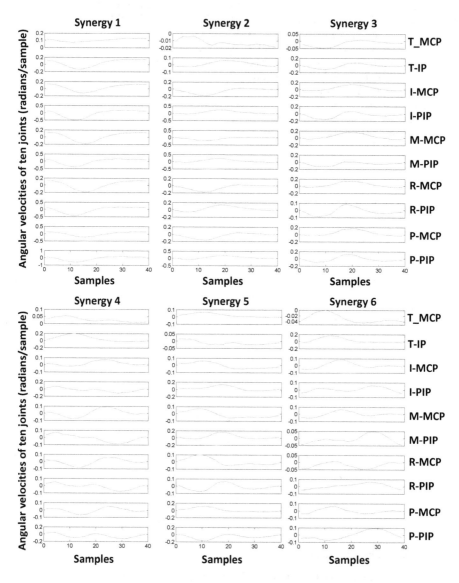

Fig. 7. Six kinematic synergies obtained for subject 1 using kPCA. Each synergy is about 0.45 s in duration (39 samples at 86 Hz). Abbreviations: T, thumb; I, index finger; M, middle finger; R, ring finger; P, pinky finger; MCP, metacarpophalangeal joint; IP, interphalangeal joint; PIP, proximal IP joint.

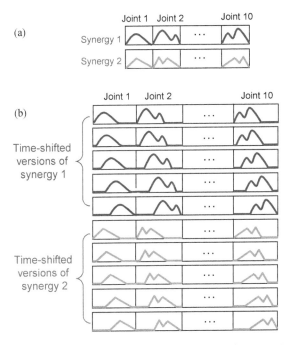

Fig. 8. (a) Synergies were rearranged as row vectors (b) A template matrix was formed by adding the time-shifted versions of synergies along the rows. Adapted from Vinjamuri et al. (2010a).

where $\mathbf{c}$ denotes

$$[0, ..., c_{11}, ..., 0, ..., c_{1K_1}, ..., 0, ..., c_{m1}, ..., 0, ..., c_{mK_m}, ..., 0]$$

with nonzero values c_{jk} appearing at the $(T - t_s + 1)(j - 1) + t_{jk}$-th elements of $\mathbf{c}$. The matrix B (shown in Fig. 8(b)) can be viewed as a bank or library of template functions with each row of B as a template. This bank can be overcomplete and contain linearly dependent subsets. Therefore, for a given movement profile $\mathbf{v}_{\text{row}}$ and an overcomplete bank of template functions B, there exists an infinite number of $\mathbf{c}$ satisfying (8).

We hypothesize that the strategy of central nervous system for dimensionality reduction in movement control is to use a small number of synergies and a small number of recruitments of these synergies for movement generation. Therefore, the coefficient vector $\mathbf{c}$ in (8) should be sparse, i.e., having a lot of zeros and only a small number of nonzero elements. Therefore, we seek the sparsest coefficient vector $\mathbf{c}$ such that $\mathbf{c}B = \mathbf{v}_{\text{row}}$.

The following was optimization problem that was used in selection of synergies in reconstruction of a particular movement.

$$\text{Minimize} \quad \| \mathbf{c} \|_1 + \frac{1}{\lambda} \| \mathbf{c}B - \mathbf{v}_{\text{row}} \|_2^2 \tag{9}$$

where $\| \cdot \|_2$ represents the l_2 norm or Euclidean norm of a vector and λ is a regulation parameter.

Using the above optimization algorithm, the synergies extracted from four methods (PCA, Isomap, LLE, and kPCA) were used in reconstruction of natural movements and ASL postural movements. The reconstruction errors were calculated using the methods in Vinjamuri et al. (2010a). Figures 9 and 10 showed the comparison between the four dimensionality reduction methods. Fig. 9 showed the reconstruction errors for 100 natural movements and Fig. 10 showed the reconstruction errors for 36 ASL postural movements for all four methods. It is observed that PCA still has the best overall performance when compared with the novel nonlinear methods. Fig. 11 showed one of the best reconstructions by all four methods. Tables 1 and 2 summarize the reconstruction results obtained for all ten subjects.

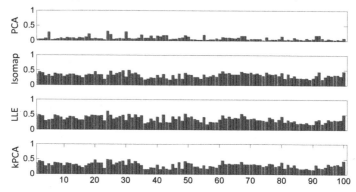

Fig. 9. The reconstruction errors for 100 natural movements with four dimensionality reduction methods PCA, Isomap, LLE, and kPCA. PCA outperformed other three nonlinear methods.

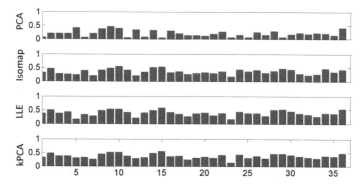

Fig. 10. The reconstruction errors for 36 ASL postural movements with four dimensionality reduction methods PCA, Isomap, LLE, and kPCA. All methods performed poorly in reconstruction of ASL movements when compared to natural movements. PCA performed better than other three nonlinear methods.

7. Summary

In this chapter we applied linear and nonlinear dimensionality reduction methods to extract movement primitives or synergies from rapid reach and grasp movements. We then used these synergies to reconstruct natural movements that were similar to activities of daily living. To broaden the applicability of synergies we also tested them on ASL postural movements which are different movements when compared to natural reach and grasp movements. We employed four types of dimensionality reduction methods: (1) PCA and MDS (2) Isomap (3) LLE (4) kPCA. PCA is a well known linear dimensionality reduction method. Two widely used PCA implementations (Covariance and SVD) were presented and relevant MATLAB codes were provided. Classical MDS is very similar to PCA but operates on a distance matrix. This was introduced as Isomap and LLE both work on a similar framework. Isomap and LLE are both neighborhood graph based nonlinear dimensionality reduction methods. The difference between Isomap and LLE is that the former is a global dimensionality reduction technique where as the former as the name suggests is a local linear technique. kPCA is similar to Isomap as it is a global dimensionality reduction method, but the uniqueness of the method is in using kernel tricks to transform the input data to higher dimensional space.

Subject	PCA	Isomap	LLE	kPCA
1	0.0797 ± 0.0596	0.3240 ± 0.0769	0.3544±0.0954	0.2954 ± 0.0851
2	0.0845 ± 0.0781	0.3780 ± 0.0519	0.3998±0.0765	0.3444 ± 0.0658
3	0.0678 ± 0.0396	0.3180 ± 0.0554	0.3234±0.0834	0.2654 ± 0.0767
4	0.0590 ± 0.0432	0.3680 ± 0.0663	0.3879±0.0645	0.3214 ± 0.0873
5	0.0997 ± 0.0831	0.2640 ± 0.0734	0.2950±0.0790	0.2230 ± 0.0667
6	0.0661 ± 0.0976	0.3240 ± 0.0545	0.3754±0.0531	0.2875 ± 0.0931
7	0.0598 ± 0.0542	0.4140 ± 0.0787	0.4344±0.0632	0.3895 ± 0.0696
8	0.0732 ± 0.0348	0.3540 ± 0.0989	0.3954±0.0854	0.3123 ± 0.0555
9	0.0814 ± 0.0212	0.2290 ± 0.0823	0.3100±0.0991	0.2074 ± 0.0651
10	0.0883 ± 0.0443	0.2490 ± 0.0665	0.2780±0.0799	0.2189 ± 0.0799

Table 1. Mean reconstruction errors (± standard deviation) in natural movements

Subject	PCA	Isomap	LLE	kPCA
1	0.2449 ± 0.0986	0.3480 ± 0.0988	0.3867±0.1052	0.3654 ± 0.0865
2	0.2782 ± 0.0915	0.3280 ± 0.0891	0.3661±0.0963	0.3291 ± 0.0745
3	0.2677 ± 0.0866	0.3110 ± 0.0764	0.3527±0.0881	0.3321 ± 0.0666
4	0.2321 ± 0.0645	0.3450 ± 0.0662	0.3667±0.1123	0.3540 ± 0.0831
5	0.2899 ± 0.0713	0.3780 ± 0.0843	0.3889±0.0792	0.3624 ± 0.0934
6	0.2569 ± 0.0456	0.3200 ± 0.0894	0.3457±0.0856	0.3654 ± 0.0678
7	0.2319 ± 0.0521	0.3890 ± 0.0663	0.3999±0.0996	0.3889 ± 0.0610
8	0.2789 ± 0.0486	0.3671 ± 0.0713	0.3897±0.1152	0.3764 ± 0.0885
9	0.2999 ± 0.0816	0.3335 ± 0.0465	0.3867±0.1672	0.3478 ± 0.0585
10	0.2569 ± 0.0936	0.3146 ± 0.0548	0.3547±0.0082	0.3354 ± 0.0635

Table 2. Mean reconstruction errors (± standard deviation) in ASL postural movements

Thus the reader was given the opportunity to sample different varieties of dimensionality reduction methods. Quantitative and qualitative comparison of the results obtained from reconstruction follows but the verdict is that PCA outperformed the nonlinear methods employed in this chapter.

The results from the reconstructions reveal that nonlinear techniques do not outperform the traditional PCA for both natural movements as well as ASL postural movements. The

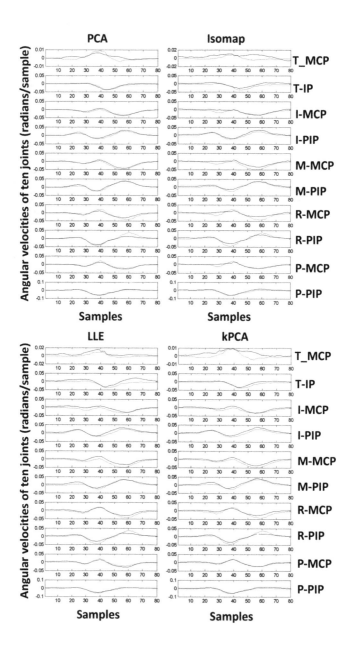

Fig. 11. An example reconstruction (in black) of a natural movement (in red) for task 24 when subject 1 was grasping an object.

reconstruction errors were more for ASL postural movements when compared to those of natural movements for all methods. The reconstruction errors were in general larger for Isomap and LLE when compared with PCA and kPCA and of course PCA had outstanding performance for more than 90% of the tasks. van der Maaten et al. (van der Maaten et al. (2009)) also found that nonlinear methods performed well on specific data sets but could not perform better than PCA for real world tasks. For example, for the Swiss roll data set that contains points that lie on a spiral like two dimensional manifold within a three dimensional space, several nonlinear techniques such as Isomap, LLE were able to find the two dimensional planar embedding, but linear techniques like PCA failed to find so. The reasons for two nonlinear methods Isomap and LLE to perform poorly in this study, might be that they relied neighborhood graphs. Moreover LLE might have been biased to local properties that do not necessarily follow the global properties of high dimensional data. It was surprising to see that Isomap, being a global dimensionality reduction technique performed poorly when compared to LLE for natural movements. kPCA performed better than Isomap and LLE, but kPCA does suffer from the limitation of selection of ideal kernel. The selection of gaussian kernel in this study might not be favorable in extracting the kinematic synergies in this study. In conclusion, although there are numerical advantages and disadvantages with both linear and nonlinear dimensionality reduction methods, PCA seemed to generalize and perform well on the real world data.

8. Acknowledgments

This work was supported by the NSF grant CMMI-0953449, NIDRR grant H133F100001. Special thanks to Laurens van der Maaten for guidance with the dimensionality reduction toolbox, and Prof. Dan Ventura (Brigham Young University) for helpful notes on comparison of LLE and Isomap. Thanks to Stephen Foldes for his suggestions with formatting. Thanks to Mr. Oliver Kurelic for his guidance and help through the preparation of the manuscript.

9. References

Bernabucci, I., Conforto, S., Capozza, M., Accornero, N., Schmid, M. & D'Alessio, T. (2007). A biologically inspired neural network controller for ballistic arm movements, *Journal of NeuroEngineering and Rehabilitation* 4(33): 1–17.

Bernstein, N. (1967). *The Co-ordination and Regulation of Movements*, Pergamon Press, Oxford, UK.

Braido, P. & Zhang, X. (2004). Quantitative analysis of finger motion coordination in hand manipulative and gestic acts, *Human Movement Science* 22: 661–678.

Cole, K. J. & Abbs, J. H. (1986). Coordination of three-joint digit movements for rapid finger-thumb grasp, *Journal of Neurophysiology* 55(6): 1407–1423.

d'Avella, A., Portone, A., Fernandez, L. & Lacquaniti, F. (2006). Control of fast-reaching movements by muscle synergy combinations, *Journal of Neuroscience* 26(30): 7791–7810.

d'Avella, A., Saltiel, P. & Bizzi, E. (2003). Combinations of muscle synergies in the contruction of a natural motor behavior, *Nature Neuroscience* 6(3): 300–308.

Flash, T. & Hochner, B. (2005). Motor primitives in vertebrates and invertebrates, *Current Opinion in Neurobiology* 15(6): 660–666.

Grinyagin, I. V., Biryukova, E. V. & Maier, M. A. (2005). Kinematic and dynamic synergies of human precision-grip movements, *Journal of Neurophysiology* 94(4): 2284–2294.

Hauser, H., Neumann, G., Ijspeert, A. J. & Maass, W. (2007). Biologically inspired kinematic synergies provide a new paradigm for balance control of humanoid robots, *Proc. IEEE-RAS 7th Intl. Conf. Humanoid Robots*, Pittsburgh, PA, USA.

Jerde, T. E., Soechting, J. F. & Flanders, M. (2003). Biological constraints simplify the recognition of hand shapes, 50(2): 565–569.

Jolliffe, I. T. (2002). *Principal Component Analysis, 2nd Ed.*, Springer, New York, NY, USA.

Mackenzie, C. L. & Iberall, T. (1994). *The Grasping Hand (Advances in Psychology)*, North-Holland, Amsterdam, Netherlands.

Mason, C. R., Gomez, J. E. & Ebner, T. J. (2001). Hand synergies during reach-to-grasp, *Journal of Neurophysiology* 86(6): 2896–2910.

Poggio, T. & Bizzi, E. (2004). Generalization in vision and motor control, *Nature* 431: 768–774.

Popovic, M. & Popovic, D. (2001). Cloning biological synergies improves control of elbow neuroprostheses, 20(1): 74–81.

Roweis, S. & Saul, L. (2000). Nonlinear dimensionality reduction by locally linear embedding, *Science* 290(5500): 2323–2326.

Santello, M., Flanders, M. & Soechting, J. F. (1998). Postural hand synergies for tool use, *Journal of Neuroscience* 18(23): 10105–10115.

Santello, M., Flanders, M. & Soechting, J. F. (2002). Patterns of hand motion during grasping and the influence of sensory guidance, *Journal of Neuroscience* 22(4): 1426–1435.

Scholkopf, B., Smola, A. . J. & Muller, K.-R. (1998). Nonlinear component analysis as a kernel eigenvalue problem., *Neural Computation* 10(5): 1299–1319.

Tenenbaum, J. . B., Silva, V. D. & Langford, J. C. (2000). A global geometric framework for nonlinear dimensionality reduction, *Science* 290: 2319–2323.

Thakur, P. H., Bastian, A. J. & Hsiao, S. S. (2008). Multidigit movement synergies of the human hand in an unconstrained haptic exploration task, *Journal of Neuroscience* 28(6): 1271–1281.

Todorov, E. & Ghahramani, Z. (2004). Analysis of the synergies underlying complex hand manipulation, *Proc. 26th Annual International Conference of the IEEE EMBS*, San Francisco, CA, USA, pp. 4637–4640.

van der Maaten, L. J. P., Postma, E. O. & van den Herik, H. J. (2009). Dimensionality reduction: A comparative review, *Tilburg University Technical Report, TiCC-TR* 2009-005.

Vinjamuri, R., Crammond, D. J., Kondziolka, D. & Mao, Z.-H. (2009). Extraction of sources of tremor in hand movements of patients with movement disorders, *IEEE Transactions on Information Technology and Biomedicine* 13(1): 49–59.

Vinjamuri, R., Mao, Z.-H., Sclabassi, R. & Sun, M. (2007). Time-varying synergies in velocity profiles of finger joints of the hand during reach and grasp, *Proc. 29th Annual International Conference of the IEEE EMBS*, Lyon, France, pp. 4846–4849.

Vinjamuri, R., Sun, M., Chang, C.-C., Lee, H.-N., Sclabassi, R. J. & Mao, Z.-H. (2010a). Dimensionality reduction in control and coordination of human hand, *IEEE Transactions on Biomedical Engineering* 57(2): 284–295.

Vinjamuri, R., Sun, M., Chang, C.-C., Lee, H.-N., Sclabassi, R. J. & Mao, Z.-H. (2010b). Temporal postural synergies of the hand in rapid grasping tasks, *IEEE Transactions on Information Technology and Biomedicine* 14(4): 986–994.

Vinjamuri, R., Sun, M., Crammond, D., Sclabassi, R. & Mao, Z.-H. (2008). Inherent bimanual postural synergies in hands, *Proc. 30th Annual International Conference of the IEEE EMBS*, Vancouver, British Columbia, Canada, pp. 5093–5096.

Vinjamuri, R., Weber, D., Mao, Z. H., Collinger, J., Degenhart, A., Kelly, J., Boninger, M., Tyler-Kabara, E. & Wang, W. (2011). Towards synergy-based brain-machine interfaces, *IEEE Transactions on Information Technology and Biomedicine* 15(5): 726–736.

5

Subset Basis Approximation of Kernel Principal Component Analysis

Yoshikazu Washizawa
The University of Electro-Communications
Japan

1. Introduction

Principal component analysis (PCA) has been extended to various ways because of its simple definition. Especially, non-linear generalizations of PCA have been proposed and used in various areas. Non-linear generalizations of PCA, such as principal curves (Hastie & Stuetzle, 1989) and manifolds (Gorban et al., 2008), have intuitive explanations and formulations comparing to the other non-linear dimensional techniques such as ISOMAP (Tenenbaum et al., 2000) and Locally-linear embedding (LLE) (Roweis & Saul, 2000).

Kernel PCA (KPCA) is one of the non-linear generalizations of PCA by using the kernel trick (Schölkopf et al., 1998). The kernel trick nonlinearly maps input samples to higher dimensional space so-called the feature space $\mathcal{F}$. The mapping is denoted by Φ, and let $\boldsymbol{x}$ be a d-dimensional input vector,

$$\Phi : \mathbb{R}^d \to \mathcal{F}, \quad \boldsymbol{x} \mapsto \Phi(\boldsymbol{x}). \tag{1}$$

Then a linear operation in the feature space is a non-linear operation in the input space. The dimension of the feature space $\mathcal{F}$ is usually much larger than the input dimension d, or could be infinite. The positive definite kernel function $k(\cdot, \cdot)$ that satisfies following equation is used to avoid calculation in the feature space,

$$k(\boldsymbol{x}_1, \boldsymbol{x}_2) = \langle \Phi(\boldsymbol{x}_1), \Phi(\boldsymbol{x}_2) \rangle \quad \forall \boldsymbol{x}_1, \boldsymbol{x}_2 \in \mathbb{R}^d, \tag{2}$$

where $\langle \cdot, \cdot \rangle$ denotes the inner product.

By using the kernel function, inner products in $\mathcal{F}$ are replaced by the kernel function $k : \mathbb{R}^d \times \mathbb{R}^d \to \mathbb{R}$. According to this replacement, the problem in $\mathcal{F}$ is reduced to the problem in $\mathbb{R}^n$, where n is the number of samples since the space spanned by mapped samples is at most n-dimensional subsapce. For example, the primal problem of Support vector machines (SVMs) in $\mathcal{F}$ is reduced to the Wolf dual problem in $\mathbb{R}^n$ (Vapnik, 1998).

In real problems, the number of n is sometimes too large to solve the problem in $\mathbb{R}^n$. In the case of SVMs, the optimization problem is reduced to the convex quadratic programming whose size is n. Even if n is too large, SVMs have efficient computational techniques such as chunking or the sequential minimal optimization (SMO) (Platt, 1999), since SVMs have sparse solutions for the Wolf dual problem. After the optimal solution is obtained, we only have to store limited number of learning samples so-called support vectors to evaluate input vectors.

In the case of KPCA, the optimization problem is reduced to an eigenvalue problem whose size is n. There are some efficient techniques for eigenvalue problems, such as the divide-and-conquer eigenvalue algorithm (Demmel, 1997) or the implicitly restarted Arnoldi method (IRAM) (Lehoucq et al., 1998) [1]. However, their computational complexity is still too large to solve when n is large, because KPCA does not have sparse solution. These algorithms require $O(n^2)$ working memory space and $O(rn^2)$ computational complexity, where r is the number of principal components. Moreover, we have to store all n learning samples to evaluate input vectors.

Subset KPCA (SubKPCA) approximates KPCA using the subset of samples for its basis, and all learning samples for the criterion of the cost function (Washizawa, 2009). Then the optimization problem for SubKPCA is reduced to the generalized eigenvalue problem whose size is the size of the subset, m. The size of the subset m defines the trade-off between the approximation accuracy and the computational complexity. Since all learning samples are utilized for its criterion, even if m is much smaller than n, the approximation error is small. The approximation error due to this subset approximation is discussed in this chapter. Moreover, after the construction, we only have to store the subset to evaluate input vectors.

An illustrative example is shown in Figure 1. Figure 1 (a) shows artificial 1000 2-dimensional samples, and contour lines of norms of transformed vectors onto one-dimensional subspace by PCA. Figure 1 (b) shows contour curves by KPCA (transformed to five-dimensional subspace in $\mathcal{F}$). This is non-linear analysis, however, it requires to solve an eigenvalue problem whose size is 1000. For an input vector, calculations of kernel function with all 1000 samples are required. Figure 1 (c) randomly selects 50 samples, and obtains KPCA. In this case, the size of the eigenvalue problem is only 50, and calculations of kernel function with only 50 samples are required to obtain the transform. However, the contour curves are rather different from (b). Figure 1 (d) shows contour curves of SubKPCA by using the 50 samples for its basis, and all 1000 samples for evaluation. The contour corves are almost that same with (b). In this case, the size of the eigenvalue problem is also only 50, and the number of calculations of kernel function is also 50.

There are some conventional approaches to reduce the computational complexity of KPCA. improved KPCA (IKPCA) (Xu et al., 2007) is similar approach to SubKPCA, however, the approximation error is much higher than SubKPCA. Experimental and theoretical difference are shown in this chapter. Comparisons with Sparse KPCAs (Smola et al., 1999; Tipping, 2001), Nyström method (Williams & Seeger, 2001), incomplete Cholesky decomposition (ICD) (Bach & Jordan, 2002) and adaptive approaches (Ding et al., 2010; Günter et al., 2007; Kim et al., 2005) are also dieeussed.

In this chapter, we denote vectors by bold-italic lower symbols $\boldsymbol{x}, \boldsymbol{y}$, and matrices by bold-italic capital symbols $\boldsymbol{A}, \boldsymbol{B}$. In kernel methods, $\mathcal{F}$ could be infinite-dimensional space up to the selection of the kernel function. If vectors could be infinite (functions), we denote them by italic lower symbols f, g. If either domain or range of linear transforms could be infinite-dimensional space, we denote the transforms by italic capital symbols X, Y. This is summarized as follows; (i) bold symbols, $\boldsymbol{x}, \boldsymbol{A}$, are always finite. (ii) non-bold symbols, f, X, could be infinite.

---

[1] IRAM is implemented as "eigs" in MATLAB

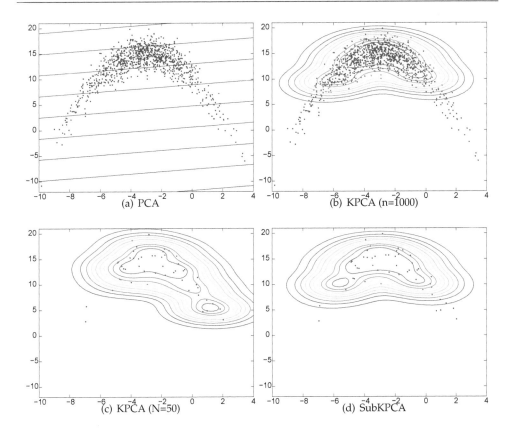

Fig. 1. Illustrative example of SubKPCA

2. Kernel PCA

This section briefly reviews KPCA, and shows some characterizations of KPCA.

2.1 Brief review of KPCA

Let $x_1, \ldots, x_n$ be d-dimensional learning samples, and $X = [x_1| \ldots |x_n] \in \mathbb{R}^{d \times n}$. Suppose that their mean is zero or subtracted. Standard PCA obtains eigenvectors of the variance-covariance matrix Σ,

$$\Sigma = \frac{1}{n} \sum_{i=1}^{n} x_i x_i^{\top} = \frac{1}{n} X X^{\top}. \tag{3}$$

Then the ith largest eigenvector corresponds to the ith principal component. Suppose $U_{\text{PCA}} = [u_1| \ldots |u_r]$. The projection and the transform of x onto r-dimensional eigenspace are $U_{\text{PCA}} U_{\text{PCA}}^{\top} x$ and $U_{\text{PCA}}^{\top} x$ respectively.

In the case of KPCA, input vectors are mapped to feature space before the operation. Let

$$S = [\Phi(x_1)| \ldots |\Phi(x_n)]$$ (4)
$$\Sigma_{\mathcal{F}} = SS^*$$ (5)
$$K = S^*S \in \mathbb{R}^{n \times n},$$ (6)

where $\cdot^*$ denotes the adjoint operator [2], and K is called the kernel Gram matrix (Schölkopf et al., 1999), and i,j-component of K is $k(x_i, x_j)$. Then the ith largest eigenvector corresponds to the ith principal component. If the dimension of $\mathcal{F}$ is large, eigenvalue decomposition (EVD) cannot be performed. Let $\{\lambda_i, u_i\}$ be the ith eigenvalue and corresponding eigenvector of $\Sigma_{\mathcal{F}}$ respectively, and $\{\lambda_i, v_i\}$ be the ith eigenvalue and eigenvector of K. Note that K and $\Sigma_{\mathcal{F}}$ have the same eigenvalues. Then the ith principal component can be obtained from the ith eigenvalue and eigenvector of K,

$$u_i = \frac{1}{\sqrt{\lambda_i}} S v_i.$$ (7)

Note that it is difficult to obtain u_i explicitly on a computer because the dimension of $\mathcal{F}$ is large. However, the inner product of a mapped input vector $\Phi(x)$ and the ith principal component is easily obtained from,

$$\langle u_i, \Phi(x) \rangle = \frac{1}{\sqrt{\lambda_i}} \langle v_i, k_x \rangle,$$ (8)

$$k_x = [k(x, x_1), \ldots, k(x, x_n)]^\top$$ (9)

k_x is an n-dimensional vector called the empirical kernel map.

Let us summarize using matrix notations. Let

$$\Lambda_{\text{KPCA}} = \text{diag}([\lambda_1, \ldots, \lambda_r])$$ (10)
$$U_{\text{KPCA}} = [u_1| \ldots |u_r]$$ (11)
$$V_{\text{KPCA}} = [v_1| \ldots |v_r].$$ (12)

Then the projection and the transform of x onto the r-dimensional eigenspace are

$$U_{\text{KPCA}} U_{\text{KPCA}}^* \Phi(x) = S V_{\text{KPCA}} \Lambda^{-1} V_{\text{KPCA}}^\top k_x,$$ (13)

$$U_{\text{KPCA}}^* \Phi(x) = \Lambda^{-1/2} V_{\text{KPCA}}^\top k_x.$$ (14)

2.2 Characterization of KPCA

There are some characterizations or definitions for PCA (Oja, 1983). SubKPCA is extended from the least mean square (LMS) error criterion [3].

$$\min_X \quad J_0(X) = \frac{1}{n} \sum_{i=1}^{n} \|x_i - X x_i\|^2$$
$$\text{Subject to rank}(X) \leq r.$$ (15)

---

[2] In real finite dimensional space, the adjoint and the transpose $\cdot^\top$ are equivalent. However, in infinite dimensional space, the transpose is not defined

[3] Since all definitions of PCA lead to the equivalent solution, SubKPCA is also defined by the other definitions. However, in this chapter, only LMS criteria is shown.

From this definition, X that minimizes the averaged distance between x_i and Xx_i over i is obtained under the rank constraint. Note that from this criterion, each principal component is not characterized, i.e., the minimum solution is $X = U_{PCA}U_{PCA}^\top$, and the transform U_{PCA} is not determined.

In the case of KPCA, the criterion is

$$\min_X \quad J_1(X) = \frac{1}{n}\sum_{i=1}^n \|\Phi(x_i) - X\Phi(x_i)\|^2 \tag{16}$$

$$\text{Subject to } \mathrm{rank}(X) \leq r, \ \mathcal{N}(X) \supset \mathcal{R}(S)^\perp,$$

where $\mathcal{R}(A)$ denotes the range or the image of the matrix or the operator A, and $\mathcal{N}(A)$ denotes the null space or the kernel of the matrix or the operator A. In linear case, we can assume that the number of samples n is sufficiently larger than r and d, and the second constraint $\mathcal{N}(X) \supset \mathcal{R}(S)^\perp$ is often ignored. However, since the dimension of the feature space is large, r could be larger than the dimension of the space spanned by mapped samples $\Phi(x_1), \ldots, \Phi(x_n)$. For such cases, the second constraint is introduced.

2.2.1 Solution to the problem (16)

Here, brief derivation of the solution to the problem (16) is shown. Since the problem is in $\mathcal{R}(S)$, X can be parameterized by $X = SAS^*$, $A \in \mathbb{R}^{n \times n}$. Accordingly, J_1 yields

$$J_1(A) = \frac{1}{n}\|S - SAS^*S\|_F^2 = \frac{1}{n}\mathrm{Trace}[K - KAK - KA^\top K + A^\top KAK]$$

$$= \frac{1}{n}\|KAK^{1/2} - K^{1/2}\|_F^2 \tag{17}$$

where $\cdot^{1/2}$ denotes the square root matrix, and $\|\cdot\|_F$ denotes the Frobenius norm. The eigenvalue decomposition of K is $K = \sum_{i=1}^n \lambda_i v_i v_i^\top$. From the Schmidt approximation theorem (also called Eckart-Young theorem) (Israel & Greville, 1973), J_1 is minimized when

$$KAK^{1/2} = \sum_{i=1}^r \sqrt{\lambda_i} v_i v_i^\top \tag{18}$$

$$A = \sum_{i=1}^r \frac{1}{\lambda_i} v_i v_i^\top = V_{KPCA}\Lambda^{-1}V_{KPCA}^\top \tag{19}$$

2.3 Computational complexity of KPCA

The procedure of KPCA is as follows;

1. Calculate K from samples. $[O(n^2)]$

2. Perform EVD for K, and obtain the r largest eigenvalues and eigenvectors, $\lambda_1, \ldots, \lambda_r$, $v_1, \ldots v_r$. $[O(rn^2)]$

3. Obtain $\Lambda^{-1/2}V_{KPCA}^\top$, and store all training samples.

4. For an input vector x, calculate the empirical kernel map k_x from Eq. (9). $[O(n)]$

5. Obtain transformed vector Eq. (14). $[O(rn)]$

The procedures 1, 2, and 3 are called the learning (training) stage, and the procedures 4 and 5 are called the evaluation stage.

The dominant computation for the learning stage is EVD. In realistic situation, n should be less than several tens of thousands. For example, if $n = 100,000$, 20Gbyte RAM is required to store K on four byte floating point system. This computational complexity is sometimes too heavy to use for real large-scale problems. Moreover, in the evaluation stage, response time of the system depends on the number of n.

3. Subset KPCA

3.1 Definition

Since the problem of KPCA in the feature space $\mathcal{F}$ is in the subspace spanned by the mapped samples, $\Phi(x_1), \ldots, \Phi(x_n)$, i.e., $\mathcal{R}(S)$, the problem in $\mathcal{F}$ is transformed to the problem in $\mathbb{R}^n$. SubKPCA seeks the optimal solution in the space spanned by smaller number of samples, $\Phi(y_1), \ldots, \Phi(y_m)$, $m \leq n$ that is called a basis set. Let $T = [\Phi(y_1), \ldots, \Phi(y_m)]$, then the optimization problem of SubKPCA is defined as

$$
\begin{aligned}
&\min_X \quad J_1(X) \\
&\text{Subject to } \operatorname{rank}(X) \leq r, \quad \mathcal{N}(X) \supset \mathcal{R}(T)^\perp, \quad \mathcal{R}(X) \subset \mathcal{R}(T).
\end{aligned}
\tag{20}
$$

The third and the fourth constraints indicate that the solution is in $\mathcal{R}(T)$. It is worth noting that SubKPCA seeks the solution in the limited space, however, the objective function is the same as that of KPCA, i.e., all training samples are used for the criterion. We call the set of all training samples the criterion set. The selection of the basis set $\{y_1, \ldots, y_m\}$ is also important problem, however, here we assume that it is given, and the selection is discussed in the next section.

3.2 Solution of SubKPCA

At first, the minimal solutions to the problem (20) are shown, then their derivations are shown. If $\mathcal{R}(T) \subset \mathcal{R}(S)$, its solution is simplified. Note that if the set $\{y_1, \ldots, y_m\}$ the subset of $\{x_1, \ldots, x_n\}$, $\mathcal{R}(T) \subset \mathcal{R}(S)$ is satisfied. Therefore, solutions for two cases are shown, ($\mathcal{R}(T) \subset \mathcal{R}(S)$ and all cases)

3.2.1 The case $\mathcal{R}(T) \subset \mathcal{R}(S)$

Let $K_y = T^*T \in \mathbb{R}^{m \times m}$, $(K_y)_{i,j} = k(y_i, y_j)$, $K_{xy} = X^*T \in \mathbb{R}^{n \times m}$, $(K_{xy})_{i,j} = k(x_i, y_j)$. Let $\kappa_1, \ldots, \kappa_r$ and $z_1, \ldots, z_r$ be sorted eigenvalues and corresponding eigenvectors of the generalized eigenvalue problem,

$$
K_{xy}^\top K_{xy} z = \kappa K_y z
\tag{21}
$$

respectively, where each eigenvector z_i is normalized by $z_i \leftarrow z_i / \sqrt{\langle z_i, K_y z_i \rangle}$, that is $\langle z_i, K_y z_j \rangle = \delta_{ij}$ (Kronecker delta). Let $Z = [z_1 | \ldots | z_r]$, then the problem (20) is minimized by

$$
P_{\text{SubKPCA}} = TZZ^\top T^*.
\tag{22}
$$

The projection and the transform of SubKPCA for an input vector x are

$$P_{\text{SubKPCA}}\Phi(x) = TZZ^\top h_x \tag{23}$$

$$U_{\text{SubKPCA}}\Phi(x) = Z^\top h_x, \tag{24}$$

where $h_x = [k(x, y_1), \dots, k(x, y_m)] \in \mathbb{R}^m$ is the empirical kernel map of x for the subset.

A matrix or an operator A that satisfies $AA = A$ and $A^\top = A$ ($A^* = A$), is called a projector (Harville, 1997). If $\mathcal{R}(T) \subset \mathcal{R}(S)$, P_{SubKPCA} is a projector since $P^*_{\text{SubKPCA}} = P_{\text{SubKPCA}}$, and

$$P_{\text{SubKPCA}} P_{\text{SubKPCA}} = TZZ^\top K_y ZZ^\top T^* = TZZ^\top T^* = P_{\text{SubKPCA}}. \tag{25}$$

3.2.2 All cases

The Moore-Penrose pseudo inverse is denoted by $\cdot^\dagger$. Suppose that EVD of $(K_y)^\dagger K_{xy}^\top K_{xy}(K_y)^\dagger$ is

$$(K_y)^\dagger K_{xy}^\top K_{xy}(K_y)^\dagger = \sum_{i=1}^m \xi_i w_i w_i^\top, \tag{26}$$

and let $W = [w_1, \dots, w_r]$. Then the problem (20) is minimized by

$$P_{\text{SubKPCA}} = T(K_y^{1/2})^\dagger WW^\top (K_y^{1/2})^\dagger (K_{xy}^\top K_{xy})(K_{xy}^\top K_{xy})^\dagger T^*. \tag{27}$$

Since the solution is rather complex, and we don't find any advantages to use the basis set $\{y_1, \dots, y_m\}$ such that $\mathcal{R}(T) \not\subset \mathcal{R}(S)$, we henceforth assume that $\mathcal{R}(T) \subset \mathcal{R}(S)$.

3.2.3 Derivation of the solutions

Since the problem (20) is in $\mathcal{R}(T)$, the solution can be parameterized as $X = TBT^*$, $B \in \mathbb{R}^{m \times m}$. Then the objective function is

$$J_1(B) = \frac{1}{n} \|S - TBT^*S\|_F^2 \tag{28}$$

$$= \frac{1}{n}\text{Trace}[BK_{xy}^\top K_{xy}B^\top K_y - B^\top K_{xy}^\top K_{xy} - BK_{xy}^\top K_{xy} + K]$$

$$= \frac{1}{n}\|K_y^{1/2}BK_{xy}^\top - (K_y^{1/2})^\dagger K_{xy}^\top\|_F^2 + \frac{1}{n}\text{Trace}[K - K_{xy}K_y^\dagger K_{xy}^\top] \tag{29}$$

where the relations $K_{xy}^\top = K_y^{1/2}(K_y^{1/2})^\dagger K_{xy}^\top$ and $K_{xy} = K_{xy}(K_y^{1/2})^\dagger K_y^{1/2}$ are used. Since the second term is a constant for B, from the Schmidt approximation theorem, The minimum solution is given by the singular value decomposition (SVD) of $(K_y^{1/2})^\dagger K_{xy}^\top$,

$$(K_y^{1/2})^\dagger K_{xy}^\top = \sum_{i=1}^m \sqrt{\xi_i} w_i \nu_i^\top. \tag{30}$$

Then the minimum solution is given by

$$K_y^{1/2}BK_{xy}^\top = \sum_{i=1}^r \sqrt{\xi_i} w_i \nu_i^\top. \tag{31}$$

From the matrix equation theorem (Israel & Greville, 1973), the minimum solution is given by Eq. (27).

Let us consider the case that $\mathcal{R}(T) \subset \mathcal{R}(S)$.

Lemma 1 (Harville (1997)). *Let A and B be non-negative definite matrices that satisfy $\mathcal{R}(A) \subset \mathcal{R}(B)$. Consider an EVD and a generalized EVD,*

$$(B^{1/2})^\dagger A (B^{1/2})^\dagger v = \lambda v$$

$$Au = \sigma Bu,$$

and suppose that $\{(\lambda_i, v_i)\}$ and $\{(\sigma_i, u_i)\}$, $i = 1, 2, \ldots$ are sorted pairs of the eigenvalues and the eigenvectors respectively. Then

$$\lambda_i = \sigma_i$$

$$u_i = \alpha (B^{1/2})^\dagger v_i, \quad \forall \alpha \in \mathbb{R}$$

$$v_i = \beta B^{1/2} u_i, \quad \forall \beta \in \mathbb{R}$$

are satisfied.

If $\mathcal{R}(T) \subset \mathcal{R}(S)$, $\mathcal{R}(K_{xy}^\top) = \mathcal{R}(K_y)$. Since $(K_{xy}^\top K_{xy})(K_{xy}^\top K_{xy})^\dagger$ is a projector onto $\mathcal{R}(K_y)$, $(K_y^{1/2})^\dagger (K_{xy}^\top K_{xy})(K_{xy}^\top K_{xy})^\dagger = (K_y^{1/2})^\dagger$ in Eq. (27). From Lemma 1, the solution Eq. (22) is derived.

3.3 Computational complexity of SubKPCA

The procedures and computational complexities of SubKPCA are as follows,

1. Select the subset from training samples (discussed in the next Section)
2. Calculate K_y and $K_{xy}^\top K_{xy}$ [$O(m^2) + O(nm^2)$]
3. Perform generalized EVD, Eq. (21). [$O(rm^2)$]
4. Store Z and the samples in the subset.
5. For an input vector x, calculate the empirical kernel map h_x. [$O(m)$]
6. Obtain transformed vector Eq. (24).

The procedures 1, 2 and 3 are the construction, and 4 and 5 are the evaluation. The dominant calculation in the construction stage is the generalized EVD. In the case of standard KPCA, the size of EVD is n, whereas for SubKPCA, the size of generalized EVD is m. Moreover, for evaluation stage, the computational complexity depends on the size of the subset, m, and required memory to store Z and the subset is also reduced. It means the response time of the system using SubKPCA for an input vector x is faster than standard KPCA.

3.4 Approximation error

It should be shown the approximation error due to the subset approximation. In the case of KPCA, the approximation error, that is the value of the objective function of the problem (16). From Eqs. (17) and (19), The value of J_1 at the minimum solution is

$$J_1 = \frac{1}{n} \sum_{i=r+1}^{n} \lambda_i. \tag{32}$$

In the case of SubKPCA, the approximation error is

$$J_1 = \frac{1}{n} \sum_{i=r+1}^{n} \xi_i + \frac{1}{n} \text{Trace}[K - K_{xy}(K_y)^{\dagger} K_{xy}^{\top}]. \tag{33}$$

The first term is due to the approximation error for the rank reduction and the second term is due to the subset approximation. Let $P_{\mathcal{R}(S)}$ and $P_{\mathcal{R}(T)}$ be orthogonal projectors onto $\mathcal{R}(S)$ and $\mathcal{R}(T)$ respectively. The second term yields that

$$\text{Trace}[K - K_{xy}(K_y)^{\dagger} K_{xy}^{\top}] = \text{Trace}[S^*(P_{\mathcal{R}(S)} - P_{\mathcal{R}(T)})S], \tag{34}$$

since $K = S^* P_{\mathcal{R}(S)} S$. Therefore, if $\mathcal{R}(S) = \mathcal{R}(T)$ (for example, the subset contains all training samples), the second term is zero. If the range of the subset is far from the range of the all training set, the second term is large.

3.5 Pre-centering

Although we have assumed that the mean of training vector in the feature space is zero so far, it is not always true in real problems. In the case of PCA, we subtract the mean vector from all training samples when we obtain the variance-covariance matrix Σ. On the other hand, in KPCA, although we cannot obtain the mean vector in the feature space, $\bar{\Phi} = \frac{1}{n} \sum_{i=1}^{n} \Phi(x_i)$, explicitly, the pre-centering can be set in the algorithm of KPCA. The pre-centering can be achieved by using subtracted vector $\bar{\Phi}(x_i)$, instead of a mapped vector $\Phi(x_i)$,

$$\bar{\Phi}(x_i) = \Phi(x_i) - \bar{\Phi}, \tag{35}$$

that is to say, S and K in Eq. (17) are respectively replaced by

$$\bar{S} = S - \bar{\Phi} 1_n^{\top} = S(I - \frac{1}{n} 1_{n,n}) \tag{36}$$

$$\bar{K} = \bar{S}^* \bar{S} = (I - \frac{1}{n} 1_{n,n}) K (I - \frac{1}{n} 1_{n,n}) \tag{37}$$

where I denotes the identify matrix, and 1_n and $1_{n,n}$ are an n-dimensional vector and an $n \times n$ matrix whose elements are all one, respectively.

For SubKPCA, following three methods to estimate the centroid can be considered,

1. $\bar{\Phi}_1 = \frac{1}{n} \sum_{i=1}^{n} \Phi(x_i)$

2. $\bar{\Phi}_2 = \frac{1}{m} \sum_{i=1}^{m} \Phi(y_i)$

3. $\bar{\Phi}_3 = \underset{\Psi \in \mathcal{R}(T)}{\text{argmin}} \|\Psi - \bar{\Phi}_1\| = \frac{1}{n} T K_y^{\dagger} K_{xy}^{\top} 1_n.$

The first one is the same as that of KPCA. The second one is the mean of the basis set. If the basis set is the subset of the criterion set, the estimation accuracy is not as good as $\bar{\Phi}_1$. The third one is the best approximation of $\bar{\Phi}_1$ in $\mathcal{R}(T)$. Since SubKPCA is discussed in $\mathcal{R}(T)$, $\bar{\Phi}_1$ and $\bar{\Phi}_3$ are equivalent. However, for the post-processing such as pre-image, they are not equivalent.

For SubKPCA, only S in Eq. (28) has to be modified for per-centering [4]. If Φ_3 is used, S and K_{xy} are replaced by

$$\bar{S} = S - \Phi_3 1_n^\top \tag{38}$$

$$\bar{K}_{xy} = \bar{S}^* T = \left(I - \frac{1}{n} 1_{n,n} \right) K_{xy}. \tag{39}$$

4. Selection of samples

Selection of samples for the basis set is an important problem in SubKPCA. Ideal criterion for the selection depends on applications such as classification accuracy or PSNR for denoising. We, here, show a simple criterion using empirical error,

$$\begin{aligned} &\min_{y_1,\ldots,y_m} \min_X J_1(X) \\ &\text{Subject to } \operatorname{rank}(X) \leq r, \ \mathcal{N}(X) \supset \mathcal{R}(T)^\perp, \ \mathcal{R}(X) \subset \mathcal{R}(T), \\ &\quad \{y_1,\ldots,y_m\} \subset \{x_1,\ldots,x_n\}, \ T = [\Phi(y_1)|\ldots|\Phi(y_m)]. \end{aligned} \tag{40}$$

This criterion is a combinatorial optimization problem for the samples, and it is hard to obtain to global solution if n and m are large. Instead of solving directly, following techniques can be introduced,

1. Greedy forward search
2. Backward search
3. Random sampling consensus (RANSAC)
4. Clustering,

and their combinations.

4.1 Sample selection methods

4.1.1 Greedy forward search

The greedy forward search adds a sample to the basis set one by one or bit by bit. The algorithm is as follows, If several samples are added at 9 and 10, the algorithm is faster, but the cost function may be larger.

4.1.2 Backward search

On the other hand, a backward search removes samples that have the least effect on the cost function. In this case, the standard KPCA using the all samples has to be constructed at the beginning, and this may have very high computational complexity. However, the backward search may be useful in combination with the greedy forward search. In this case, the size of the temporal basis set does not become large, and the value of the cost function is monotonically decreasing.

Sparse KPCA (Tipping, 2001) is a kind of backward procedures. Therefore, the kernel Gram matrix K using all training samples and its inverse have to be calculated in the beginning.

---

[4] Of course, for KPCA, we can also consider the criterion set and the basis set, and perform pre-centering only for the criterion set. It produces the equivalent result.

Algorithm 1 Greedy forward search (one-by-one version)

1: Set initial basis set $\mathcal{T} = \phi$, size of current basis set $\tilde{m} = 0$, residual set $\mathcal{S} = \{x_1, \ldots, x_n\}$, size of the residual set $\tilde{n} = n$.
2: **while** $\tilde{m} < m$, **do**
3:    **for** $i = 1, \ldots, \tilde{n}$ **do**
4:       Let temporal basis set be $\tilde{\mathcal{T}} = \mathcal{T} \cup \{x_i\}$
5:       Obtain SubKPCA using the temporal basis set
6:       Store the empirical error $E_i = J_1(X)$.
7:    **end for**
8:    Obtain the smallest E_i, $k = \operatorname{argmin}_i E_i$.
9:    Add x_k to the current basis set, $\mathcal{T} \leftarrow \mathcal{T} \cup \{x_k\}$, $\tilde{m} \leftarrow \tilde{m} + 1$
10:   Remove x_k from the residual set, $\mathcal{S} \leftarrow \mathcal{S} \backslash \{x_k\}$, $\tilde{n} \leftarrow \tilde{n} - 1$.
11: **end while**

4.1.3 Random sampling consensus

RANSAC is a simple sample (or parameter) selection technique. The best basis set is chosen from many random sampling trials. The algorithm is simple to code.

4.1.4 Clustering

Clustering techniques also can be used for sample selection. When the subset is used for the basis set, i) a sample that is the closest to each centroid should be used, or ii) centroids should be included to the criterion set. Clustering in the feature space $\mathcal{F}$ is also proposed (Girolami, 2002).

5. Comparison with conventional methods

This section compares SubKPCA with related conventional methods.

5.1 Improved KPCA

Improved KPCA (IKPCA) (Xu et al., 2007) directly approximates $u_i \simeq T\tilde{v}_i$ in Eq. (7). From $SS^* u_i = \lambda_i u_i$, the approximated eigenvalue problem is

$$SS^* T\tilde{v} = \lambda_i T\tilde{v}_i. \tag{41}$$

By multiplying T^* from left side, one gets the approximated generalized EVD, $K_{xy}^\top K_{xy}\tilde{v} = \lambda_i K_y \tilde{v}_i$. The parameter vector v_i is substituted to the relation $\tilde{u}_i = T\tilde{v}_i$, hence, the transform of an input vector x is

$$U_{\text{IKPCA}}^* \Phi(x) = \left(\operatorname{diag}([\frac{1}{\sqrt{\kappa_1}}, \ldots, \frac{1}{\sqrt{\kappa_r}}]) \right) Z^\top h_x, \tag{42}$$

where κ_i is the ith largest eigenvalue of (21).

This approximation has no guarantee to be good approximation of u_i. In our experiments in the next section, IKPCA showed worse performance than SubKPCA. In so far as feature extraction, each dimension of the feature vector is multiplied by $\frac{1}{\sqrt{\kappa_i}}$ comparing to SubKPCA. If the classifier accepts such linear transforms, the classification accuracy of feature vectors may be the same with SubKPCA. Indeed, (Xu et al., 2007) uses IKPCA only for feature extraction of a classification problem, and IKPCA shows good performance.

5.2 Sparse KPCA

Two methods to obtain a sparse solution to KPCA are proposed (Smola et al., 1999; Tipping, 2001). Both approaches focus on reducing the computational complexity in the evaluation stage, and do not consider that in the construction stage. In addition, the degree of sparsity cannot be tuned directly for these sparse KPCAs, where as the number of the subset m can be tuned for SubKPCA.

As mentioned in Section 4.1.2, (Tipping, 2001) is based on a backward search, therefore, it requires to calculate the kernel Gram matrix using all training samples, and its inverse. These procedures have high computational complexity, especially, when n is large.

(Smola et al., 1999) utilizes l_1 norm regularization to make the solution sparse. The principal components are represented by linear combinations of mapped samples, $u_i = \sum_{j=1}^{n} \alpha_i^j \Phi(x_j)$. The coefficients α_i^j have many zero entry due to l_1 norm regularization. However, since α_i^j has two indeces, even if each principal component u_i is represented by a few samples, it may not be sparse for many i.

5.3 Nyström approximation

Nyström approximation is a method to approximate EVD, and it is applied to KPCA (Williams & Seeger, 2001). Let $\tilde{u}_i$ and u_i be the ith eigenvectors of K_y and K respectively. Nyström approximation approximates

$$\tilde{v}_i = \sqrt{\frac{m}{n}} \frac{1}{\lambda_i} K_{xy} v_i, \tag{43}$$

where λ_i is the ith eigenvalue of K_y. Since the eigenvector of K_x is approximated by the eigenvector of K_y, the computational complexity in the construction stage is reduced, but that in the evaluation stage is not reduced. In our experiments, SubKPCA shows better performance than Nyström approximation.

5.4 Iterative KPCA

There are some iterative approaches for KPCA (Ding et al., 2010; Günter et al., 2007; Kim et al., 2005). They update the transform matrix $\Lambda^{-1/2} V_{\text{KPCA}}^\top$ in Eq. (14) for incoming samples.

Iterative approaches are sometimes used for reduction of computational complexities. Even if optimization step does not converge to the optimal point, early stopping point may be a good approximation of the optimal solution. However, Kim et al. (2005) and Günter et al. (2007) do not compare their computational complexity with standard KPCA. In the next section, comparisons of run-times show that iterative KPCAs are not faster than batch approaches.

5.5 Incomplete Cholesky decomposition

ICD can also be used for reduction of computational complexity of KPCA. ICD approximates the kernel Gram matrix K by

$$K \simeq GG^\top, \tag{44}$$

where $G \in \mathbb{R}^{n \times m}$ whose upper triangle part is zero, and m is a parameter that specifies the trade-off between approximation accuracy and computational complexity. Instead of performing EVD of K, eigenvectors of K is obtained from EVD of $G^\top G \in \mathbb{R}^{m \times m}$ using the relation Eq. (7) approximately. Along with Nyström approximation, ICD reduces computational complexity in the construction stage, but not in evaluation stage, and all training samples have to be stored for the evaluation.

In the next section, our experimental results indicate that ICD is slower than SubKPCA for very large dataset, n is more than several thousand.

ICD can also be applied to SubKPCA. In Eq. (21), $K_{xy}^\top K_{xy}$ is approximated by

$$K_{xy}^\top K_{xy} \simeq GG^\top. \tag{45}$$

Then approximated z is obtained from EVD of $G^\top K_y G$.

6. Numerical examples

This section presents numerical examples and numerical comparisons with the other methods.

6.1 Methods and evaluation criteria

At first, methods to be compared and evaluation criteria are described. Following methods are compared,

1. SubKPCA [SubKp]

2. Full KPCA [FKp]
   Standard KPCA using all training samples.

3. Reduced KPCA [RKp]
   Standard KPCA using subset of training samples.

4. Improved KPCA (Xu et al., 2007) [IKp]

5. Sparse KPCA (Tipping, 2001) [SpKp]

6. Nyström approximation (Williams & Seeger, 2001) [Nys]

7. ICD (Bach & Jordan, 2002) [ICD]

8. Kernel Hebbian algorithm with stochastic meta-decent (Günter et al., 2007) [KHA-SMD]

Abbreviations in [] are used in Figures and Tables.

For evaluation criteria, the empirical error that is J_1, is used.

$$E_{\mathrm{emp}}(X) = J_1(X) = \frac{1}{n} \sum_{i=1}^{n} \|\Phi(x_i) - X\Phi(x_i)\|^2, \tag{46}$$

where X is replaced by each operator. Note that full KPCA gives the minimum values for $E_{\mathrm{emp}}(X)$ under the rank constraint. Since $E_{\mathrm{emp}}(X)$ depends on the problem, normalized by that of full KPCA is also used, $E_{\mathrm{emp}}(X)/E_{\mathrm{emp}}(P_{\mathrm{Fkp}})$, where P_{FKp} is a projector of full KPCA. Validation error E_{val} that uses validation samples instead of training samples in the empirical error is also used.

Sample selection	SubKPCA	Reduced KPCA	Improved KPCA	Nyström method
Random	1.0025±0.0019	1.1420±0.0771	4.7998±0.0000	2.3693±0.6826
K-means	1.0001±0.0000	1.0282±0.0114	4.7998±0.0000	1.7520±0.2535
Forward	1.0002±0.0001	1.3786±0.0719	4.7998±0.0000	14.3043±9.0850
Random	0.0045±0.0035	0.2279±0.1419	0.9900±0.0000	0.3583±0.1670
K-means	0.0002±0.0001	0.0517±0.0318	0.9900±0.0000	0.1520±0.0481
Forward	0.0002±0.0001	0.5773±0.0806	0.9900±0.0000	1.7232±0.9016

Table 1. Mean values and standard deviations over 10 trials of E_{emp} and D in Experiment 1: Upper rows are $E_{emp}(X)/E_{emp}(P_{FKp})$; lower rows are D; Sparse KPCA does not require sample selection.

The alternative criterion is operator distance from full KPCA. Since these methods are approximation of full KPCA, an operator that is closer to that of full KPCA is the better one. In the feature space, the distance between projectors is measured by the Frobenius distance,

$$D(X, P_{FKp}) = \|X - P_{FKp}\|_F. \tag{47}$$

For example, if $X = P_{SubKPCA} = TZZ^\top T^*$ (Eq. (27)),

$$D^2(P_{SubKPCA}, P_{FKp}) = \|TZZ^\top T^* - SV_{KPCA}\Lambda^{-1}V_{KPCA}^\top\|_F^2$$
$$= \text{Trace}[Z^\top K_y Z + V_{KPCA}^\top K V_{KPCA}\Lambda^{-1}$$
$$- 2Z^\top K_{xy}^\top V_{KPCA}\Lambda^{-1}V_{KPCA}^\top K_{xy}Z].$$

6.2 Artificial data

Two-dimensional artificial data described in Introduction is used again with more comparisons and quantitative evaluation. Gaussian kernel function $k(x_1, x_2) = \exp(-0.1\|x_1 - x_2\|^2)$ and the number of principal components, $r = 5$ are chosen. Training samples of Reduced KPCA and the basis set of SubKPCA, Nyström approximation, and IKPCA are identical, and chosen randomly. For Sparse KPCA (SpKp), a parameter σ is chosen to have the same sparsity level with SubKPCA. Figure 2 shows contour curves and values of evaluation criteria. From evaluation criteria E_{emp} and D, SubKp shows the best approximation accuracy among these methods.

Table 1 compares sample selection methods. The values in the table are the mean values and standard deviations over 10 trials using different random seeds or initial point. SubKPCA performed better than the other methods. Regarding sample selection, K-means and forward search give almost the same results for SubKPCA.

6.3 Open dataset

Three open benchmark datasets, "concrete," "housing," and "tic" from UCI (University of California Irvine) machine learning repository are used [5] (Asuncion & Newman, 2007). Table 2 shows properties of the datasets.

---

[5] As of Oct. 2011, the datasets are available from http://archive.ics.uci.edu/ml/index.html

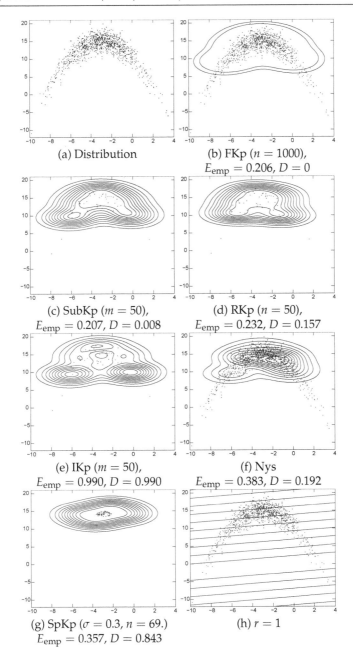

(a) Distribution

(b) FKp ($n = 1000$),
$E_{\mathrm{emp}} = 0.206, D = 0$

(c) SubKp ($m = 50$),
$E_{\mathrm{emp}} = 0.207, D = 0.008$

(d) RKp ($n = 50$),
$E_{\mathrm{emp}} = 0.232, D = 0.157$

(e) IKp ($m = 50$),
$E_{\mathrm{emp}} = 0.990, D = 0.990$

(f) Nys
$E_{\mathrm{emp}} = 0.383, D = 0.192$

(g) SpKp ($\sigma = 0.3, n = 69.$)
$E_{\mathrm{emp}} = 0.357, D = 0.843$

(h) $r = 1$

Fig. 2. Contour curves of projection norms

Gaussian kernel $k(\boldsymbol{x}_1.\boldsymbol{x}_2) = \exp(-\|\boldsymbol{x}_1 - \boldsymbol{x}_2\|^2/(2\sigma^2))$ whose σ^2 is set to be the variance of for all elements of each dataset is used for the kernel function. The number of principal

dataset	no. of dim.	no. of samples
concrete	9	1030
housing	14	506
tic	85	9822

Table 2. Open dataset

components, r, is set to be the input dimension of each dataset. 90% of samples are used for training, and the remaining 10% of samples are used for validation. The division of the training and the validation sets is repeated 50 times randomly.

Figures 3-(a) and (b) show the averaged squared distance from KPCA using all samples. SubKPCA shows better performance than Reduced KPCA and the Nyström method, especially SubKPCA with a forward search performed the best of all. In both datasets, even if the number of basis is one of tenth that of all samples, the distance error of SubKPCA is less than 1%.

Figures 3-(c) and (d) show the average normalized empirical error, and Figures (e) and (f) show the averaged validation error. SubKPCA with K-means or forward search performed the best, and its performance did not change much with 20% more basis. The results for the Nyström method are outside of the range illustrated in the figures.

Figures 4-(a) and (b) show the calculation times for construction. The simulation was done on the system that has an Intel Core 2 Quad CPU 2.83GHz and an 8Gbyte RAM. The routines dsygvx and dsyevx in the Intel math kernel library (MKL) were respectively used for the generalized eigenvalue decomposition of SubKPCA and the eigenvalue decomposition of KPCA. The figures indicate that SubKPCA is faster than Full KPCA if the number of basis is less than 80%.

Figure 5 shows the relation between runtime [s] and squared distance from Full KPCA. In this figure, "kmeans" includes runtime for K-means clustering. The vertical dotted line stands for run-time of full KPCA. For (a) concrete and (b) housing, incomplete Cholesky decomposition is faster than our method. However, for a larger dataset, (c) tic, incomplete Cholesky decomposition is slower than our method. KHA-SMD Günter et al. (2007) is slower than full KPCA in these three methods.

6.4 Classification

PCA and KPCA are also used for classifier as subspace methods (Maeda & Murase, 1999; Oja, 1983; Tsuda, 1999). Subspace methods obtain projectors onto subspaces that correspond with classes. Let P_i be a projector onto the subspace of the class i. In the class feature information compression (CLAFIC) that is one of the subspace methods, P_i is a projector of PCA for each class. Then an input sample x is classified to a class k whose squared distance is the largest, that is,

$$k = \underset{i=1,\dots,c}{\operatorname{argmax}} \|x - P_i x\|^2, \tag{48}$$

where c is the number of classes. Binary classifiers such as SVM cannot be applied to multi-class problems directly, therefore, some extentions such as one-against-all strategy have to be used. However, subspace methods can be applied to many-class problems

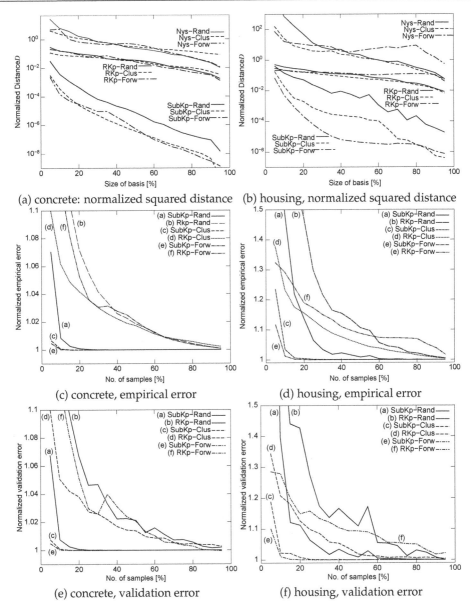

(a) concrete: normalized squared distance (b) housing, normalized squared distance

(c) concrete, empirical error

(d) housing, empirical error

(e) concrete, validation error

(f) housing, validation error

Fig. 3. Results for open datasets. Rand: random, Clus: Clustering (K-means), Forw: Forward search

easily. Furthermore, subspace methods are easily to be applied to multi-label problems or class-addition/reduction problems. CLAFIC is easily extended to KPCA (Maeda & Murase, 1999; Tsuda, 1999).

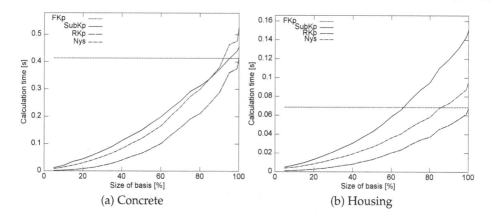

(a) Concrete   (b) Housing

Fig. 4. Calculation time

	No. of basis			
Method	10%	30%	50%	100%
SubKp (nc)	3.89±0.82	2.87±0.71	2.55±0.64	2.03±0.56
SubKp (c)	3.93±0.82	2.83±0.70	2.55±0.64	2.02±0.56
RKp (nc)	4.92±0.95	3.17±0.71	2.65±0.60	2.03±0.56
RKp (c)	4.91±0.92	3.16±0.70	2.65±0.62	2.02±0.56
CLAFIC (nc)	5.24±0.92	3.64±0.79	3.25±0.70	2.95±0.73
CLAFIC (c)	5.24±0.89	3.71±0.76	3.38±0.68	3.06±0.74

Table 3. Minimum validation errors [%] and standard deviations I; random selection, nc: non-centered, c: centered

A handwritten digits database, USPS (U.S. postal service database), is used for the demonstration. The database has 7291 images for training, and 2001 images for testing. Each image is 16x16 pixel gray-scale, and has a label $(0, \ldots, 9)$.

10% of samples (729 samples) from training set are extracted for validation, and rest 90% (6562 samples) are used for training. This division is repeated 100 times, and obtained the optimal parameters from several picks, width of Gaussian kernel $c \in \{10^{-4.0}, 10^{-3.8}, \ldots, 10^{0.0}\}$, the number of principal components $r \in \{10, 20, \ldots, 200\}$.

Tables 3 and 4 respectively show the validation errors [%] and standard deviations over 100 validations when the samples of the basis are selected randomly and by k-means respectively. SubKPCA has lower error rate than reduced KPCA when the number of basis is small. Tables 5 and 6 show the test errors when the optimum parameters are given by the validation.

6.5 Denoising using a huge dataset

KPCA is also used for image denoising (Kim et al., 2005; Mika et al., 1999). This subsection demonstrate image denoising by KPCA using MNIST database. The database has 60000 images for training, and 10000 samples for testing. Each image is a 28x28 pixel gray-scale image of a handwritten digit. Each pixel value of the original image is scaled from 0 to 255.

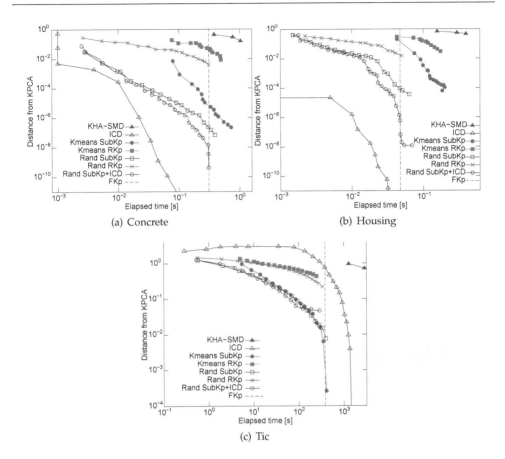

Fig. 5. Relation between runtime [s] and squared distance from Full KPCA

	No. of basis			
Method	10%	30%	50%	100%
SubKp (nc)	2.69±0.58	2.30±0.61	2.18±0.58	2.03±0.56
SubKp (c)	2.68±0.60	2.29±0.61	2.15±0.55	2.02±0.56
RKp (nc)	2.75±0.61	2.35±0.60	2.22±0.57	2.03±0.56
RKp (c)	2.91±0.63	2.40±0.59	2.24±0.58	2.02±0.56
PCA (nc)	3.60±0.66	3.38±0.60	3.21±0.56	3.03±0.60

Table 4. Minimum validation errors [%] and standard deviations II; K-means, nc: non-centered, c: centered

Before the demonstration of image denoising, comparisons of computational complexities are presented since the database has rather large data. The Gaussian kernel function $k(x_1, x_2) = \exp(-10^{-5.1}\|x_1 - x_2\|^2)$ and the number of principal components $r = 145$ are used because these parameters show the best result in latter denoising experiment. The random selection

	No. of basis			
Method	10%	30%	50%	100%
SubKp (nc)	6.50±0.36	5.69±0.15	5.20±0.14	4.78±0.00
SubKp (c)	6.54±0.36	5.52±0.16	5.20±0.14	4.83±0.00
RKp (nc)	7.48±0.44	5.71±0.31	5.26±0.20	4.78±0.00
RKp (c)	7.50±0.43	5.76±0.31	5.28±0.20	4.83±0.00

Table 5. Test errors [%] and standard deviations; random selection, nc: non-centered, c: centered

	No. of basis			
Method	10%	30%	50%	100%
SubKp (nc)	5.14±0.17	4.99±0.14	4.97±0.13	4.78±0.00
SubKp (c)	5.14±0.18	4.99±0.14	4.87±0.14	4.83±0.00
RKp (nc)	5.18±0.21	5.01±0.16	4.89±0.15	4.78±0.00
RKp (c)	5.36±0.24	5.07±0.17	4.93±0.15	4.83±0.00

Table 6. Test errors [%] and standard deviations; K-means, nc: non-centered, c: centered

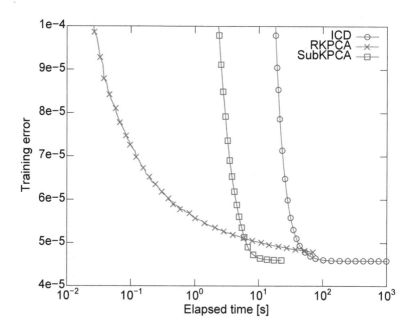

Fig. 6. Relation between training error and elapsed time in MNIST dataset

is used for basis of the SubKPCA. Figure 6 shows relation between run-time and training error. SubKPCA achieves lower training error $E_{\text{emp}} = 4.57 \times 10^{-5}$ in 28 seconds, whereas ICD achieves $E_{\text{emp}} = 4.59 \times 10^{-5}$ in 156 seconds.

Denoising is done by following procedures,

1. Rescale each pixel value from 0 to 1.
2. Obtain the subset using K-means clustering from 60000 training samples.
3. Obtain operators,
   (a) Obtain centered SubKPCA using 60000 training samples and the subset.
   (b) Obtain centered KPCA using the subset.
4. Prepare noisy images using 10000 test samples;
   (a) Add Gaussian noise whose variance is σ^2.
   (b) Add salt-and-pepper noise with a probability of p (a pixel flips white (1) with probability $p/2$, and flips black (0) with probability $p/2$).
5. Obtain each transformed vector and pre-image using the method in (Mika et al., 1999).
6. Rescale each pixel value from 0 to 255, and truncate values if the values less than 0 or grater than 255.

The evaluation criterion is the mean squared error

$$E_{\text{MSE}} = \frac{1}{10000} \sum_{i=1}^{10000} \|f_i - \hat{f}_i\|^2, \tag{49}$$

where f_i is the ith original test image, and $\hat{f}$ is its denoising image. The optimal parameters, r: the number of principal components, and c: parameter of the Gaussian kernel, are chosen to show the best performance in several picks $r \in \{5, 10, 15, \ldots, m\}$ and $c \in \{10^{-6.0}, 10^{-5.9}, \ldots, 10^{-2.0}\}$.

Tables 7 and 8 are denoising results. SubKPCA shows always lower errors than errors of Reduced KPCA. Figures 7 show the original images, noisy images, and de-noised images. Fields of experts (FoE) Roth & Black (2009) and block-matching and 3D filtering (BM3D) Dabov et al. (2007) are state-of-the-art denoising methods for natural images [6]. FoE and BM3D

σ	20	50	80	100
SubKp (100)	3.38±1.37	4.64±1.49	6.73±1.70	8.33±2.17
RKp (100)	3.48±1.42	4.71±1.51	6.80±1.74	8.55±2.11
SubKp (500)	0.99±0.24	3.64±0.82	6.22±1.43	7.95±1.91
RKp (500)	1.01±0.27	3.73±0.81	6.39±1.51	8.14±2.01
SubKp (1000)	0.93±0.22	3.20±0.83	5.11±1.67	6.18±2.02
RKp (1000)	0.94±0.20	3.27±0.87	5.49±1.60	7.18±1.88
WF	0.88±0.24	3.14±0.81	5.49±1.43	7.01±1.84
FoE	1.15±2.08	8.48±0.78	23.29±1.90	36.53±2.81
BM3D	1.07±1.80	7.17±1.09	17.39±2.96	25.49±4.17

Table 7. Denoising results for Gaussian noise , mean and SD of squared errors, values are divided by 10^5; the numbers in brackets denote the numbers of basis

---

[6] MATLAB codes were downloaded from http://www.gris. tu-darmstadt.de/~sroth/research/foe/downloads.html.   and http://www.cs.tut.fi/~foi/GCF-BM3D/index.html

p	0.05	0.10	0.20	0.40
SubKp (100)	4.73±1.51	6.45±1.73	10.07±2.16	18.11±3.06
RKp (100)	4.79±1.54	6.52±1.72	10.51±2.30	18.80±3.30
SubKp (500)	3.61±1.00	5.73±1.34	9.66±2.02	17.87±2.95
RKp (500)	3.72±1.00	6.00±1.39	10.04±2.13	18.31±3.05
SubKp (1000)	3.22±0.98	4.99±1.46	7.93±2.22	13.58±4.72
RKp (1000)	3.25±1.00	5.15±1.53	8.78±2.22	18.21±2.96
WF	3.15±0.87	5.15±1.17	8.93±1.72	17.07±2.61
Median	3.26±1.44	4.34±1.70	7.36±2.45	21.78±5.62

Table 8. Denoising results for salt-and-pepper noise , mean and SD of squared errors, values are divided by 10^5; the numbers in brackets denote the numbers of basis

are assumed that the noise is Gaussian whose mean is zero and variance is known. Thus these two methods are compared only in Gaussian noise case. Since the datasets is not natural images, these methods are not better than SubKPCA. "WF" and "Median" denote Wiener filter and median filter respectively. When noise is relatively small, ($\sigma = 20 \sim 50$ in Gaussian or $p = 0.05 \sim 0.10$), these classical methods show better performance. On the other hand, when noise is large, our method shows better performance. Note that Wiener filter is known to be the optimal filter in terms of the mean squared error among linear operators. From different point of view, Wiener filter is optimal among all linear and non-linear operators if both signal and noise are Gaussian. However, KPCA is non-linear because of non-linear mapping Φ, and pixel values of images and salt-and-pepper noise are not Gaussian in this case.

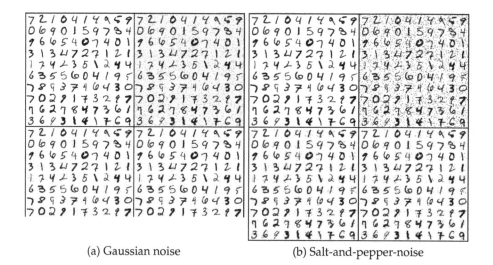

(a) Gaussian noise   (b) Salt-and-pepper-noise

Fig. 7. Results of denoising (first 100 samples), top-left: original image, top-right: noisy image (Gaussian, $\sigma = 50$), bottom-left: image de-noised by SubKPCA, bottom-right: image de-noised by KPCA.

7. Conclusion

Theories, properties, and numerical examples of SubKPCA have been presented in this chapter. SubKPCA has a simple solution form Eq. (22) and no constraint for its kernel functions. Therefore, SubKPCA can be applied to any applications of KPCA. Furthermore, it should be emphasized that SubKPCA is always better than reduced KPCA in the sense of the empirical errors if the subset is the same.

8. References

Asuncion, A. & Newman, D. (2007). UCI machine learning repository.
  URL: *http://www.ics.uci.edu/~mlearn/MLRepository.html*

Bach, F. R. & Jordan, M. I. (2002). Kernel independent component analysis, *Journal of Machine Learning Research* 3: 1–48.

Dabov, K., Foi, A., Katkovnik, V. & Egiazarian, K. (2007). Image denoising by sparse 3D transform-domain collaborative filtering, *IEEE Trans. on Image Processing* 16(8): 2080–2095.

Demmel, J. (1997). *Applied Numerical Linear Algebra*, Society for Industrial Mathematics.

Ding, M., Tian, Z. & Xu, H. (2010). Adaptive kernel principal component analysis, *Signal Processing* 90(5): 1542–1553.

Girolami, M. (2002). Mercer kernel-based clustering in feature space, *IEEE Trans. on Neural Networks* 13(3): 780–784.

Gorban, A., Kégl, B., Wunsch, D. & (Eds.), A. Z. (2008). *Principal Manifolds for Data Visualisation and Dimension Reduction*, LNCSE 58, Springer.

Günter, S., Schraudolph, N. N. & Vishwanathan, S. V. N. (2007). Fast iterative kernel principal component analysis, *Journal of Machine Learning Research* 8: 1893–1918.

Harville, D. A. (1997). *Matrix Algebra From a Statistician's Perspective*, Springer-Verlag.

Hastie, T. & Stuetzle, W. (1989). Principal curves, *Journal of the American Statistical Association* Vol. 84(No. 406): 502–516.

Israel, A. B. & Greville, T. N. E. (1973). *Generalized inverses, Theorey and applications*, Springer.

Kim, K., Franz, M. O. & Schölkopf, B. (2005). Iterative kernel principal component analysis for image modeling, *IEEE Trans. Pattern Analysis and Machine Intelligence* 27(9): 1351–1366.

Lehoucq, R. B., Sorensen, D. C. & Yang, C. (1998). *ARPACK Users' Guide: Solution of Large-Scale Eigenvalue Problems with Implicitly Restarted Arnoldi Methods*, Software, Environments, and Tools 6, SIAM.

Maeda, E. & Murase, H. (1999). Multi-category classification by kernel based nonlinear subspace method, *IEEE International Conference On Acoustics, speech, and signal processing (ICASSP)*, Vol. 2, IEEE press., pp. 1025–1028.

Mika, S., Schölkopf, B. & Smola, A. (1999). Kernel PCA and de-noising in feature space, *Advances in Neural Information Processing Systems (NIPS)* 11: 536–542.

Oja, E. (1983). *Subspace Methods of Pattern Recognition*, Wiley, New-York.

Platt, J. C. (1999). Fast training of support vector machines using sequential minimal optimization, *in* B. Scholkopf, C. Burges & A. J. Smola (eds), *Advances in Kernel Methods - Support Vector Learning*, MIT press, pp. 185–208.

Roth, S. & Black, M. J. (2009). Fields of experts, *International Journal of Computer Vision* 82(2): 205–229.

Roweis, S. T. & Saul, L. K. (2000). Nonlinear dimensionality reduction by locally linear embedding, *Science* 290: 2323–2326.

Schölkopf, B., Mika, S., Burges, C., Knirsch, P., Müller, K.-R., Rätsch, G. & Smola, A. (1999). Input space vs. feature space in kernel-based methods, *IEEE Trans. on Neural Networks* 10(5): 1000–1017.

Schölkopf, B., Smola, A. & Müller, K.-R. (1998). Nonlinear component analysis as a kernel eigenvalue problem, *Neural Computation* 10(5): 1299–1319.

Smola, A. J., Mngasarian, O. L. & Schölkopf, B. (1999). Sparse kernel feature analysis, *Technical report 99-04, University of Wisconsin* .

Tenenbaum, J. B., de Silva, V. & Langford, J. C. (2000). A global geometric framework for nonlinear dimensionality reduction, *Science* 290: 2319–2323.

Tipping, M. E. (2001). Sparse kernel principal component analysis, *Advances in Neural Information Processing Systems (NIPS)* 13: 633–639.

Tsuda, K. (1999). Subspace classifier in the Hilbert space, *Pattern Recognition Letters* 20: 513–519.

Vapnik, V. (1998). *Statistical Learning Theory*, Wiley, New-York.

Washizawa, Y. (2009). Subset kernel principal component analysis, *Proceedings of 2009 IEEE International Workshop on Machine Learning for Signal Processing*, IEEE, pp. 1–6.

Williams, C. K. I. & Seeger, M. (2001). Using the Nyström method to speed up kernel machines, *Advances in Neural Information Processing Systems (NIPS)* 13: 682–688.

Xu, Y., Zhang, D., Song, F., Yang, J., Jing, Z. & Li, M. (2007). A method for speeding up feature extraction based on KPCA, *Neurocomputing* pp. 1056–1061.

6

Multilinear Supervised Neighborhood Preserving Embedding Analysis of Local Descriptor Tensor

Xian-Hua Han and Yen-Wei Chen
Ritsumeikan University
Japan

1. Introduction

Subspace learning based pattern recognition methods have attracted considerable interests in recent years, including Principal Component Analysis (PCA), Independent Component Analysis (ICA), Linear Discriminant Analysis (LDA), and some extensions for 2D analysis. However, a disadvantage of all these approaches is that they perform subspace analysis directly on the reshaped vector or matrix of pixel-level intensity, which is usually unstable under appearance variance. In this chapter, we propose to represent an image as a local descriptor tensor, which is a combination of the descriptor of local regions (K*K-pixel patch) in the image, and is more efficient than the popular Bag-Of-Feature (BOF) model for local descriptor combination. As we know that the idea of BOF is to quantize local invariant descriptors, e.g., obtained using some interest-point detector techniques by Harris & Stephens (1998), and a description with SIFT by Lowe (2004) into a set of visual words by Lazebnik et al. (2006). The frequency vector of the visual words then represents the image, and an inverted file system is used for efficient comparison of such BOFs. However, the BOF model approximately represents each local descriptor feature as a predefined visual word, and vectorizes the local descriptors of an image into a orderless histogram, which may lose some important (discriminant) information of local features and spatial information hold in the local regions of the image. Therefore, this paper proposes to combine the local features of an image as a descriptor tensor. Because the local descriptor tensor retains all information of local features, it will be more efficient for image representation than the BOF model and then can use a moderate amount of local regions to extract the descriptor for image representation, which will be more effective in computational time than the BOF model. For feature representation of image regions, SIFT proposed by Lowe (2004) is improved to be a powerful local descriptor by Lazebnik et al. (2006) for object or scene recognition, which is somewhat invariant to small illumination change. However, in some benchmark database such as YALE and PIE face data sets by Belhumeur et al. (1997), the illumination variance is very large. Then, in order to extract robust features invariant to large illumination, we explore an improved gradient (intensity-normalized gradient) of the image and use histogram of orientation weighed with the improved gradient for local region representation.

With the local descriptor tensor of image representation, we propose to use a tensor subspace analysis algorithm, which is called as multilinear Supervised Neighborhood Preserving Embedding (MSNPE), for discriminant feature extraction, and then use it for object or scene recognition. As we know, subspace learning approaches, such as PCA and LDA by Belhumeur et al. (1997), have widely used in computer vision research filed for feature extraction or selection and have been proven to be efficient for modeling or classification.

Recently there are considerable interests in geometrically motivated approaches to visual analysis. Therein, the most popular ones include locality preserving projection by He et al. (2005), neighborhood preserving embedding, and so on, which cannot only preserve the local structure between samples but also obtain acceptable recognition rates for face recognition. In real applications, all these subspace learning methods need to firstly reshape the multilinear data into a 1D vector for analysis, which usually suffers an overfitting problem. Therefore, some researchers proposed to solve the curse-of-dimension problem with 2D subspace learning such as 2-D PCA and 2-D LDA by ming Wang et al. (2009) for analyzing directly on a 2D image matrix, which was proven to be suitable in some extend. However, all of the conventional methods usually perform subspace analysis directly on the reshaped vector or matrix of pixel-level intensity, which would be unstable under illumination and background variance. In this paper, we propose MSNPE for discriminant feature extraction on the local descriptor tensor. Unlike tensor discriminant analysis by Wang (2006), which equally deals with the samples in the same category, the proposed MSNPE uses neighbor similarity in the same category as a weight of minimizing the cost function for N^{th} order tensor analysis, which is able to estimate geometrical and topological properties of the sub-manifold tensor from random points ("scattered data") lying on this unknown sub-manifold. In addition, compared with TensorFaces by Casilescu & D.Terzopoulos (2002) method, which also directly analyzes multi-dimensional data, the proposed multilinear supervised neighborhood preserving embedding uses supervised strategy and thus can extract more discriminant features for distinguishing different objects and, at the same time, can preserve samples' relationship of inner object instead of only dimension reduction in TensorFaces. We validate our proposed algorithm on different benchmark databases such as view-based object data sets (Coil-100 and Eth-70) and Facial image data sets (YALE and CMU PIE) by Belhumeur et al. (1997) and Sim et al. (2001).

2. Related work

In this section, we firstly briefly introduce the tensor algebra and then review subspace-based feature extraction approaches such as PCA, LPP.

Tensors are arrays of numbers which transform in certain ways under coordinate transformations. The order of a tensor $\mathcal{X} \in R^{N_1 \times N_2 \times \cdots \times N_M}$, represented by a multi-dimensional array of real numbers, is M. An element of $\mathcal{X}$ is denoted as $\mathcal{X}_{i_1,i_2,\cdots,i_M}$, where $1 \leq i_j \leq N_j$ and $1 \leq j \leq M$. In the tensor terminology, the mode-j vectors of the nth-order tensor $\mathcal{X}$ are the vectors in R^{N_j} obtained from $\mathcal{X}$ by varying the index i_j while keeping the other indices fixed. For example, the column vectors in a matrix are the mode-1 vectors and the row vectors in a matrix are the mode-2 vectors.

Definition. (**Modeproduct**). The tensor product $\mathcal{X}_{\times d}\mathbf{U}$ of tensor $\mathcal{X} \in R^{N_1 \times N_2 \times \cdots \times N_M}$ and a matrix $\mathbf{U} \in R^{N_d \times N'}$ is the $N_1 \times N_2 \times \cdots \times N_{d-1} \times N' \times N_{d+1} \times \cdots \times N_M$ tensor:

$$(\mathcal{X}_{\times d}\mathbf{U})_{i_1,i_2,\cdots,i_{d-1},j,i_{d+1},\cdots,i_M} = \sum_{i_d}(\mathcal{X}_{i_1,i_2,\cdots,i_{d-1},i_d,i_{d+1},\cdots,i_M}\mathbf{U}_{i_d,j}) \tag{1}$$

for all index values. $\mathcal{X}_{\times d}\mathbf{U}$ means the mode d's product of the tensor $\mathcal{X}$ with the matrix $\mathbf{U}$. The mode product is a special case of a contraction, which is defined for any two tensors not just for a tensor and a matrix. In this paper, we follow the definitions in Lathauwer (1997) and avoid the use of the term "contraction".

In tensor analysis, Principal Component Analysis (PCA) is used to extract the basis for each mode. The proposed MSNPE approach is based on the basis idea of Locality Preserving Projection (LPP). Therefore, we simply introduce PCA, LPP and a 2D extension of LPP as the following.

(1) Principal component analysis extracts the principal eigen-space associated with a set (matrix) $\mathbf{X} = [\mathbf{x}_i|_{i=1}^N]$ of training samples ($\mathbf{x}_i \in R^n$ with $1 \leq i \leq N$; N: sample number; n: dimension of the samples). Let $\mathbf{m}$ be the mean of the N training samples, and $\mathbf{C} = \frac{1}{N}\sum_{i=1}^N (\mathbf{x}_i - \mathbf{m})(\mathbf{x}_i - \mathbf{m})^T$ be the covariance matrix of the $\mathbf{x}_i$. One solves the eigenvalue equation $\lambda \mathbf{u}_i = \mathbf{C}\mathbf{u}_i$ for eigenvalues $\lambda_i \geq 0$. The principal eigenspace $\mathbf{U}$ is spanned by the first K eigenvectors with the largest eigenvalues, $\mathbf{U} = [\mathbf{u}_i|_{i=1}^K]$. If $\mathbf{x}_t$ is a new feature vector, then it is projected to eigenspace $\mathbf{U}$: $\mathbf{y}_t = \mathbf{U}^T(\mathbf{x}_t - \mathbf{m})$. The vector $\mathbf{y}_t$ is used in place of $\mathbf{x}_t$ for representation and classification.

(2)Locality Preserving Projection: LPP seeks a linear transformation $\mathbf{P}$ to project high-dimensional data into a low-dimensional sub-manifold that preserves the local Structure of the data. Let $\mathbf{X} = [\mathbf{x}_1, \mathbf{x}_2, \cdots, \mathbf{x}_N]$ denotes the set representing features of N training image samples, and $\mathbf{Y} = [\mathbf{y}_1, \mathbf{y}_2, \cdots, \mathbf{y}_N] = [\mathbf{P}^T\mathbf{x}_1, \mathbf{P}^T\mathbf{x}_2, \cdots, \mathbf{P}^T\mathbf{x}_N]$ denotes the samples feature in transformed subspace. Then, the linear transformation $\mathbf{P}$ can be obtained by solving the following minimization problem with some constraints, which will be given later:

$$\min_{\mathbf{P}} \sum_{ij} ||\mathbf{y}_i - \mathbf{y}_j||^2 W_{ij} = \min_{\mathbf{P}} \sum_{ij} ||\mathbf{P}^T\mathbf{x}_i - \mathbf{P}^T\mathbf{x}_j||^2 W_{ij} \qquad (2)$$

where W_{ij} evaluate the local structure of the image space. It can be simply defined as follows:

$$W_{ij} = \begin{cases} 1 \text{ if } \mathbf{x}_i \text{ is among the } k \text{ nearest neighbors of } \mathbf{x}_j \\ 0 \text{ otherwise} \end{cases} \qquad (3)$$

By simple algebra formulation, the objective function can be reduced to:

$$\frac{1}{2}\sum_{ij}(\mathbf{P}^T\mathbf{x}_i - \mathbf{P}^T\mathbf{x}_j)^2 W_{ij} = \sum_i \mathbf{P}^T\mathbf{x}_i D_{ii}\mathbf{x}_i^T\mathbf{P} - \sum_{ij}\mathbf{P}^T\mathbf{x}_i W_{ij}\mathbf{x}_i^T\mathbf{P}$$
$$= \mathbf{P}^T\mathbf{X}(\mathbf{D} - \mathbf{W})\mathbf{X}^T\mathbf{P} = \mathbf{P}^T\mathbf{X}\mathbf{L}\mathbf{X}^T\mathbf{P} \qquad (4)$$

where each column $\mathbf{P}_i$ of the LPP linear transformation matrix $\mathbf{P}$ can not be zero vector, and a constraint is imposed as follows:

$$\mathbf{Y}^T\mathbf{D}\mathbf{Y} = \mathbf{I} \Rightarrow \mathbf{P}^T\mathbf{X}\mathbf{D}\mathbf{X}^T\mathbf{P} = \mathbf{I} \qquad (5)$$

where $\mathbf{I}$ in constraint term $\mathbf{P}^T\mathbf{X}\mathbf{D}\mathbf{X}^T\mathbf{P} = \mathbf{I}$ or $\mathbf{Y}^T\mathbf{D}\mathbf{Y} = \mathbf{I}$ is an identity matrix. $\mathbf{D}$ is a diagonal matrix; its entries are column (or row, since $\mathbf{W}$ is symmetric) sums of $\mathbf{W}$, $D_{ii} = \sum_j W_{ij}$; $\mathbf{L} = \mathbf{D} - \mathbf{W}$ is the Laplacian matrix [5]. Matrix D provides a natural measure on the data samples. The bigger the value D_{ii} (corresponding to $\mathbf{y}_i$) is, the more importance is $\mathbf{y}_i$. The constraint for the sample $\mathbf{y}_i$ in $\mathbf{Y}^T\mathbf{D}\mathbf{Y} = \mathbf{I}$ is $D_{ii} * \mathbf{y}_i^T\mathbf{y}_i = 1$, which means that the more importance (D_{ii} is larger) the sample $\mathbf{y}_i$ is, the smaller the value of $\mathbf{y}_i^T\mathbf{y}_i$ is. Therefor, the constraint $\mathbf{Y}^T\mathbf{D}\mathbf{Y} = \mathbf{I}$ will try to make the important point (has density distribution around the important point) near the origin of the projected subspace. Then, the density region near the origin of the projected subspace includes most of the samples, which can make the objecrive function in Eq. (2) as small as possible, and at same time, can avoid the trivial solution $||\mathbf{P}_i||^2 = 0$ for the transformation matrix $\mathbf{P}$.

Then, The linear transformation $\mathbf{P}$ can be obtained by minimizing the objective function under constraint $\mathbf{P}^T\mathbf{X}\mathbf{D}\mathbf{X}^T\mathbf{P} = \mathbf{I}$:

$$\underset{\mathbf{P}^T\mathbf{X}\mathbf{D}\mathbf{X}^T\mathbf{P}=\mathbf{I}}{\text{argmin}} \ \mathbf{P}^T\mathbf{X}(\mathbf{D} - \mathbf{W})\mathbf{X}^T\mathbf{P} \tag{6}$$

Finally, the minimization problem can be converted to solve a generalized eigenvalue problem as follows:

$$\mathbf{X}\mathbf{L}\mathbf{X}^T\mathbf{P} = \lambda\mathbf{X}\mathbf{D}\mathbf{X}^T\mathbf{P} \tag{7}$$

In Face recognition application, He et al [8] extended LPP method into 2D dimension analysis, named as Tensor Subspace Analysis (TSA). TSA can directly deal with 2D gray images, and achieved better recognition results than the conventional 1D subspace learning methods such as PCA, LDA and LPP. However, for object recognition, color information also plays an important role for distinguishing different objects. Then, in this paper, we extend LPP to ND tensor analysis, which can directly deal with not only 3D Data but also ND data structure. At the same time, in order to obtain stable transformation tensor basis, we regularize a term in the proposed MSNPE objective function for abject recognition, which is introduced in Sec. 3 in detail.

3. Local descriptor tensor for image representation

In computer vision, local descriptors (i.e., features computed over limited spatial support) have been proven to be well-adapted for matching and recognition tasks as they are robust to partial visibility and clutter. The current popular one for a local descriptor is the SIFT feature, which is proposed by Lowe (2004). With the local SIFT descriptor, usually there are two types of algorithms for object recognition. One is to match the local points with SIFT features in two images, and the other one is to use the popular BOF model, which forms a frequency histogram of a predefined visual-words for all sampled region features by Belhumeur et al. (1997). For a matching algorithm, it is usually not enough to recognize the unknown image even if there are several points that are well matched. The popular BOF model usually can achieve good recognition performance in most applications such as scene and object recognition. However, in BOF model, in order to achieve an acceptable recognition rate, it is necessary to sample a lot of points for extracting SIFT features (usually more than 1000 in an image) and to compare the extracted local SIFT feature with the predefined visual words (usually more than 1000) to obtain the visual-word occurrence histogram. Therefore, BOF model needs a lot of computing time to extract visual-words occurrence histogram. In addition, BOF model just approximately represents each local region feature as a predefined visual-word; then, it may lose a lot of information and will be not efficient for image representation. Therefore, in this paper, we propose to represent a color or gray image as a combined local descriptor tensor, which can use different features (such as SIFT or other descriptors) for local region representation.

In order to extract the local descriptor tensor for image representation, we firstly grid-segment an image into K regions with some overlapping, and in each region, we extract some descriptors (can be consider tensor) for local region representation. For a gray image, a M-dimensional feature vector, which can be considered as a 1D tensor, is extracted from the local gray region. For a color image, a M-dimensional feature vector can be extracted from each color channel such as R, G and B color channels. With the feature vectors of the three color channels, a combined 2D $M \times 3$ tensor can represent the local color region. Furthermore we combine the K 1D or 2D local tensor (M-dimensional vector or $M \times 3$ 2D tensor) into a 2D or 3D tensor with of size $M \times K \times L$ (L: 1 or 3). The tensor feature extraction procedure of a color image is shown in Fig. 1(a). For feature representation of the local regions such as the red, orange and green rectangles in Fig. 1 (a), the popular SIFT proposed by Lowe (2004) is proved to be a powerful one for object recognition, which is somewhat invariant to small illumination change. However, in some benchmark database such as YALE and CMU PIE face datasets, the illumination variance is very large. Then, in order to extract robust feature invariant to large illumination, we explore an normalized gradient (intensity-normalized gradient) of the image, and use Histogram of Orientation weighed with Normalized Gradient (NHOG) for local region representation. Therefore, for the benchmark databases without large illumination variance such as COIL-100 dataset or where the illumination information is also useful for recognition such as scene dataset, we use the popular SIFT for local region representation. However, for the benchmark database with large illumination variation, which will be harmful for subject recognition such as YALE and CMU PIE facial datasets, we use Histogram of Orientation weighed with Normalized Gradient (NHOG) for local region representation.

(1) SIFT: The SIFT descriptor computes a gradient orientation histogram within the support region. For each of 8 orientation planes, the gradient image is sampled over a 4 by 4 grid of locations, thus resulting in a 128-dimensional feature vector for each region. A Gaussian window function is used to assign a weight to the magnitude of each sample point. This makes the descriptor less sensitive to small changes in the position of the support region and puts more emphasis on the gradients that are near the center of the region. To obtain robustness to illumination changes, the descriptors are made invariant to illumination transformations of the form $aI(x) + b$ by scaling the norm of each descriptor to unity [8]. For representing the local region of a color image, we extract SIFT feature in each color component (R, G and B color components), and then can achieve a $128 * 3$ 2D tensor for each local region.

(2) Histogram of Orientation weighed with the Normalized Gradient (NHOG): Given an image $\mathbf{I}$, we calculate the improved gradient (Intensity-normalized gradient) using the following Eq.:

$$
\mathbf{I}_x(i,j) = \frac{\mathbf{I}(i+1,j) - \mathbf{I}(i-1,j)}{\mathbf{I}(i+1,j) + \mathbf{I}(i-1,j)}
$$
$$
\mathbf{I}_y(i,j) = \frac{\mathbf{I}(i,j+1) - \mathbf{I}(i,j-1)}{\mathbf{I}(i,j+1) + \mathbf{I}(i,j-1)} \tag{8}
$$
$$
\mathbf{I}_{xy}(i,j) = \sqrt{\mathbf{I}_x(i,j)^2 + \mathbf{I}_y(i,j)^2}
$$

where $\mathbf{I}_x(i,j)$ and $\mathbf{I}_y(i,j)$ mean the horizontal and vertical gradient in pixel position i,j, respectively, $\mathbf{I}_{xy}(i,j)$ means the global gradient in pixel position i,j. The idea of the normalized gradient is from χ^2 distance: a normalized Euclidean distance. For x-direction, the gradient is normalized by summation of the upper one and the bottom one pixel centered by the focused pixel; for y-direction, the gradient is normalized by that of the right and left one. With the intensity-normalized gradient, we can extract robust and invariant features to illumination changing in a local region of an image. Some examples with the intensity-normalized and conventional gradients are shown in Fig. 2

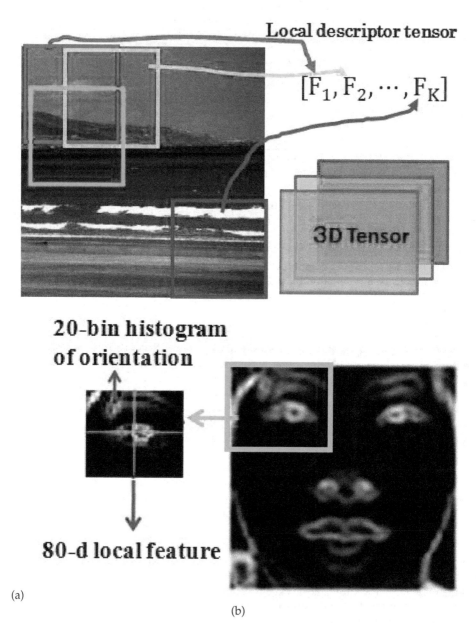

Local descriptor tensor

$$[F_1, F_2, \cdots, F_K]$$

3D Tensor

20-bin histogram of orientation

80-d local feature

(a)

(b)

Fig. 1. (a) Extraction of local descriptor tensor for color image representation; (b)NHOG feature extraction from a gray region.

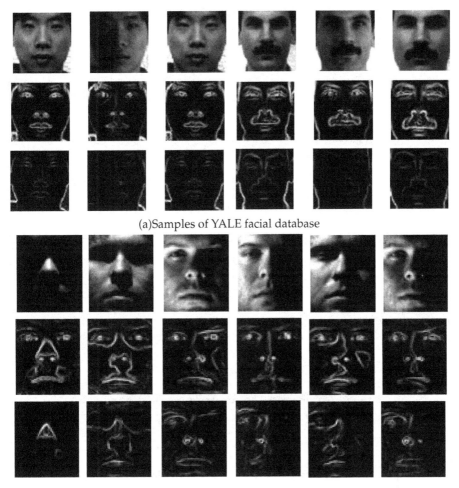

(a)Samples of YALE facial database

(b)Samples of PIE facial database

Fig. 2. Gradient image samples. Top row: Original face images; Middle row: the intensity-normalized gradient images; Bottom row: the conventional gradient images.

For feature extraction of a local region I^R in the normalized gradient image shown in Fig. 1(b), we firstly segment the region into 4 (2×2) patches, and then in each patch extract a 20-bin histogram of orientation weighted by global gradient $\mathbf{I}_{xy}^R$ calculated using the intensity-normalized gradients $\mathbf{I}_x^R$, $\mathbf{I}_y^R$. Therefore, each region in a gray image can be represented by 80-bin (20×4) histogram as shown in Fig. 1(b).

4. Multilinear supervised neighborhood preserving embedding

In order to model N-Dimensional data without rasterization, tensor representation is proposed and analyzed for feature extraction or modeling. In this section, we propose a multilinear supervised neighborhood preserving embedding by Han et al. (2011) Han et al.

(2011) to not only extract discriminant feature but also preserve the local geometrical and topological properties in same category for recognition. The proposed approach decompose each mode of tensor with objective function, which consider neighborhood relation and class label of training samples.

Suppose we have ND tensor objects $\mathcal{X}$ from C classes. The cth class has n^c tensor objects and the total number of tensor objects is n. Let $\mathcal{X}_{i_c} \in R^{N_1 \times N_2 \times \cdots \times N_L}(i_c = 1, 2, \cdots, n^c)$ be the ith object in the cth class. For color object image tensor, L is 3, N_1 is the row number, N_2 is the column number, and N_3 is the color space components (N_3=3). We can build a nearest neighbor graph $\mathcal{G}$ to model the local geometrical structure and label information of $\mathcal{X}$. Let **W** be the weight matrix of $\mathcal{G}$. A possible definition of **W** is as follows:

$$W_{ij} = \begin{cases} exp^{-\frac{\|\mathcal{X}_i - \mathcal{X}_j\|^2}{t}} & \text{if sample } i \text{ and } j \text{ is in same class} \\ 0 & \text{otherwise} \end{cases} \tag{9}$$

where $\|\mathcal{X}_i - \mathcal{X}_j\|^2$ means Euclidean distance of two tensor, which is the summation square root of all corresponding elements between $\mathcal{X}_i$ and $\mathcal{X}_j$, and $\| \bullet \|$ means l_2 norm in our paper.

Let $\mathbf{U}_d$ be the d-mode transformation matrices (Dimension: $N_d \times N'_d$). A reasonable transformation respecting the graph structure can be obtained by solving the following objective functions:

$$\min_{\mathbf{U}_1, \mathbf{U}_2, \cdots, \mathbf{U}_L} \frac{1}{2} \sum_{ij} \|\mathcal{X}_i \times_1 \mathbf{U}_1 \times_2 \mathbf{U}_2 \cdots \times_L \mathbf{U}_L - \mathcal{X}_j \times_1 \mathbf{U}_1 \times_2 \mathbf{U}_2 \cdots \times_L \mathbf{U}_L\|_2 W_{ij} \tag{10}$$

Algorithm 1: ND tensor supervised neighborhood embedding

Input: Tensor objects $\mathcal{X}_i^c$ from C classes, $\mathcal{X}_i^c$ denots the ith tensor object in the c_{th} class

Graph-based weights: Building nearest neighbor graph in same class and calculate the graph weight **W** according to Eq. 9 and **D** from **W**

Initialize: Randomly initialize $\mathbf{U}_r^d \in R^{N_d}$ for d =1,2,$\cdots$, L

for t=1:T (Iteration steps) or until converge **do**

    for d=1:L (Iteration steps) **do**

        • Calculate $\mathbf{D}_d$ and $\mathbf{S}_d$ assuming $\mathbf{U}_i(i = 1, 2, \cdots, d-1, d+1, \cdots, L)$ fixed.

        • Solve the minimizing problem:

        $\min_{\mathbf{U}_d} tr(\mathbf{U}_d^T(\mathbf{D}_d - \mathbf{S}_d)U_d)$ with eigenspace analysis

    end for

end for

output: the MSNPE tensor $\mathcal{T}_j = \mathbf{U}_1 \times \mathbf{U}_2 \times \cdots \times \mathbf{U}_L, j = 1, 2, \cdots, (N'_1 \times N'_2 \times \cdots \times N'_L)$.

Table 1. The flowchart of multilinear supervised neighborhood preserving embedding (MSNPE).

where $\mathcal{X}_i$ is the tensor representation of the ith sample; $\mathcal{X}_{i\times 1}\mathbf{U}_1$ means the mode 1's product of the tensor $\mathcal{X}_i$ with the matrix $\mathbf{U}_1$, and $\mathcal{X}_{i\times 1}\mathbf{U}_{1\times 2}\mathbf{U}_2$ means the mode 2's product of the tensor $\mathcal{X}_{i\times 1}\mathbf{U}_1$ with the matrix $\mathbf{U}_2$, and so on. The above objective function incurs a heavy penalty if neighboring points of same class $\mathcal{X}_i$ and $\mathcal{X}_j$ are mapped far apart. Therefore, minimizing it is an attempt to ensure that if $\mathcal{X}_i$ and $\mathcal{X}_j$ are "close", then $\mathcal{X}_{i\times 1}\mathbf{U}_{1\times 2}\mathbf{U}_2\cdots_{\times L}\mathbf{U}_L$ and $\mathcal{X}_{j\times 1}\mathbf{U}_{1\times 2}\mathbf{U}_2\cdots_{\times L}\mathbf{U}_L$ are "close" as well. Let $\mathcal{Y}_i = \mathcal{X}_{i\times 1}\mathbf{U}_{1\times 2}\mathbf{U}_2\cdots_{\times L}\mathbf{U}_L$ with dimension $N_1\times N_2\times\cdots\times N_L$, and $(\mathbf{Y}_i)^d = (\mathcal{X}_{i\times 1}\mathbf{U}_{1\times 2}\mathbf{U}_2\cdots_{\times d-1}\mathbf{U}_{d-1\times d+1}\mathbf{U}_{d+1}\cdots_{\times L}\mathbf{U}_L)^d$ with dimension:$N_d\times(N_1\times N_2\times\cdots\times N_{d-1}\times N_{d+1}\times\cdots\times N_L)$ is the d-mode extension of tensor $\mathcal{Y}_i$, which is a 2D matrix. Let D be a diagonal matrix,$D_{ii}=\sum_j W_{ij}$. Since $\|\mathbf{A}\|^2 = tr(\mathbf{A}\mathbf{A}^T)$, we see that

$$\frac{1}{2}\sum_{ij}\|\mathcal{X}_{i\times 1}\mathbf{U}_1\cdots_{\times L}\mathbf{U}_L - \mathcal{X}_{j\times 1}\mathbf{U}_1\cdots_{\times L}\mathbf{U}_L\|^2 W_{ij}$$

$$=\frac{1}{2}\sum_{ij}tr(((\mathbf{Y}_i)^d - (\mathbf{Y}_j)^d)((\mathbf{Y}_i)^d - (\mathbf{Y}_j)^d)^T)W_{ij}$$

$$=tr(\sum_i D_{ii}(\mathbf{Y}_i)^d((\mathbf{Y}_i)^d)^T - \sum_{ij}W_{ij}(\mathbf{Y}_i)^d((\mathbf{Y}_j)^d)^T)$$

$$=tr(\sum_i D_{ii}(\mathbf{U}_d^T(\mathcal{X}_{i\times 1}\mathbf{U}_1\cdots_{\times d-1}\mathbf{U}_{d-1\times d+1}\mathbf{U}_{d+1}\cdots_{\times L}\mathbf{U}_L)^d$$

$$((\mathcal{X}_{i\times 1}\mathbf{U}_1\cdots_{\times d-1}\mathbf{U}_{d-1\times d+1}\mathbf{U}_{d+1}\cdots_{\times L}\mathbf{U}_L)^d)^T\mathbf{U}_d$$

$$-\sum_{ij}W_{ij}(\mathbf{U}_d^T(\mathcal{X}_{i\times 1}\mathbf{U}_1\cdots_{\times d-1}\mathbf{U}_{d-1\times d+1}\mathbf{U}_{d+1}\cdots_{\times L}\mathbf{U}_L)^d \tag{11}$$

$$((\mathcal{X}_{j\times 1}\mathbf{U}_1\cdots_{\times d-1}\mathbf{U}_{d-1\times d+1}\mathbf{U}_{d+1}\cdots_{\times L}\mathbf{U}_L)^d)^T\mathbf{U}_d)$$

$$=tr(\mathbf{U}_d^T(\sum_i D_{ii}((\mathcal{X}_{i\times 1}\mathbf{U}_1\cdots_{\times d-1}\mathbf{U}_{d-1\times d+1}\mathbf{U}_{d+1}\cdots_{\times L}\mathbf{U}_L)^d$$

$$((\mathcal{X}_{i\times 1}\mathbf{U}_1\cdots_{\times d-1}\mathbf{U}_{d-1\times d+1}\mathbf{U}_{d+1}\cdots_{\times L}\mathbf{U}_L)^d)^T$$

$$-\sum_{ij}W_{ij}((\mathcal{X}_{i\times 1}\mathbf{U}_1\cdots_{\times d-1}\mathbf{U}_{d-1\times d+1}\mathbf{U}_{d+1}\cdots_{\times L}\mathbf{U}_L)^d$$

$$((\mathcal{X}_{j\times 1}\mathbf{U}_1\cdots_{\times d-1}\mathbf{U}_{d-1\times d+1}\mathbf{U}_{d+1}\cdots_{\times L}\mathbf{U}_L)^d)^T)\mathbf{U}_d)$$

$$=tr(\mathbf{U}_d^T(\mathbf{D}_d - \mathbf{S}_d)\mathbf{U}_d)$$

where $\mathbf{D}_d = \sum_i D_{ii}(\mathcal{X}_{i\times 1}\mathbf{U}_1\cdots_{\times d-1}\mathbf{U}_{d-1\times d+1}\mathbf{U}_{d+1}\cdots_{\times L}\mathbf{U}_L)^d((\mathcal{X}_{i\times 1}\mathbf{U}_1\cdots_{\times d-1}\mathbf{U}_{d-1\times d+1}$ $\mathbf{U}_{d+1}\cdots_{\times L}\mathbf{U}_L)^d)^T$ and $\mathbf{S}_d = \sum_{ij}W_{ij}(\mathcal{X}_{i\times 1}\mathbf{U}_1\cdots_{\times d-1}\mathbf{U}_{d-1\times d+1}\mathbf{U}_{d+1}\cdots_{\times L}\mathbf{U}_L)^d((\mathcal{X}_{j\times 1}\mathbf{U}_1$ $\cdots_{\times d-1}\mathbf{U}_{d-1\times d+1}\mathbf{U}_{d+1}\cdots_{\times L}\mathbf{U}_L)^d)^T$. In optimization procedure of each mode, we also impose a constraint to achieve the transformation matrix (such as $\mathbf{U}_d$ in mode d) as the following:

$$\mathbf{U}_d^T\mathbf{Y}^d\mathbf{D}(\mathbf{Y}^d)^T\mathbf{U}_d = \mathbf{I} \Rightarrow \mathbf{U}^T\mathbf{D}_d\mathbf{U} = \mathbf{I} \tag{12}$$

For the optimization problem of all modes, we adopt an alternative least square (ALS) approach. In ALS, we can obtain the optimal base vectors on one mode by fixing the base vectors on the other modes and cycle for the remaining variables. The d-mode transformation matrix $\mathbf{U}_d$ can be achieved by minimizing the following cost function:

$$\underset{\mathbf{U}_d^T\mathbf{D}_d\mathbf{U}_d=\mathbf{I}}{\textbf{argmin}}\ \mathbf{U}_d^T(\mathbf{D}_d - \mathbf{S}_d)\mathbf{U}_d \tag{13}$$

In order to achieve the stable solution, we firstly regularize the symmetric matrix $\mathbf{D}_d$ as $\mathbf{D}_d = \mathbf{D}_d + \alpha \mathbf{I}$ (α is a small value, $\mathbf{I}$ is an identity matrix of same size with the matrix $\mathbf{D}_d$). Then, the minimization problem for obtaining d-mode matrix can be converted to solve a generalized eigenvalue problem as follows:

$$(\mathbf{D}_d - \mathbf{S}_d)\mathbf{U}_d = \lambda \mathbf{D}_d \mathbf{U}_d \tag{14}$$

We can select the corresponding generalized eigenvectors with the first N'_d smaller eigenvalues in Eq.(14), which can minimize the objective function in Eq.(13). However, the eigenvectors with the smallest eigenvalues are usually unstable. Therefore, we convert Eq. (14) into:

$$\mathbf{S}_d \mathbf{U}_d = (1 - \lambda)\mathbf{D}_d \mathbf{U}_d \Rightarrow \mathbf{S}_d \mathbf{U}_d = \beta \mathbf{D}_d \mathbf{U}_d \tag{15}$$

The corresponding generalized eigenvectors with the first N'_d smaller eigenvalues λ in Eq. (14) means those with the first N'_d larger eigenvalues $\beta(1 - \lambda)$ in Eq. (15). Therefore, the corresponding generalized eigenvectors with the first N'_d larger eigenvalues can be selected for minimizing the objective function in Eq.(13). The details algorithm of MSNPE are listed in Algorithm 1. In MSNPE algorithm, we need to decide the retained number of the generalized eigenvectors (mode dimension) for each mode. Usually, the dimension numbers in most discriminant tensor analysis methods are decided empirically or according to applications. In our experiments, we retain different dimension numbers for different modes, and do recognition for objects or scene categories. The recognition accuracy with varied dimensions in different modes are also given in the experiment part. The dimension numbers is decided empirically in the compared results with the state-of-art algorithms.

After obtaining the MSNPE basis of each mode, we can project each tensor object into these MSNPE tensors. For classification, the projection coefficients can represent the extracted feature vectors and can be inputted into any other classification algorithm. In our work, beside Euclidean distance as KNN (k=1) classifier, we also use Random Forest (RF) for recognition.

5. Experiments

5.1 Database

We evaluated our proposed framework on two different types of datasets.

(i) View-based object datasets, which includes two datasets: The first one is the Columbia COIL-100 image library by Nene et al. (1996). It consists of color images of 72 different views of 100 objects. The images were obtained by placing the objects on a turntable and taking a view every $5°$. The objects have a wide variety of complex geometric and reflectance characteristics. Fig. 3(a) shows some sample images from COIL-100. The second one is the ETH Zurich CogVis ETH-80 dataset by Leibe & Schiele (2003a). This dataset was setup by Leibe and Schiele to explore the capabilities of different features for object class recognition. In this dataset, eight object categories including apple, pear, tomato, cow, dog, horse, cup and car have been collected. There are 10 different objects spanned large intra-class variance in each category. Each object has 41 images from viewpoints spaced equally over the upper viewing hemisphere. On the whole we have 3280 images, 41 images for each object and 10 object for each category. Fig.3(b) shows some sample images from ETH-80.

(a)COIL-100 dataset;

(b) ETH80 dataset;

Fig. 3. Sample images from view-based object data sets.

(ii) Facial dataset: We use two facial datasets for evaluating the tensor representation with the proposed NHOG for image representation. One is Yale databae which includes 15 people and 11 facial images of each individual with different illuminations and expressions. Some sample facial images are shown in the top row of Fig. 2(a). The other one is CMU PIE, which includes 68 people and about 170 facial images for each individual with 13 different poses, 43 different illumination conditions, and with 4 different expressions. Some sample facial images are shown in the top row of Fig. 2(b).

5.2 Methodology

The recognition task is to assign each test image to one of a number of categories or objects. The performance is measured using recognition rates.

For view-based object databases, we take different experimental setup in COIL-100 and ETH80 datasets. For COIL-100, the objective is to discriminate between the 100 individual objects. In most previous experiments on object recognition using COIL-100, the number of views used as training set for each object varied from 36 to 4. When 36 views are used for training, the recognition rate using SVM was reported approaching 100% by Pontil & Verri (1998). In practice, however, only very few views of an object are available. In our experiment, in order to compare experimental results with those by Wang (2006), we follows the experiment setup, which used only 4 views of each object for training and the rest 68 views for testing. In total it is equivalent to 400 images for training and 6800 images for testing. The error rate is the overall error rate over 100 objects. The 4 training viewpoints are sampled evenly from the 72 viewpoints, which can capture enough variance on the change of viewpoints for tensor learning. For ETH-80, it aims to discriminate between the 8 object categories. Most previous experiments using ETH-80 dataset all adopted leave-one-object-out cross-validation. The training set consists of all views from 9 objects from each category. The testing set consists of all views from the remaining object from each category. In this setting, objects in the testing set have not appeared in the training set, but those belonging to the same category have. Classification of a test image is a process of labeling the image by one of the categories. Reported results are based on average error rate over all 80 possible test objects by Leibe & Schiele (2003b). Similar to the above, instead of taking all possible views of each object in the training set, we take only 5 views of each object as training data. By doing so we have decreased the number of the training data to $1/8$ of that used by Leibe & Schiele (2003b), Marrr et al. (2005). The testing set consists of all the views of an object. The recognition rate with the proposed scheme is compared to those of different conventional approaches by Wang (2006) and those with MSNPE analysis directly on pixel-level intensity tensor.

For facial dataset, which has large illumination variance in images, we validate that the tensor representation with the proposed NHOG for image representation will be much more efficient for face recognition than that with the popular SIFT descriptor, which only is somewhat robust to small illumination variance. In experiments Yale dataset, we randomly select 2, 3, 4 and 5 facial images from each individual for training, and the remainders for test. For CMU PIE dataset, we randomly select 5 and 10 facial images from each individual for training, and the remainder for test. We do 20 runs for different training number and average recognition rate in all experiments. The recognitions with our proposed approach are compared to those by the state-of-art algorithm by Cai et al. (2007a), Cai et al. (2007b).

6. Experimental results

(1) View-based object data sets

We investigate the performance of the proposed MSNPE tensor learning compared with conventional tensor analysis such as tensor LDA by Wang (2006), which is also used in view-base object recognition, and the efficiency of the proposed tensor representation compared to the pixel-level intensity tensor, which directly consider a whole image as a tensor, on COIL-100 and ETH80 datasets. In these experiments, all samples are also color images, and SIFT descriptor for local region representation is used. Therefore, the pixel-level intensity tensor is 3rd tensor with dimension $R1 \times C1 \times 3$, where $R1$ and $C1$ is row and column number of the image, and the local descriptor tensor is with $128 \times K \times 3$, where K is the segmented region number of an image (here $K=128$). In order to compare with the state-of-art works by Wang (2006), simple KNN method ($k=1$ in our experiments) is also used for recognition. Experimental setup was given in Sec. 5, and we did 18 runs so that all samples can be as test. Figure 6(a) shows the compared results of MSNPE using pixel-level tensor and local descriptor tensor (denoted MSNPE-PL and MSNPE with KNN classifier, respectively, MSNPE-RF-PL and MSNPE-RF with random forest) and traditional methods by Wang (2006) on COIL-100

Methods	DTROD	DTROD+AdaB	RSW	LS
Rate(%)	70.0	76.0	75.0	65.0
Methods	MSNPE-PL		RF-PL	MSNPE	RF
Rate(%)	76.83		77.74	83.54	85.98

Table 2. The compared recognition rates on ETH-80. RSW denotes random subwindow method Marrr et al. (2005) and LS denotes the results from Leibe and Schiele Leibe & Schiele (2003b) with 2925 samples for training and 328 for testing. The others are with 360 samples for training. MSNPE-PL and RF-PL mean MSNPE analysis on pixel-level intensity tensor using simple Euclidean distance and random forest classifier, respectively; MSNPE and RF mean the proposed MSNPE analysis on local SIFT tensor using simple Euclidean distance random forest classifier, respectively.

dataset . The best result with the same experiment setup (400 training samples and 6800 test samples) on COIL-100 is reported by Wang (2006), in which the average recognition rate using tensor LDA and AdBoost classifier (DTROD+AdaBoost) is 84.5%, and the recognition rate of the tensor LDA and simple Euclidean distance(DTROD) by Wang (2006) (same as KNN method with $k=1$) is 79.7. However, The MSNPE approach with pixel-intensity tensor can achieve about 85.28% with same classifier (KNN), and 90% average recognition rate with random forest classifier. Furthermore, the MSNPE approach with local SIFT tensor achieved 93.68% average recognition rate. The compared recognition rate results with the state-of-art approaches are shown in Fig. 4 (a). Figure 4(b) shows the compared recognition rates of one run on different mode dimension of MSNPE between using pixel-level intensity and local SIFT tensor with random forest classifier. It is obvious that the recognition rates by using pixel-level tensor have very large variance with differen mode dimension changing. Therefore, we must select a optimized row and column mode dimension to achieve better recognition rate. However, it is usually difficult to decide the optimized dimension number of different modes automatically. If we just shift the mode dimension number a little from the optimized mode dimension, the recognition rate can be decreased significantly shown in Fig. 4(b) when using pixel-level tensor. For the local SIFT tensor representing an object image, the average recognition rates in lager mode dimension changing (Both row and column mode dimension numbers are from 3 to 8; color mode dimension is 3) are very stable.

For ETH-80 dataset, we also do similar experiments to COIL-100 using the proposed MSNPE analysis with pixel-level and local SIFT tensor, respectively. The compared results with the state of the art approach are shown in Table 2. From Table 2, it can be seen that our proposed approach can greatly improve the overall recognition rate compared with the state of the art method (from 60-80% to about 86%).

(2) Facial Datasets: With the two used facial datasets, we investigate the efficiency of the proposed local NHOG feature on large illumination variance dataset compared with local SIFT descriptor. We do 20 runs for different training number and average recognition rate. For comparison, we also do experiments using the proposed MSNPE analysis directly on the gray face image (pixel-level intensity, denoted MSNPE-PL), local feature tensor with SIFT descriptor (denoted MSNPE-SIFT) and our proposed intensity-normalized histogram of orientation (denoted MSNPE-NHOG). Table 3 gives the compared results using MSNPE analysis with different tensors using KNN classifier ($k=1$) and other subspace learning methods by Cai et al. (2007a), Cai et al. (2007b), Cai (2009) and Cai (n.d.) on YALE dataset, and the compared results on CMU PIE dataset are shown in Table 4 with our proposed framework and the conventional ones by Cai et al. (2007a) Cai et al. (2007b) Cai (2009) Cai

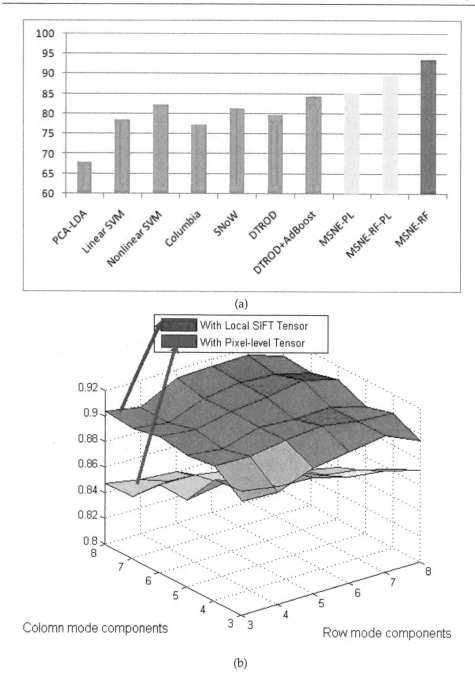

(a)

(b)

Fig. 4. (a) The compared recognition rates on COIL-100 between the proposed framework and the state-of-art approaches Wang (2006). (b) Average recognition rate with different mode dimension using random forest classifier.

Method	2 Train	3 Train	4 Train	5 Train
PCA	56.5	51.1	57.8	45.6
LDA	54.3	35.5	27.3	22.5
Laplacianface	43.5	31.5	25.4	21.7
O-Laplacianface	44.3	29.9	22.7	17.9
TensorLPP	54.5	42.8	37	32.7
R-LDA	42.1	28.6	21.6	17.4
S-LDA	37.5	25.6	19.7	14.9
MSNPE	41.89	31.67	24.86	23.06
MSNPE-SIFT	35.22	26.33	22.19	20.83
MSNPE-NHOG	**29.74**	**22.87**	**18.52**	**17.44**

Table 3. Average recognition error rates (%) on YALE dataset with different training number.

Method	5 Train	10 Train
PCA	75.33	65.5
LDA	42.8	29.7
LPP	38	29.6
MSNPE	37.66	23.57
MSNPE-NHOG	**33.85**	**22.06**

Table 4. Average recognition error rates (%) on PIE dataset with different training number.

(n.d.). From Table 3 and 4, it is obvious that our proposed algorithm can achieve the best recognition performances for all most cases, and the recognition rate improvements become greater when the training sample number is small compared to those by the conventional subspace learning methods by Cai et al. (2007a), Cai et al. (2007b), Cai (2009) and Cai (n.d.). In addition, as we have shown in the previous section, our proposed strategy can be applied not only for recognition of face with small variance (such as mainly frontal face database), but also for recognition of generic object with large variance. With generic object dataset with large variance, the recognition rates are also improved greatly compared with using pixel-level tensor.

7. Conclusion

In this paper, we proposed to represent an image as a local descriptor tensor, which is a combination of the descriptor of local regions ($K * K$-pixel patch) in the image, and more efficient than the popular Bag-Of-Feature (BOF) model for local descriptor combination, and at the same time, we explored a local descriptor for region representation for databases with large illumination variance, Which is improved to be more efficient than the popular SIFT descriptor. Furthermore, we proposed to use Multilinear Supervised Neighborhood Preserving Embedding (MSNPE) for discriminant feature extraction from the local descriptor tensor of different images, which can preserve local sample structure in feature space. We validate our proposed algorithm on different Benchmark databases such as view-based and facial datasets, and experimental results show recognition rate with our method can be greatly improved compared conventional subspace analysis methods.

8. References

Belhumeur, P. N., Hepanha, J. P. & Kriegman, D. J. (1997). Eigenfaces vs. fisherfaces: recognition using class specific linear projection, *IEEE Transactions on Pattern Analysis and Machine Intelligence* (7): 711–720.

Cai, D. (2009). Spectral regression: A regression framework for efficient regularized subspace learning.

Cai, D. (n.d.). *http://www.zjucadcg.cn/dengcai/Data/Yale/results_new.html* .

Cai, D., He, X., Hu, Y., Han, J. & Huang, T. (2007a). Learning a spatially smooth subspace for face recognition, *CVPR* .

Cai, D., He, X., Hu, Y., Han, J. & Huang, T. (2007b). Spectral regression for efficient regularized subspace learning, *ICCV* .

Casilescu, M. & D.Terzopoulos (2002). Multilinear analysis of image ensembles: Tensorfaces, *ECCV* .

Han, X.-H., Qiao, X. & wei Chen, Y. (2011). Multilinear supervised neighborhood embedding with local descriptor tensor for face recognition, *IEICE Trans. Inf. & Syst.* .

Han, X.-H., wei Chen, Y. & Ruan, X. (2011). Multilinear supervised neighborhood embedding of local descriptor tensor for scene/object recognition, *IEEE Transaction on Image Processing* .

Harris, C. & Stephens, M. (1998). A combined corner and edge detector, *In Proc. Alvey Vision Conference* .

He, X., Yan, S., Hu, Y., Niyogi, P. & Zhang, H.-J. (2005). Face recognition using laplacianfaces, *IEEE Transactions on Pattern Analysis and Machine Intelligence* (3): 328–340.

Lazebnik, S., Schmid, C. & Ponce, J. (2006). Beyond bags of features: Spatial pyramid matching for recognizing natural scene categories, *CVPR* pp. 2169–2178.

Leibe, B. & Schiele, B. (2003a). Analyzing appearance and contour based methods for object categorization, *CVPR* .

Leibe, B. & Schiele, B. (2003b). Analyzing appearance and contour based methods for object categorization, *CVPR* .

Lowe, D. (2004). Distinctive image features from scale-invariant keypoints, *International Journal of Computer Vision* (2): 91–110.

Marrr, R., Geurts, P., Piater, J. & Wehenkel, L. (2005). Random subwindows for robust image classification, *CVPR* pp. 34–40.

ming Wang, X., Huang, C., ying Fang, X. & gao Liu, J. (2009). 2dpca vs. 2dlda: Face recognition using two-dimensional method, *International Conference on Artificial Intelligence and Computational Intelligence* pp. 357–360.

Nene, S. A., Nayar, S. K. & Murase, H. (1996). Columbia object image library (coil-100), *Technical Report CUCS-006-96* .

Pontil, M. & Verri, A. (1998). Support vector machines for 3d object recognition, *PAMI* pp. 637–646.

Sim, T., Baker, S. & Bsat, M. (2001). The cmu pose, illumination, and expression (pie) database of human faces, *Robotics Institute, CMU-RI-TR-01-02, Pittsburgh, PA,* .

Wang, Y. & Gong, S. (2006). Tensor discriminant analysis for view-based object recognition, *ICPR* pp. 439–454.

L.D. Lathauwer. Signal processing based on multilinear algebra, *Ph.D. Thesis*, Katholike Universiteit Leu- ven.

7

Acceleration of Convergence of the Alternating Least Squares Algorithm for Nonlinear Principal Components Analysis

Masahiro Kuroda[1], Yuichi Mori[1], Masaya Iizuka[2] and Michio Sakakihara[1]
[1]*Okayama University of Science*
[2]*Okayama University*
Japan

1. Introduction

Principal components analysis (PCA) is a popular descriptive multivariate method for handling quantitative data. In PCA of a mixture of quantitative and qualitative data, it requires quantification of qualitative data to obtain optimal scaling data and use ordinary PCA. The extended PCA including such quantification is called *nonlinear PCA*, see Gifi [Gifi, 1990]. The existing algorithms for nonlinear PCA are PRINCIPALS of Young et al. [Young et al., 1978] and PRINCALS of Gifi [Gifi, 1990] in which the alternating least squares (ALS) algorithm is utilized. The algorithm alternates between quantification of qualitative data and computation of ordinary PCA of optimal scaling data.

In the application of nonlinear PCA for very large data sets and variable selection problems, many iterations and much computation time may be required for convergence of the ALS algorithm, because its speed of convergence is linear. Kuroda et al. [Kuroda et al., 2011] proposed an acceleration algorithm for speeding up the convergence of the ALS algorithm using the vector ε (vε) algorithm of Wynn [Wynn, 1962]. During iterations of the vε accelerated ALS algorithm, the vε algorithm generates an accelerated sequence of optimal scaling data estimated by the ALS algorithm. Then the vε accelerated sequence converges faster than the original sequence of the estimated optimal scaling data. In this paper, we use PRINCIPALS as the ALS algorithm for nonlinear PCA and provide the vε acceleration for PRINCIPALS (vε-PRINCIPALS). The computation steps of PRINCALS are given in Appendix A. As shown in Kuroda et al. [Kuroda et al., 2011], the vε acceleration is applicable to PRINCALS.

The paper is organized as follows. We briefly describe nonlinear PCA of a mixture of quantitative and qualitative data in Section 2, and describe PRINCIPALS for finding least squares estimates of the model and optimal scaling parameters in Section 3. Section 4 presents the procedure of vε-PRINCIPALS that adds the vε algorithm to PRINCIPALS for speeding up convergence and demonstrate the performance of the vε acceleration using numerical experiments. In Section 5, we apply vε-PRINCIPALS to variable selection in nonlinear PCA. Then we utilize modified PCA (M.PCA) approach of Tanaka and Mori [Tanaka and Mori, 1997] for variable selection problems and give the variable selection procedures in M.PCA of qualitative data. Numerical experiments examine the the performance and properties of vε-PRINCIPALS. In Section 6, we present our concluding remarks.

2. Nonlinear principal components analysis

PCA transforms linearly an original data set of variables into a substantially smaller set of uncorrelated variables that contains much of the information in the original data set. The original data matrix is then replaced by an estimate constructed by forming the product of matrices of component scores and eigenvectors.

Let $\mathbf{X} = (\mathbf{X}_1 \, \mathbf{X}_2 \, \cdots \, \mathbf{X}_p)$ be an $n \times p$ matrix of n observations on p variables and be columnwise standardized. In PCA, we postulate that $\mathbf{X}$ is approximated by the following bilinear form:

$$\hat{\mathbf{X}} = \mathbf{Z}\mathbf{A}^\top, \tag{1}$$

where $\mathbf{Z} = (\mathbf{Z}_1 \, \mathbf{Z}_2 \, \cdots \, \mathbf{Z}_r)$ is an $n \times r$ matrix of n component scores on r $(1 \leq r \leq p)$ components, and $\mathbf{A} = (\mathbf{A}_1 \, \mathbf{A}_2 \, \cdots \, \mathbf{A}_r)$ is a $p \times r$ matrix consisting of the eigenvectors of $\mathbf{X}^\top\mathbf{X}/n$ and $\mathbf{A}^\top\mathbf{A} = \mathbf{I}_r$. Then we determine model parameters $\mathbf{Z}$ and $\mathbf{A}$ such that

$$\theta = \text{tr}(\mathbf{X} - \hat{\mathbf{X}})^\top(\mathbf{X} - \hat{\mathbf{X}}) = \text{tr}(\mathbf{X} - \mathbf{Z}\mathbf{A}^\top)^\top(\mathbf{X} - \mathbf{Z}\mathbf{A}^\top) \tag{2}$$

is minimized for the prescribed r components.

Ordinary PCA assumes that all variables are measured with interval and ratio scales and can be applied only to quantitative data. When the observed data are a mixture of quantitative and qualitative data, ordinary PCA cannot be directly applied to such data. In such situations, optimal scaling is used to quantify the observed qualitative data and then ordinary PCA can be applied.

To quantify $\mathbf{X}_j$ of qualitative variable j with K_j categories, the vector is coded by using an $n \times K_j$ indicator matrix $\mathbf{G}_j$ with entries $g_{(j)ik} = 1$ if object i belongs to category k, and $g_{(j)ik'} = 0$ if object i belongs to some other category $k'(\neq k)$, $i = 1,\ldots,n$ and $k = 1,\ldots,K_j$. Then the optimally scaled vector $\mathbf{X}_j^*$ of $\mathbf{X}_j$ is given by $\mathbf{X}_j^* = \mathbf{G}_j\alpha_j$, where α_j is a $K_j \times 1$ score vector for categories of $\mathbf{X}_j$. Let $\mathbf{X}^* = (\mathbf{X}_1^* \, \mathbf{X}_2^* \, \cdots \, \mathbf{X}_p^*)$ be an $n \times p$ matrix of optimally scaled observations to satisfy restrictions

$$\mathbf{X}^{*\top}\mathbf{1}_n = \mathbf{0}_p \quad \text{and} \quad \text{diag}\left[\frac{\mathbf{X}^{*\top}\mathbf{X}^*}{n}\right] = \mathbf{I}_p, \tag{3}$$

where $\mathbf{1}_n$ and $\mathbf{0}_p$ are vectors of ones and zeros of length n and p respectively. In the presence of nominal and/or ordinal variables, the optimization criterion (2) is replaced by

$$\theta^* = \text{tr}(\mathbf{X}^* - \hat{\mathbf{X}})^\top(\mathbf{X}^* - \hat{\mathbf{X}}) = \text{tr}(\mathbf{X}^* - \mathbf{Z}\mathbf{A}^\top)^\top(\mathbf{X}^* - \mathbf{Z}\mathbf{A}^\top). \tag{4}$$

In nonlinear PCA, we determine the optimal scaling parameter $\mathbf{X}^*$, in addition to estimating $\mathbf{Z}$ and $\mathbf{A}$.

3. Alternating least squares algorithm for nonlinear principal components analysis

A possible computational algorithm for estimating simultaneously $\mathbf{Z}$, $\mathbf{A}$ and $\mathbf{X}^*$ is the ALS algorithm. The algorithm involves dividing an entire set of parameters of a model into the model parameters and the optimal scaling parameters, and finds the least squares estimates for these parameters. The model parameters are used to compute the predictive values of the model. The optimal scaling parameters are obtained by solving the least squares regression problem for the predictive values. Krijnen [Krijnen, 2006] gave sufficient conditions for convergence of the ALS algorithm and discussed convergence properties in its application to several statistical models. Kiers [Kiers, 2002] described setting up the ALS and iterative majorization algorithms for solving various matrix optimization problems.

PRINCIPALS

PRINCIPALS proposed by Young et al. [Young et al., 1978] is a method for utilizing the ALS algorithm for nonlinear PCA of a mixture of quantitative and qualitative data. PRINCIPALS alternates between ordinary PCA and optimal scaling, and minimizes θ^* defined by Equation (4) under the restriction (3). Then θ^* is to be determined by model parameters $\mathbf{Z}$ and $\mathbf{A}$ and optimal scaling parameter $\mathbf{X}^*$, by updating each of the parameters in turn, keeping the others fixed.

For the initialization of PRINCIPALS, we determine initial data $\mathbf{X}^{*(0)}$. The observed data $\mathbf{X}$ may be used as $\mathbf{X}^{*(0)}$ after it is standardized to satisfy the restriction (3). For given initial data $\mathbf{X}^{*(0)}$ with the restriction (3), PRINCIPALS iterates the following two steps:

- *Model parameter estimation step*: Obtain $\mathbf{A}^{(t)}$ by solving

$$\left[\frac{\mathbf{X}^{*(t)\top}\mathbf{X}^{*(t)}}{n} \right] \mathbf{A} = \mathbf{A}\mathbf{D}_r, \tag{5}$$

  where $\mathbf{A}^\top\mathbf{A} = \mathbf{I}_r$ and $\mathbf{D}_r$ is an $r \times r$ diagonal matrix of eigenvalues, and the superscript (t) indicates the t-th iteration. Compute $\mathbf{Z}^{(t)}$ from $\mathbf{Z}^{(t)} = \mathbf{X}^{*(t)}\mathbf{A}^{(t)}$.

- *Optimal scaling step*: Calculate $\hat{\mathbf{X}}^{(t+1)} = \mathbf{Z}^{(t)}\mathbf{A}^{(t)\top}$ from Equation (1). Find $\mathbf{X}^{*(t+1)}$ such that

$$\mathbf{X}^{*(t+1)} = \arg\min_{\mathbf{X}^*} \operatorname{tr}(\mathbf{X}^* - \hat{\mathbf{X}}^{(t+1)})^\top(\mathbf{X}^* - \hat{\mathbf{X}}^{(t+1)})$$

  for fixed $\hat{\mathbf{X}}^{(t+1)}$ under measurement restrictions on each of the variables. Scale $\mathbf{X}^{*(t+1)}$ by columnwise centering and normalizing.

4. The $v\varepsilon$ acceleration of the ALS algorithm

We briefly introduce the $v\varepsilon$ algorithm of Wynn [Wynn, 1962] used in the acceleration of the ALS algorithm. The $v\varepsilon$ algorithm is utilized to speed up the convergence of a slowly convergent vector sequence and is very effective for linearly converging sequences. Kuroda and Sakakihara [Kuroda and Sakakihara, 2006] proposed the ε-accelerated EM algorithm that speeds up the convergence of the EM sequence via the $v\varepsilon$ algorithm and demonstrated that its speed of convergence is significantly faster than that of the EM algorithm. Wang et al. [Wang et al., 2008] studied the convergence properties of the ε-accelerated EM algorithm.

Let $\{\mathbf{Y}^{(t)}\}_{t\geq 0} = \{\mathbf{Y}^{(0)}, \mathbf{Y}^{(1)}, \mathbf{Y}^{(2)}, \ldots\}$ be a linear convergent sequence generated by an iterative computational procedure and let $\{\dot{\mathbf{Y}}^{(t)}\}_{t\geq 0} = \{\dot{\mathbf{Y}}^{(0)}, \dot{\mathbf{Y}}^{(1)}, \dot{\mathbf{Y}}^{(2)}, \ldots\}$ be the accelerated sequence of $\{\mathbf{Y}^{(t)}\}_{t\geq 0}$. Then the $v\varepsilon$ algorithm generates $\{\dot{\mathbf{Y}}^{(t)}\}_{t\geq 0}$ by using

$$\dot{\mathbf{Y}}^{(t-1)} = \mathbf{Y}^{(t)} + \left[\left[(\mathbf{Y}^{(t-1)} - \mathbf{Y}^{(t)}) \right]^{-1} + \left[(\mathbf{Y}^{(t+1)} - \mathbf{Y}^{(t)}) \right]^{-1} \right]^{-1}, \qquad (6)$$

where $[\mathbf{Y}]^{-1} = \mathbf{Y} / ||\mathbf{Y}||^2$ and $||\mathbf{Y}||$ is the Euclidean norm of $\mathbf{Y}$. For the detailed derivation of Equation (6), see Appendix B. When $\{\mathbf{Y}^{(t)}\}_{t\geq 0}$ converges to a limit point $\mathbf{Y}^{(\infty)}$ of $\{\mathbf{Y}^{(t)}\}_{t\geq 0}$, it is known that, in many cases, $\{\dot{\mathbf{Y}}^{(t)}\}_{t\geq 0}$ generated by the $v\varepsilon$ algorithm converges to $\mathbf{Y}^{(\infty)}$ faster than $\{\mathbf{Y}^{(t)}\}_{t\geq 0}$.

We assume that $\{\mathbf{X}^{*(t)}\}_{t\geq 0}$ generated by PRINCIPALS converges to a limit point $\mathbf{X}^{*(\infty)}$. Then $v\varepsilon$-PRINCIPALS produces a faster convergent sequence $\{\dot{\mathbf{X}}^{*(t)}\}_{t\geq 0}$ of $\{\mathbf{X}^{*(t)}\}_{t\geq 0}$ by using the $v\varepsilon$ algorithm and enables the acceleration of convergence of PRINCIPALS. The general procedure of $v\varepsilon$-PRINCIPALS iterates the following two steps:

- *PRINCIPALS step*: Compute model parameters $\mathbf{A}^{(t)}$ and $\mathbf{Z}^{(t)}$ and determine optimal scaling parameter $\mathbf{X}^{*(t+1)}$.

- *Acceleration step*: Calculate $\dot{\mathbf{X}}^{*(t-1)}$ using $\{\mathbf{X}^{*(t-1)}, \mathbf{X}^{*(t)}, \mathbf{X}^{*(t+1)}\}$ from the $v\varepsilon$ algorithm:

$$\mathrm{vec}\dot{\mathbf{X}}^{*(t-1)} = \mathrm{vec}\mathbf{X}^{*(t)} + \left[\left[\mathrm{vec}(\mathbf{X}^{*(t-1)} - \mathbf{X}^{*(t)}) \right]^{-1} + \left[\mathrm{vec}(\mathbf{X}^{*(t+1)} - \mathbf{X}^{*(t)}) \right]^{-1} \right]^{-1},$$

where $\mathrm{vec}\mathbf{X}^* = (\mathbf{X}_1^{*\top} \mathbf{X}_2^{*\top} \cdots \mathbf{X}_p^{*\top})^{\top}$, and check the convergence by

$$\left\| \mathrm{vec}(\dot{\mathbf{X}}^{*(t-1)} - \dot{\mathbf{X}}^{*(t-2)}) \right\|^2 < \delta,$$

where δ is a desired accuracy.

Before starting the iteration, we determine initial data $\mathbf{X}^{*(0)}$ satisfying the restriction (3) and execute the *PRINCIPALS step* twice to generate $\{\mathbf{X}^{*(0)}, \mathbf{X}^{*(1)}, \mathbf{X}^{*(2)}\}$.

$v\varepsilon$-PRINCIPALS is designed to generate $\{\dot{\mathbf{X}}^{*(t)}\}_{t\geq 0}$ converging to $\mathbf{X}^{*(\infty)}$. Thus the estimate of $\mathbf{X}^*$ can be obtained from the final value of $\{\dot{\mathbf{X}}^{*(t)}\}_{t\geq 0}$ when $v\varepsilon$-PRINCIPALS terminates. The estimates of $\mathbf{Z}$ and $\mathbf{A}$ can then be calculated immediately from the estimate of $\mathbf{X}^*$ in the *Model parameter estimation step* of PRINCIPALS.

Note that $\dot{\mathbf{X}}^{*(t-1)}$ obtained at the t-th iteration of the *Acceleration step* is not used as the estimate $\mathbf{X}^{*(t+1)}$ at the $(t+1)$-th iteration of the *PRINCIPALS step*. Thus $v\varepsilon$-PRINCIPALS speeds up the convergence of $\{\mathbf{X}^{*(t)}\}_{t\geq 0}$ without affecting the convergence properties of ordinary PRINCIPALS.

Numerical experiments 1: Comparison of the number of iterations and CPU time

We study how much faster $v\varepsilon$-PRINCIPALS converges than ordinary PRINCIPALS. All computations are performed with the statistical package R [R Development Core Team, 2008] executing on Intel Core i5 3.3 GHz with 4 GB of memory. CPU times (in seconds) taken are measured by the function proc.time[1]. For all experiments, δ for convergence

---

[1] Times are typically available to 10 msec.

of vε-PRINCIPALS is set to 10^{-8} and PRINCIPALS terminates when $|\theta^{(t+1)} - \theta^{(t)}| < 10^{-8}$, where $\theta^{(t)}$ is the t-th update of θ calculated from Equation (4). The maximum number of iterations is also set to 100,000.

We apply these algorithms to a random data matrix of 100 observations on 20 variables with 10 levels and measure the number of iterations and CPU time taken for $r = 3$. The procedure is replicated 50 times.

Table 1 is summary statistics of the numbers of iterations and CPU times of PRINCIPALS and vε-PRINCIPALS from 50 simulated data. Figure 1 shows the scatter plots of the number of iterations and CPU time. The values of the second to fifth columns of the table and the figure show that PRINCIPALS requires more iterations and takes a longer computation time than vε-PRINCIPALS. The values of the sixth and seventh columns in the table are summary statistics of the iteration and CPU time speed-ups for comparing the speed of convergence of PRINCIPALS with that of vε-PRINCIPALS. The iteration speed-up is defined as the number of iterations required for PRINCIPALS divided by the number of iterations required for vε-PRINCIPALS. The CPU time speed-up is calculated similarly to the iteration speed-up. We can see from the values of the iteration and CPU time speed-ups that vε-PRINCIPALS converges 3.23 times in terms of the mean number of iterations and 2.92 times in terms of the mean CPU time faster than PRINCIPALS. Figure 2 shows the boxplots of the iteration and CPU time speed-ups. Table 1 and Figure 2 show that vε-PRINCIPALS well accelerates the convergence of $\{\mathbf{X}^{*(t)}\}_{t \geq 0}$.

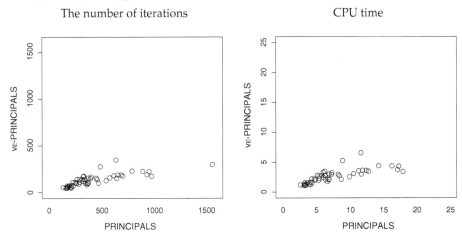

Fig. 1. Scatter plots of the number iterations and CPU time from 50 simulated data.

Figure 3 is the scatter plots of iteration and CPU time speed-ups for the number of iterations of PRINCIPALS. The figure demonstrates that the vε acceleration speeds up greatly the convergence of $\{\mathbf{X}^{*(t)}\}_{t \geq 0}$ and its speed of convergence is faster for the larger number of iterations of PRINCIPALS. For more than 400 iterations of PRINCIPALS, the speed of the vε acceleration is faster 3 times more than that of PRINCIPALS and the maximum values of both speed-ups are for around 1,000 iterations of PRINCIPALS. The advantage of the vε acceleration is very obvious.

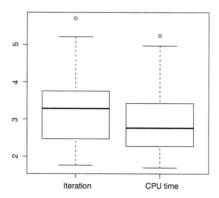

Fig. 2. Boxplots of iteration and CPU time speed-ups from 50 simulated data.

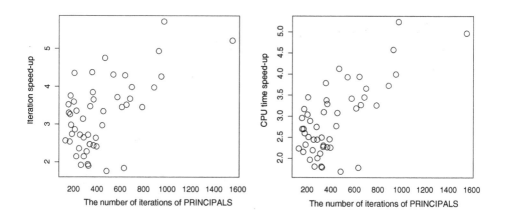

Fig. 3. Scatter plots of iteration and CPU time speed-ups for the number of iterations of PRINCIPALS from 50 simulated data.

	PRINCIPALS		vε-PRINCIPALS		Speed-up	
	Iteration	CPU time	Iteration	CPU time	Iteration	CPU time
Minimum	136.0	2.64	46.0	1.07	1.76	1.69
1st Quartile	236.5	4.44	85.0	1.81	2.49	2.27
Median	345.5	6.37	137.0	2.72	3.28	2.76
Mean	437.0	8.02	135.0	2.70	3.23	2.92
3rd Quartile	573.2	10.39	171.2	3.40	3.74	3.41
Maximum	1564.0	28.05	348.0	6.56	5.71	5.24

Table 1. Summary statistics of the numbers of iterations and CPU times of PRINCIPALS and vε-PRINCIPALS and iteration and CPU time speed-ups from 50 simulated data.

Numerical experiments 2: Studies of convergence

We introduce the result of studies of convergence of vε-PRINCIPALS from Kuroda et al. [Kuroda et al., 2011]. The data set used in the experiments is obtained in teacher evaluation by students and consists of 56 observations on 13 variables with 5 levels each; the lowest evaluation level is 1 and the highest 5.

The rates of convergence of these algorithms are assessed as

$$\tau = \lim_{t \to \infty} \tau^{(t)} = \lim_{t \to \infty} \frac{\|\mathbf{X}^{*(t)} - \mathbf{X}^{*(t-1)}\|}{\|\mathbf{X}^{*(t-1)} - \mathbf{X}^{*(t-2)}\|} \quad \text{for PRINCIPALS,}$$

$$\dot{\tau} = \lim_{t \to \infty} \dot{\tau}^{(t)} = \lim_{t \to \infty} \frac{\|\dot{\mathbf{X}}^{*(t)} - \dot{\mathbf{X}}^{*(t-1)}\|}{\|\dot{\mathbf{X}}^{*(t-1)} - \dot{\mathbf{X}}^{*(t-2)}\|} \quad \text{for vε-PRINCIPALS.}$$

If the inequality $0 < \dot{\tau} < \tau < 1$ holds, we say that $\{\dot{\mathbf{X}}^{*(t)}\}_{t \geq 0}$ converges faster than $\{\mathbf{X}^{*(t)}\}_{t \geq 0}$. Table 2 provides the rates of convergence τ and $\dot{\tau}$ for each r. We see from the table that $\{\dot{\mathbf{X}}^{*(t)}\}_{t \geq 0}$ converges faster than $\{\mathbf{X}^{*(t)}\}_{t \geq 0}$ in comparison between τ and $\dot{\tau}$ for each r and thus conclude that vε-PRINCIPALS significantly improves the rate of convergence of PRINCIPALS. The speed of convergence of vε-PRINCIPALS is investigate by

r	τ	$\dot{\tau}$
1	0.060	0.001
2	0.812	0.667
3	0.489	0.323
4	0.466	0.257
5	0.493	0.388
6	0.576	0.332
7	0.473	0.372
8	0.659	0.553
9	0.645	0.494
10	0.678	0.537
11	0.592	0.473
12	0.648	0.465

Table 2. Rates of convergence τ and $\dot{\tau}$ of PRINCIPALS to vε-PRINCIPALS.

$$\dot{\rho} = \lim_{t \to \infty} \dot{\rho}^{(t)} = \lim_{t \to \infty} \frac{\|\dot{\mathbf{X}}^{*(t)} - \mathbf{X}^{*(\infty)}\|}{\|\mathbf{X}^{*(t+2)} - \mathbf{X}^{*(\infty)}\|} = 0. \tag{7}$$

If $\{\dot{X}^{*(t)}\}_{t\geq 0}$ converges to th same limit point $X^{*(\infty)}$ as $\{X^{*(t)}\}_{t\geq 0}$ and Equation (7) holds, we say that $\{\dot{X}^{*(t)}\}_{t\geq 0}$ accelerates the convergence of $\{X^{*(t)}\}_{t\geq 0}$. See Brezinski and Zaglia [Brezinski and Zaglia, 1991]. In the experiments, $\{\dot{X}^{*(t)}\}_{t\geq 0}$ converges to the final value of $\{X^{*(t)}\}_{t\geq 0}$ and $\dot{\rho}$ is reduced to zero for all r. We see from the results that vε-PRINCIPALS accelerates the convergence of $\{X^{*(t)}\}$.

5. Variable selection in nonlinear PCA: Modified PCA approach

In the analysis of data with large numbers of variables, a common objective is to reduce the dimensionality of the data set. PCA is a popular dimension-reducing tool that replaces the variables in the data set by a smaller number of derived variables. However, for example, in PCA of a data set with a large number of variables, the result may not be easy to interpret. One way to give a simple interpretation of principal components is to select a subset of variables that best approximates all the variables. Various variable selection criteria in PCA has been proposed by Jolliffe [Jolliffe, 1972], McCabe [McCabe, 1984], Robert and Escoufier [Robert and Escoufier, 1976], Krzanowski [Krzanowski, 1987]. Al-Kandari et al. [Al-Kandari et al., 2001; Al-Kandari et al., 2005] gave guidelines as to the types of data for which each variable selection criteria is useful. Cadima et al. [Cadima et al., 2004] reported computational experiments carried out with several heuristic algorithms for the optimization problems resulting from the variable selection criteria in PCA found in the above literature.

Tanaka and Mori [Tanaka and Mori, 1997] proposed modified PCA (M.PCA) for deriving principal components which are computed by using only a selected subset of variables but which represent all the variables including those not selected. Since M.PCA includes variable selection procedures in the analysis, its criteria can be used directly to find a reasonable subset of variables. Mori et al. [Mori et al., 1997] extended M.PCA to qualitative data and provided variable selection procedures, in which the ASL algorithm is utilized.

5.1 Formulation of modified PCA

M.PCA derives principal components which are computed as linear combinations of a subset of variables but which can reproduce all the variables very well. Let X be decomposed into an $n \times q$ submatrix X_{V_1} and an $n \times (p-q)$ remaining submatrix X_{V_2}. Then M.PCA finds r linear combinations $Z = X_{V_1}A$. The matrix A consists of the eigenvectors associated with the largest r eigenvalues $\lambda_1 \geq \lambda_2 \geq \cdots \geq \lambda_r$ and is obtained by solving the eigenvalue problem:

$$[(S_{11}^2 + S_{12}S_{21}) - DS_{11}]A = 0, \tag{8}$$

where $S = \begin{pmatrix} S_{11} & S_{12} \\ S_{21} & S_{22} \end{pmatrix}$ is the covariance matrix of $X = (X_{V_1}, X_{V_2})$ and D is a $q \times q$ diagonal matrix of eigenvalues. A best subset of q variables has the largest value of the proportion $P = \sum_{j=1}^{r} \lambda_j / \text{tr}(S)$ or the RV-coefficient $RV = \left\{ \sum_{j=1}^{r} \lambda_j^2 / \text{tr}(S^2) \right\}^{1/2}$. Here we use P as variable selection criteria.

5.2 Variable selection procedures

In order to find a subset of q variables, we employ Backward elimination and Forward selection of Mori et al. [Mori et al., 1998; Mori et al., 2006] as cost-saving stepwise selection procedures in which only one variable is removed or added sequentially.

[Backward elimination]

Stage A: *Initial fixed-variables stage*

    A-1 Assign q variables to subset $\mathbf{X}_{V_1}$, usually $q := p$.

    A-2 Solve the eigenvalue problem (8).

    A-3 Look carefully at the eigenvalues, determine the number r of principal components to be used.

    A-4 Specify kernel variables which should be involved in $\mathbf{X}_{V_1}$, if necessary. The number of kernel variables is less than q.

Stage B: *Variable selection stage (Backward)*

    B-1 Remove one variable from among q variables in $\mathbf{X}_{V_1}$, make a temporary subset of size $q - 1$, and compute P based on the subset. Repeat this for each variable in $\mathbf{X}_{V_1}$, then obtain q values on P. Find the best subset of size $q - 1$ which provides the largest P among these q values and remove the corresponding variable from the present $\mathbf{X}_{V_1}$. Put $q := q - 1$.

    B-2 If P or q is larger than preassigned values, go to **B-1**. Otherwise stop.

[Forward selection]

Stage A: *Initial fixed-variables stage*

    A-1 $\sim$ **A-3** Same as A-1 to A-3 in Backward elimination.

    A-4 Redefine q as the number of kernel variables (here, $q \geq r$). If you have kernel variables, assign them to $\mathbf{X}_{V_1}$. If not, put $q := r$, find the best subset of q variables which provides the largest P among all possible subsets of size q and assign it to $\mathbf{X}_{V_1}$.

Stage B: *Variable selection stage (Forward)*

    B-1 Adding one of the $p - q$ variables in $\mathbf{X}_{V_2}$ to $\mathbf{X}_{V_1}$, make a temporary subset of size $q + 1$ and obtain P. Repeat this for each variable in $\mathbf{X}_{V_2}$, then obtain $p - q$ Ps. Find the best subset of size $q + 1$ which provides the largest (or smallest) P among the $p - q$ Ps and add the corresponding variable to the present subset of $\mathbf{X}_{V_1}$. Put $q := q + 1$.

    B-2 If the P or q are smaller (or larger) than preassigned values, go back to **B-1**. Otherwise stop.

In Backward elimination, to find the best subset of $q - 1$ variables, we perform M.PCA for each of q possible subsets of the $q - 1$ variables among q variables selected in the previous selection step. The total number of estimations for M.PCA from $q = p - 1$ to $q = r$ is therefore large, i.e., $p + (p - 1) + \cdots + (r + 1) = (p - r)(p + r + 1)/2$. In Forward selection, the total number of estimations for M.PCA from $q = r$ to $q = p - 1$ is $_pC_r + (p - r) + (p - (r + 1)) + \cdots + 2 = {}_pC_r + (p - r - 1)(p - r + 2)/2$.

Numerical experiments 3: Variable selection in M.PCA for simulated data

We apply PRINCIPALS and vε-PRINCIPALS to variable selection in M.PCA of qualitative data using simulated data consisting of 100 observations on 10 variables with 3 levels.

Table 3 shows the number of iterations and CPU time taken by two algorithms for finding a subset of q variables based on 3 ($= r$) principal components. The values of the second to fifth columns in the table indicate that the number of iterations of PRINCIPALS is very large and a long computation time is taken for convergence, while vε-PRINCIPALS converges considerably faster than PRINCIPALS. We can see from the sixth and seventh columns in the table that vε-PRINCIPALS requires the number of iterations 3 - 5 times smaller and CPU time 2 - 5 times shorter than vε-PRINCIPALS. In particular, the vε acceleration effectively works to accelerate the convergence of $\{\mathbf{X}^{*(t)}\}_{t\geq 0}$ for the larger number of iterations of PRINCIPALS.

(a) Backward elimination

q	PRINCIPALS		vε-PRINCIPALS		Speed-up	
	Iteration	CPU time	Iteration	CPU time	Iteration	CPU time
10	141	1.70	48	0.68	2.94	2.49
9	1,363	17.40	438	6.64	3.11	2.62
8	1,620	20.19	400	5.98	4.05	3.37
7	1,348	16.81	309	4.80	4.36	3.50
6	4,542	53.72	869	11.26	5.23	4.77
5	13,735	159.72	2,949	35.70	4.66	4.47
4	41,759	482.59	12,521	148.13	3.34	3.26
3	124	1.98	44	1.06	2.82	1.86
Total	64,491	752.40	17,530	213.57	3.68	3.52

(b) Forward selection

q	PRINCIPALS		vε-PRINCIPALS		Speed-up	
	Iteration	CPU time	Iteration	CPU time	Iteration	CPU time
3	4,382	67.11	1442	33.54	3.04	2.00
4	154,743	1,786.70	26,091	308.33	5.93	5.79
5	13,123	152.72	3,198	38.61	4.10	3.96
6	3,989	47.02	1,143	14.24	3.49	3.30
7	1,264	15.27	300	4.14	4.21	3.69
8	340	4.38	108	1.70	3.15	2.58
9	267	3.42	75	1.17	3.56	2.93
10	141	1.73	48	0.68	2.94	2.54
Total	178,249	2,078.33	32,405	402.40	5.50	5.16

Table 3. The numbers of iterations and CPU times of PRINCIPALS and vε-PRINCIPALS and their speed-ups in application to variable selection for finding a subset of q variables using simulated data.

The last row in Table 3 shows the total number of iterations and total CPU time for selecting 8 subsets for $q = 3, \ldots, 10$. When searching the best subset for each q, PRINCIPALS requires 64,491 iterations in Backward elimination and 178,249 iterations in Forward selection, while vε-PRINCIPALS finds the subsets after 17,530 and 32,405 iterations, respectively. These values show that the computation times by vε-PRINCIPALS are reduced to only 28%$(= 1/3.52)$ and 19% $= (1/5.16)$ of those of ordinary PRINCIPALS. The iteration and CPU time speed-ups given in the sixth and seventh columns of the table demonstrate that the vε acceleration works well to speed up the convergence of $\{\mathbf{X}^{*(t)}\}_{t\geq 0}$ and consequently results in greatly reduced computation times in variable selection problems.

Numerical experiments 4: Variable selection in M.PCA for real data

We consider the variable selection problems in M.PCA of qualitative data to mild distribution of consciousness (MDOC) data from Sano et al. [Sano et al. 1977]. MDOC is the data matrix of 87 individuals on 23 variables with 4 levels. In the variable selection problem, we select a suitable subset based on 2 $(= r)$ principal components.

Table 4 summarizes the results of variable selection using Backward elimination and Forward selection procedures for finding a subset of q variables. We see from the last row of the table

(a) Backward elimination

q	PRINCIPALS		$v\varepsilon$-PRINCIPALS		Speed-up		P
	Iteration	CPU time	Iteration	CPU time	Iteration	CPU time	
23	36	1.39	10	0.65	3.60	2.13	0.694
22	819	32.42	231	15.40	3.55	2.11	0.694
21	779	30.79	221	14.70	3.52	2.10	0.693
20	744	29.37	212	14.05	3.51	2.09	0.693
19	725	28.43	203	13.41	3.57	2.12	0.692
18	705	27.45	195	12.77	3.62	2.15	0.692
17	690	26.67	189	12.25	3.65	2.18	0.691
16	671	25.73	180	11.61	3.73	2.22	0.690
15	633	24.26	169	10.85	3.75	2.24	0.689
14	565	21.79	153	10.02	3.69	2.17	0.688
13	540	20.69	147	9.48	3.67	2.18	0.687
12	498	19.09	132	8.64	3.77	2.21	0.686
11	451	17.34	121	7.95	3.73	2.18	0.684
10	427	16.29	117	7.46	3.65	2.18	0.682
9	459	16.99	115	7.05	3.99	2.41	0.679
8	419	15.43	106	6.42	3.95	2.40	0.676
7	382	14.02	100	5.89	3.82	2.38	0.673
6	375	13.51	96	5.41	3.91	2.50	0.669
5	355	12.58	95	5.05	3.74	2.49	0.661
4	480	16.11	117	5.33	4.10	3.02	0.648
3	2,793	86.55	1,354	43.48	2.06	1.99	0.620
2	35	1.92	10	1.34	3.50	1.43	0.581
Total	13,581	498.82	4,273	229.20	3.18	2.18	

(b) Forward selection

q	PRINCIPALS		$v\varepsilon$-PRINCIPALS		Speed-up		P
	Iteration	CPU time	Iteration	CPU time	Iteration	CPU time	
2	3,442	176.76	1,026	119.07	3.35	1.48	0.597
3	5,389	170.82	1,189	44.28	4.53	3.86	0.633
4	1,804	60.96	429	20.27	4.21	3.01	0.650
5	1,406	48.53	349	17.41	4.03	2.79	0.662
6	1,243	43.25	305	15.75	4.08	2.75	0.668
7	1,114	39.03	278	14.61	4.01	2.67	0.674
8	871	31.35	221	12.39	3.94	2.53	0.677
9	789	28.57	202	11.52	3.91	2.48	0.680
10	724	26.32	187	10.74	3.87	2.45	0.683
11	647	23.69	156	9.39	4.15	2.52	0.685
12	578	21.30	142	8.60	4.07	2.48	0.687
13	492	18.39	125	7.76	3.94	2.37	0.688
14	432	16.23	110	6.94	3.93	2.34	0.689
15	365	13.91	95	6.13	3.84	2.27	0.690
16	306	11.80	80	5.30	3.83	2.22	0.691
17	267	10.32	71	4.66	3.76	2.21	0.691
18	226	8.77	60	3.96	3.77	2.21	0.692
19	193	7.48	51	3.39	3.78	2.21	0.692
20	152	5.91	40	2.65	3.80	2.23	0.693
21	108	4.26	30	2.00	3.60	2.13	0.693
22	72	2.85	20	1.33	3.60	2.14	0.694
23	36	1.39	10	0.66	3.60	2.11	0.694
Total	20,656	771.88	5,176	328.81	3.99	2.35	

Table 4. The numbers of iterations and CPU times of PRINCIPALS and $v\varepsilon$-PRINCIPALS, their speed-ups and P in application to variable selection for finding a subset of q variables using MDOC.

that the iteration speed-ups are 3.18 in Backward elimination and 3.99 in Forward selection and thus vε-PRINCIPALS well accelerates the convergence of $\{X^{*(t)}\}_{t\geq 0}$. The CPU time speed-ups are 2.18 in Backward elimination and 2.35 in Forward selection, and are not as large as the iteration speed-ups. The computation time per iteration of vε-PRINCIPALS is greater than that of PRINCIPALS due to computation of the *Acceleration step*. Therefore, for the smaller number of iterations, the CPU time of vε-PRINCIPALS is almost same as or may be longer than that of PRINCIPALS. For example, in Forward selection for $q = 2$, PRINCIPALS converges in almost cases after less than 15 iterations and then the CPU time speed-up is 1.48.

The proportion P in the eighth column of the table indicates the variation explained by the first 2 principal components for the selected q variables. Iizuka et al. [Iizuka et al., 2003] selected the subset of 6 variables found by either procedures as a best subset, since P slightly changes until $q = 6$ in Backward elimination and after $q = 6$ in Forward selection.

6. Concluding remarks

In this paper, we presented vε-PRINCIPALS that accelerates the convergence of PRINCIPALS by using the vε algorithm. The algorithm generates the vε accelerated sequence $\{\dot{X}^{*(t)}\}$ using $\{X^{*(t)}\}_{t\geq 0}$ but it does not modify the estimation equations in PRINCIPALS. Therefore the algorithm enables an acceleration of the convergence of PRINCIPALS, while still preserving the stable convergence property of PRINCIPALS. The vε algorithm in itself is a fairly simple computational procedure and, at each iteration, it requires only $O(np)$ arithmetic operations. For each iteration, the computational complexity of the vε algorithm may be less expensive than that for computing a matrix inversion and for solving the eigenvalue problem in PRINCIPALS.

The most appealing points of the vε algorithm are that, if an original sequence converges to a limit point then the accelerated sequence converges to the same limit point as the original sequence and its speed of convergence is faster than the original sequence. In all the numerical experiments, the vε accelerated sequence $\{\dot{X}^{*(t)}\}_{t\geq 0}$ converges to the final value of $\{X^{*(t)}\}_{t\geq 0}$ after the significantly fewer number of iterations than that of PRINCIPALS.

The numerical experiments employing simulated data in Section 4 demonstrated that vε acceleration for PRINCIPALS significantly speeds up the convergence of $\{X^{*(t)}\}_{t\geq 0}$ in terms of the number of iterations and the computation time. In particular, the vε acceleration effectively works to speed up the convergence for the larger number of iterations of PRINCIPALS. Furthermore, we evaluate the performance of the vε acceleration for PRINCIPALS by applying to variable selection in M.PCA of qualitative data. Numerical experiments using simulated and real data showed that vε-PRINCIPALS improves the speed of convergence of ordinary PRINCIPALS and enables greatly the reduction of computation times in the variable selection for finding a suitable variable set using Backward elimination and Forward selection procedures. The results indicate that the vε acceleration well works in saving the computational time in variable selection problems.

The computations of variable selection in M.PCA of qualitative data are partially performed by the statistical package VASpca(VAriable Selection in principal component analysis) that was developed by Mori, Iizuka, Tarumi and Tanaka in 1999 and can be obtained from Mori's website in Appendix C. We will provide VASpca using vε-PRINCIPALS as the iterative algorithm for PCA and M.PCA of qualitative data.

7. Acknowledgment

The authors would like to thank the editor and two referees whose valuable comments and kind suggestions that led to an improvement of this paper. This research is supported by the Japan Society for the Promotion of Science (JSPS), Grant-in-Aid for Scientific Research (C), No 20500263.

8. Appendix A: PRINCALS

PRINCALS by Gifi [Gifi, 1990] can handle multiple nominal variables in addition to the single nominal, ordinal and numerical variables accepted in PRINCIPALS. We denote the set of multiple variables by $\mathcal{J}_M$ and the set of single variables with single nominal and ordinal scales and numerical measurements by $\mathcal{J}_S$. For $\mathbf{X}$ consisting of a mixture of multiple and single variables, the algorithm alternates between estimation of $\mathbf{Z}$, $\mathbf{A}$ and $\mathbf{X}^*$ subject to minimizing

$$\theta^* = \mathrm{tr}(\mathbf{Z} - \mathbf{X}^*\mathbf{A})^\top(\mathbf{Z} - \mathbf{X}^*\mathbf{A})$$

under the restriction

$$\mathbf{Z}^\top \mathbf{1}_n = \mathbf{0}_r \quad \text{and} \quad \mathbf{Z}^\top\mathbf{Z} = n\mathbf{I}_p. \tag{9}$$

For the initialization of PRINCALS, we determine initial data $\mathbf{Z}^{(0)}$, $\mathbf{A}^{(0)}$ and $\mathbf{X}^{*(0)}$. The values of $\mathbf{Z}^{(0)}$ are initialized with random numbers under the restriction (9). For $j \in \mathcal{J}_M$, the initial value of $\mathbf{X}_j^*$ is obtained by $\mathbf{X}_j^{*(0)} = \mathbf{G}_j(\mathbf{G}_j^\top\mathbf{G}_j)^{-1}\mathbf{G}_j^\top\mathbf{Z}^{(0)}$. For $j \in \mathcal{J}_S$, $\mathbf{X}_j^{*(0)}$ is defined as the first K_j successive integers under the normalization restriction, and the initial value of $\mathbf{A}_j$ is calculated as the vector $\mathbf{A}_j^{(0)} = \mathbf{Z}^{(0)\top}\mathbf{X}_j^{*(0)}$. Given these initial values, PRINCALS as provided in Michailidis and de Leeuw [Michailidis and Leeuw, 1998] iterates the following two steps:

- *Model parameter estimation step*: Calculate $\mathbf{Z}^{(t+1)}$ by

$$\mathbf{Z}^{(t+1)} = p^{-1}\left(\sum_{j\in\mathcal{J}_M}\mathbf{X}_j^{*(t)} + \sum_{j\in\mathcal{J}_S}\mathbf{X}_j^{*(t)}\mathbf{A}_j^{(t)}\right).$$

Columnwise center and orthonormalize $\mathbf{Z}^{(t+1)}$. Estimate $\mathbf{A}_j^{(t+1)}$ for the single variable j by $\mathbf{A}_j^{(t+1)} = \mathbf{Z}^{(t+1)\top}\mathbf{X}_j^{*(t)}/\mathbf{X}_j^{*(t)\top}\mathbf{X}_j^{*(t)}$.

- *Optimal scaling step*: Estimate the optimally scaled vector for $j \in \mathcal{J}_M$ by

$$\mathbf{X}_j^{*(t+1)} = \mathbf{G}_j(\mathbf{G}_j^\top\mathbf{G}_j)^{-1}\mathbf{G}_j^\top\mathbf{Z}^{(t+1)}$$

and for $j \in \mathcal{J}_S$ by

$$\mathbf{X}_j^{*(t+1)} = \mathbf{G}_j(\mathbf{G}_j^\top\mathbf{G}_j)^{-1}\mathbf{G}_j^\top\mathbf{Z}^{(t+1)}\mathbf{A}_j^{(t+1)}/\mathbf{A}_j^{(t+1)\top}\mathbf{A}_j^{(t+1)}$$

under measurement restrictions on each of the variables.

9. Appendix B: The $v\varepsilon$ algorithm

Let $\mathbf{Y}^{(t)}$ denote a vector of dimensionality d that converges to a vector $\mathbf{Y}^{(\infty)}$ as $t \to \infty$. Let the inverse $[\mathbf{Y}]^{-1}$ of a vector $\mathbf{Y}$ be defined by

$$[\mathbf{Y}]^{-1} = \frac{\mathbf{Y}}{\|\mathbf{Y}\|^2},$$

where $\|\mathbf{Y}\|$ is the Euclidean norm of $\mathbf{Y}$.

In general, the $v\varepsilon$ algorithm for a sequence $\{\mathbf{Y}^{(t)}\}_{t \geq 0}$ starts with

$$\varepsilon^{(t,-1)} = 0, \qquad \varepsilon^{(t,0)} = \mathbf{Y}^{(t)},$$

and then generates a vector $\varepsilon^{(t,k+1)}$ by

$$\varepsilon^{(t,k+1)} = \varepsilon^{(t+1,k-1)} + \left[\varepsilon^{(t+1,k)} - \varepsilon^{(t,k)}\right]^{-1}, \qquad k = 0, 1, 2, \ldots \tag{10}$$

For practical implementation, we apply the $v\varepsilon$ algorithm for $k = 1$ to accelerate the convergence of $\{\mathbf{Y}^{(t)}\}_{t \geq 0}$. From Equation (10), we have

$$\varepsilon^{(t,2)} = \varepsilon^{(t+1,0)} + \left[\varepsilon^{(t+1,1)} - \varepsilon^{(t,1)}\right]^{-1} \quad \text{for } k = 1,$$

$$\varepsilon^{(t,1)} = \varepsilon^{(t+1,-1)} + \left[\varepsilon^{(t+1,0)} - \varepsilon^{(t,0)}\right]^{-1} = \left[\varepsilon^{(t+1,0)} - \varepsilon^{(t,0)}\right]^{-1} \quad \text{for } k = 0.$$

Then the vector $\varepsilon^{(t,2)}$ becomes as follows:

$$\varepsilon^{(t,2)} = \varepsilon^{(t+1,0)} + \left[\left[\varepsilon^{(t,0)} - \varepsilon^{(t+1,0)}\right]^{-1} + \left[\varepsilon^{(t+2,0)} - \varepsilon^{(t+1,0)}\right]^{-1}\right]^{-1}$$

$$= \mathbf{Y}^{(t+1)} + \left[\left[\mathbf{Y}^{(t)} - \mathbf{Y}^{(t+1)}\right]^{-1} + \left[\mathbf{Y}^{(t+2)} - \mathbf{Y}^{(t+1)}\right]^{-1}\right]^{-1}.$$

10. Appendix C: VASpca

URL of VASpca
http://mo161.soci.ous.ac.jp/vaspca/indexE.html

11. References

Al-Kandari, N.M. and Jolliffe, I.T. (2001). Variable selection and interpretation of covariance principal components. *Communications in Statistics. Simulation and Computation*, 30, 339-354.

Al-Kandari, N.M. and Jolliffe, I.T. (2005). Variable selection and interpretation in correlation principal components. *Environmetrics*, 16, 659-672.

Brezinski, C. and Zaglia, M. (1991). *Extrapolation methods: theory and practice*. Elsevier Science Ltd. North-Holland, Amsterdam.

Cadima, J., Cerdeira, J.O. and Manuel, M. (2004). Computational aspects of algorithms for variable selection in the context of principal components. *Computational Statistics and Data Analysis*, 47, 225-236.

Gifi, A. (1990). *Nonlinear multivariate analysis.* John Wiley & Sons, Ltd., Chichester.

Iizuka, M., Mori, Y., Tarumi, T. and Tanaka, Y. (2003). Computer intensive trials to determine the number of variables in PCA. *Journal of the Japanese Society of Computational Statistics*, 15, 337-345.

Jolliffe, I.T. (1972). Discarding variables in a principal component analysis. I. Artificial data. *Applied Statistics*, 21, 160-173.

Kiers, H.A.L. (2002). Setting up alternating least squares and iterative majorization algorithm for solving various matrix optimization problems. *Computational Statistics and Data Analysis*, 41, 157-170.

Krijnen, W.P. (2006). Convergence of the sequence of parameters generated by alternating least squares algorithms. *Computational Statistics and Data Analysis*, 51, 481-489.

Krzanowski, W.J. (1987). Selection of variables to preserve multivariate data structure using principal components. *Applied Statistics*, 36, 22-33.

Kuroda, M. and Sakakihara, M. (2006). Accelerating the convergence of the EM algorithm using the vector epsilon algorithm. *Computational Statistics and Data Analysis*, 51, 1549-1561.

Kuroda, M., Mori, Y., Iizuka, M. and Sakakihara, M. (2011). Accelerating the convergence of the EM algorithm using the vector epsilon algorithm. *Computational Statistics and Data Analysis*, 55, 143-153.

McCabe, G.P. (1984). Principal variables. *Technometrics*, 26, 137-144.

Michailidis, G. and de Leeuw, J. (1998). The Gifi system of descriptive multivariate analysis. *Statistical Science*, 13, 307-336.

Mori, Y., Tanaka, Y. and Tarumi, T. (1997). Principal component analysis based on a subset of variables for qualitative data. *Data Science, Classification, and Related Methods (Proceedings of IFCS-96)*, 547-554, Springer-Verlag.

Mori, Y., Tarumi, T. and Tanaka, Y. (1998). Principal component analysis based on a subset of variables - Numerical investigation on variable selection procedures -. *Bulletin of the Computational Statistics Society of Japan*, 11, 1-12 (in Japanese).

Mori, Y., Iizuka, M., Tanaka, Y. and Tarumi, T. (2006). Variable Selection in Principal Component Analysis. Härdle, W., Mori, Y. and Vieu, P. (eds), *Statistical Methods for Biostatistics and Related Fields*, 265-283, Springer.

R Development Core Team (2008). *R: A language and environment for statistical computing.* R Foundation for Statistical Computing, Vienna, Austria. ISBN 3-900051-07-0, URL http://www.R-project.org.

Robert, P. and Escoufier, Y. (1976). A unifying tool for linear multivariate statistical methods: the RV-coefficient. *Applied Statistics*, 25, 257-265.

Sano, K., Manaka, S., Kitamura, K., Kagawa M., Takeuchi, K., Ogashiwa, M., Kameyama, M., Tohgi, H. and Yamada, H. (1977). Statistical studies on evaluation of mild disturbance of consciousness - Abstraction of characteristic clinical pictures by cross-sectional investigation. *Sinkei Kenkyu no Shinpo* 21, 1052-1065 (in Japanese).

Tanaka, Y. and Mori, Y. (1997). Principal component analysis based on a subset of variables: Variable selection and sensitivity analysis. *The American Journal of Mathematical and Management Sciences*, 17, 61-89.

Young, F.W., Takane, Y., and de Leeuw, J. (1978). Principal components of mixed measurement level multivariate data: An alternating least squares method with optimal scaling features. *Psychometrika*, 43, 279-281.

Wang, M., Kuroda, M., Sakakihara, M. and Geng, Z. (2008). Acceleration of the EM algorithm using the vector epsilon algorithm. *Computational Statistics*, 23, 469-486.

Wynn, P. (1962). Acceleration techniques for iterated vector and matrix problems. *Mathematics of Computation*, 16, 301-322.

8

The Maximum Non-Linear Feature Selection of Kernel Based on Object Appearance

Mauridhi Hery Purnomo[1], Diah P. Wulandari[1],
I. Ketut Eddy Purnama[1] and Arif Muntasa[2]
*[1]Electrical Engineering Department – Industrial Engineering Faculty,
Institut Teknologi Sepuluh Nopember, Surabaya
[2]Informatics Engineering Department – Engineering Faculty,
Universitas Trunojoyo Madura
Indonesia*

1. Introduction

Principal component analysis (PCA) is linear method for feature extraction that is known as Karhonen Loove method. PCA was first proposed to recognize face by Turk and Pentland, and was also known as eigenface in 1991 [Turk, 1991]. However, PCA has some weaknesses. The first, it cannot capture the simplest invariance of the face image [Arif et al., 2008b] , when this information is not provided in the training data. The last, the result of feature extraction is global structure [Arif, 2008]. The PCA is very simple, has overcome curse of dimensionality problem, this method have been known and expanded by some researchers to recognize face such as Linear Discriminant Analysis (LDA)[Yambor, 2000; A.M. Martinez, 2003; J.H.P.N. Belhumeur 1998], Linear Preserving Projection that known Lapalacianfaces [Cai, 2005; Cai et al, 2006; Kokiopoulou, 2004; X. He et al., 2005], Independent Component Analysis, Kernel Principal Component Analysis [Scholkopf et al., 1998; Sch¨olkopf 1999], Kernel Linear Discriminant Analysis (KLDA) [Mika, 1999] and maximum feature value selection of nonlinear function based on Kernel PCA [Arif et al., 2008b]. As we know, PCA is dimensionality reduction method based on object appearance by projecting an original n-*dimensional* (row*column) image into k eigenface where $k<<n$. Although PCA have been developed into some methods, but in some cases, PCA can outperform LDA, LPP and ICA when it uses small sample size.

This chapter will explain some theoretical of modified PCA that derived from Principal Component Analysis. The first, PCA transforms input space into feature space by using three non-linear functions followed by selection of the maximum value of kernel PCA. The feature space is called as kernel of PCA [Arif et al., 2008b]. The function used to transform is the function that qualifies Mercer Kernel and generates positive semi-definite matrix. Kernel PCA as been implemented to recognize face image[Arif et al., 2008b] and has been compared with some method such as Principal Component Analysis, Principal Linear Discriminant Analysis, and Linear Preserving Projection. The last, the maximum value selection has been enhanced and implemented to classify smiling stage by using Kernel Laplacianlips [Mauridhi et al., 2010]. Kernel Laplacianlips transform from input space into feature space on the lips data, followed by PCA and LPP process on feature space. Kernel Laplacianlips yield local structure in feature space. Local structure is more important than global structure. The experimental results show that, Kernel Laplacianlips using selection of non-linear function maximum value outperforms another methods [Mauridhi et al., 2010], such as Two Dimensional Principal Component Analysis (2D-PCA) [Rima et al., 2010], PCA+LDA+Support Vector Machine [Gunawan et al., 2009]. This chapter is composed as follows:

1. Principal Component Analysis in input space
2. Kernel Principal Component Analysis
3. Maximum Value Selection of Kernel Principal Component Analysis as Feature Extraction in Feature Space
4. Experimental Results of Face Recognition by Using Maximum Value Selection of Kernel Principal Component Analysis as Feature Extraction in Feature Space
5. The Maximum Value Selection of Kernel Linear Preserving Projection as Extension of Kernel Principal Component Analysis
6. Experimental Results of Simile Stage Classification Based on Maximum Value Selection of Kernel Linear Preserving Projection
7. Conclusions

2. Principal component analysis in input space

Over the last two decades, many subspace algorithms have been developed for feature extraction. One of the most popular is Principal Component Analysis (PCA) [Arif et al., 2008a, Jon, 2003; A.M. Martinez and A.C. Kak, 2001; M. Kirby and L. Sirovich, 1990; M. Turk and A. Pentland, 1991]. PCA has overcome Curse of Dimensionality in object recognition, where it has been able to reduce the number of object characteristics fantastically. Therefore, until now PCA is still used as a reference to develop a feature extraction.

Suppose a set of training image containing m training image $X^{(k)}$, $\forall k$, $k \in 1$. .m, each training image has $h \times w$ size where $\forall H$, $H \in 1..h$ and $\forall W$, $W \in 1..w$. Each training image is represented as:

$$X = \begin{pmatrix} X_{1,1}^{(k)} & X_{1,2}^{(k)} & X_{1,3}^{(k)} & \cdots\cdots & X_{1,w-1}^{(k)} & X_{1,w}^{(k)} \\ X_{2,1}^{(k)} & X_{2,2}^{(k)} & X_{2,3}^{(k)} & \cdots\cdots & X_{2,w-1}^{(k)} & X_{2,w}^{(k)} \\ X_{3,1}^{(k)} & X_{3,2}^{(k)} & X_{3,3}^{(k)} & \cdots\cdots & X_{3,w-1}^{(k)} & X_{3,w}^{(k)} \\ X_{4,1}^{(k)} & X_{4,2}^{(k)} & X_{4,3}^{(k)} & \cdots\cdots & X_{4,w-1}^{(k)} & X_{4,w}^{(k)} \\ \cdots\cdots & \cdots\cdots & \cdots\cdots & \cdots\cdots & \cdots\cdots & \cdots\cdots \\ X_{h-1,1}^{(k)} & X_{h-1,2}^{(k)} & X_{h-1,3}^{(k)} & \cdots\cdots & X_{h-1,w-1}^{(k)} & X_{h-1,w}^{(k)} \\ X_{h,1}^{(k)} & X_{h,2}^{(k)} & X_{h,3}^{(k)} & \cdots\cdots & X_{h,w-1}^{(k)} & X_{h,w}^{(k)} \end{pmatrix} \qquad (1)$$

Equation (1) can be transformed into one dimensional matrix form, by placing (t+1) [th] row to t[th]. If $\forall N$, $N \in 1..n$ and $n = h \times w$, then Equation (1) can be changed into the following equation

$$X = \begin{pmatrix} X_{1,1}^{(k)} & X_{1,2}^{(k)} & \cdots & X_{1,w}^{(k)} & X_{1,w+1}^{(k)} & \cdots & X_{1,2w}^{(k)} & \cdots & X_{1,n-1}^{(k)} & X_{1,n}^{(k)} \end{pmatrix} \qquad (2)$$

To expresses m training image set, it is necessary to composed Equation (2) in the following equation:

$$X = \begin{pmatrix} X_{1,1}^{(1)} & X_{1,2}^{(1)} & \cdots & X_{1,w}^{(1)} & X_{1,w+1}^{(1)} & \cdots & X_{1,2w}^{(1)} & \cdots & X_{1,n-1}^{(1)} & X_{1,n}^{(1)} \\ X_{1,1}^{(2)} & X_{1,2}^{(2)} & \cdots & X_{1,w}^{(2)} & X_{1,w+1}^{(2)} & \cdots & X_{1,2w}^{(2)} & \cdots & X_{1,n-1}^{(2)} & X_{1,n}^{(2)} \\ X_{1,1}^{(3)} & X_{1,2}^{(3)} & \cdots & X_{1,w}^{(3)} & X_{1,w+1}^{(3)} & \cdots & X_{1,2w}^{(3)} & \cdots & X_{1,n-1}^{(3)} & X_{1,n}^{(3)} \\ X_{1,1}^{(4)} & X_{1,2}^{(4)} & \cdots & X_{1,w}^{(4)} & X_{1,w+1}^{(4)} & \cdots & X_{1,2w}^{(4)} & \cdots & X_{1,n-1}^{(4)} & X_{1,n}^{(4)} \\ \cdots & \cdots & \cdots & \cdots & \cdots & & \cdots & \cdots & \cdots & \cdots \\ X_{1,1}^{(m-1)} & X_{1,2}^{(m-1)} & \cdots & X_{1,w}^{(m-1)} & X_{1,w+1}^{(m-1)} & \cdots & X_{1,2w}^{(m-1)} & \cdots & X_{1,n-1}^{(m-1)} & X_{1,n}^{(m-1)} \\ X_{1,1}^{(m)} & X_{1,2}^{(m)} & \cdots & X_{1,w}^{(m)} & X_{1,w+1}^{(m)} & \cdots & X_{1,2w}^{(m)} & \cdots & X_{1,n-1}^{(m)} & X_{1,n}^{(m)} \end{pmatrix} \qquad (3)$$

The average of training image set of (Equation (3)) can be obtained by column-wise summation. It can be formulated by using the following equation

$$\overline{X} = \frac{\sum_{k=1}^{m} X_{1,N}^{(k)}}{m} \qquad (4)$$

And $\forall N, N \in 1..n$.

The result of Equation (4) is in the row vector form, it has $1 \times N$ dimension. It can be re-written in the following equation

$$\overline{X} = [\overline{X}_{1,1} \ \overline{X}_{1,2} \ \overline{X}_{1,3} \ \overline{X}_{1,4} \ \dots\dots\dots\dots\dots\dots\dots\dots \ \overline{X}_{1,n-1} \ \overline{X}_{1,n}] \qquad (5)$$

The result of Equation (5) can be re-formed as original training image. To illustrate Equation (1), (2), (3) and (4), it is important to give an example of image average of the Olivetty Research Laboratory (ORL) face image database as seen In Figure 1

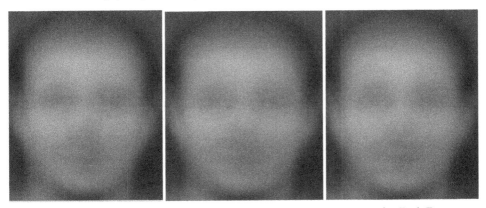

Fig. 1. average of ORL Face Image Database Using 3, 5 and 7 Face Image for Each Person

The zero mean matrix can be calculated by subtracting the face image values of training set with Equation (5). In order to perform the subtraction, both face image and Equation (5) must have the same size.

Therefore, Equation (5) can be replicated as many as m row size. The zero mean matrix can be formulated by using the following equation

$$\Phi_M = X_M - \overline{X} \tag{6}$$

$\forall M$, $M \in 1..m$. Furthermore, the covariance value can be computed by using the following equation

$$C = (X_M - \overline{X}).\left(X_M - \overline{X}\right)^T \tag{7}$$

As shown in Equation (7), C has mxm size and the value of $m<<n$. To obtain the principal components, the eigenvalues and eigenvectors can be computed by using the following equation:

$$
\begin{aligned}
C.\Lambda &= \lambda.\Lambda \\
C.\Lambda &= \lambda.I.\Lambda \\
(\lambda I - C).\Lambda &= 0 \\
Det(\lambda I - C) &= 0
\end{aligned}
\tag{8}
$$

The values of λ and Λ represent eigenvalues and eigenvectors of C respectively.

$$
\lambda = \begin{pmatrix}
\lambda_{1,1} & 0 & 0 & 0 & 0 \\
0 & \lambda_{2,2} & 0 & 0 & 0 \\
0 & 0 & .... & 0 & 0 \\
0 & 0 & 0 & \lambda_{M-1,M-1} & 0 \\
0 & 0 & 0 & 0 & \lambda_{M,M}
\end{pmatrix}
\tag{9}
$$

$$
\Lambda = \begin{pmatrix}
\Lambda_{1,1} & \Lambda_{1,2} & ..... & \Lambda_{1,m-1} & \Lambda_{1,m} \\
\Lambda_{2,1} & \lambda_{2,2} & .... & \Lambda_{2,m-1} & \Lambda_{2,m} \\
...... & ...... & .... & ...... & ...... \\
\Lambda_{m-1,1} & \Lambda_{m-1,2} & ...... & \Lambda_{m-1,m-1} & \Lambda_{m-1,m} \\
\Lambda_{m,1} & \Lambda_{m,2} & ...... & \Lambda_{m,m-1} & \lambda_{m,m}
\end{pmatrix}
\tag{10}
$$

Equation (9) can be changed into row vector as seen in the following equation

$$\lambda = \begin{bmatrix} \lambda_{1,1} & \lambda_{2,2} & \lambda_{3,3} & \cdots\cdots\cdots\cdots & \lambda_{m-1,m-1} & \lambda_{m,m} \end{bmatrix} \tag{10}$$

To obtain the most until the less dominant features, the eigenvalues were sorted in descending order $(\lambda_1 > \lambda_2 > \lambda_3 > \cdots\cdots\cdots\cdots > \lambda_{m-1} > \lambda_m)$ and followed by corresponding eigenvectors.

3. Kernel principal component analysis

Principal Component Analysis has inspired some researchers to develop it. Kernel Principal Component Analysis (KPCA) is Principal Component Analysis in feature space [Sch"olkopf et al., 1998; Sch"olkopf et al., 1999; Arif et al., 2008b; Mauridhi et al., 2010]. Principally, KPCA works in feature space [Arif et al., 2008b]. Input space of training set is transformed into feature space by using Mercer Kernel that yields positive semi definite matrix as seen in the Kernel Trick [Sch"olkopf et al., 1998; Sch"olkopf et al., 1999]

$$k(X,Y) = (\phi(X), \phi(Y)) \tag{11}$$

Functions that can be used to transform are *Gaussian, Polynomial,* and *Sigmoidal* as seen in the following equation

$$k(X,Y) = \exp(\frac{-||X-Y||^2}{\sigma}) \tag{12}$$

$$k(X,Y) = (a(X.Y)+b)^d \tag{13}$$

$$k(X,Y) = \tanh(a(X.Y)+b)^d \tag{14}$$

4. Maximum value selection of kernel principal component analysis as feature extraction in feature space

The results of Equation (12), (13) and (14) will be selected as object feature candidates [Arif et al., 2008b, Mauridhi, 2010]. The biggest value of them will be employed as feature space in the next stage, as seen in the following equation

$$F = \max(\phi_i : R_i \xrightarrow[k(x,y)]{} F_i) \tag{15}$$

For each kernel function has yielded one matrix feature, so we have 3 matrix of feature space from 3 kernel functions. For each corresponding matrix position will be compared and will be selected the maximum value (the greatest value). The maximum value will be used as feature candidate. It can be represented by using the following equation

$$\phi(X) = \begin{pmatrix} \phi\left(X_{1,1}^{(k)}\right) & \phi\left(X_{1,2}^{(k)}\right) & \phi\left(X_{1,3}^{(k)}\right) & \cdots & \phi\left(X_{1,w-1}^{(k)}\right) & \phi\left(X_{1,m}^{(k)}\right) \\ \phi\left(X_{2,1}^{(k)}\right) & \phi\left(X_{2,2}^{(k)}\right) & \phi\left(X_{2,3}^{(k)}\right) & \cdots & \phi\left(X_{2,w-1}^{(k)}\right) & \phi\left(X_{2,m}^{(k)}\right) \\ \phi\left(X_{3,1}^{(k)}\right) & \phi\left(X_{3,2}^{(k)}\right) & \phi\left(X_{3,3}^{(k)}\right) & \cdots & \phi\left(X_{3,w-1}^{(k)}\right) & \phi\left(X_{3,m}^{(k)}\right) \\ \phi\left(X_{4,1}^{(k)}\right) & \phi\left(X_{4,2}^{(k)}\right) & \phi\left(X_{4,3}^{(k)}\right) & \cdots & \phi\left(X_{4,w-1}^{(k)}\right) & \phi\left(X_{4,m}^{(k)}\right) \\ \cdots & \cdots & \cdots & \cdots & \cdots & \cdots \\ \phi\left(X_{m-1,1}^{(k)}\right) & \phi\left(X_{m-1,2}^{(k)}\right) & \phi\left(X_{m-1,3}^{(k)}\right) & \cdots & \phi\left(X_{m-1,m-1}^{(k)}\right) & \phi\left(X_{m-1,m}^{(k)}\right) \\ \phi\left(X_{m,1}^{(k)}\right) & \phi\left(X_{m,2}^{(k)}\right) & \phi\left(X_{m,3}^{(k)}\right) & \cdots & \phi\left(X_{m,m-1}^{(k)}\right) & \phi\left(X_{m,m}^{(k)}\right) \end{pmatrix} \tag{16}$$

The biggest value of feature space is the most dominant feature value. As we know, feature space as seen on equation (16) is yielded by using kernel (in this case, training set is transformed into feature space using equation (12), (13) and (14) and followed by selection of the biggest value at the same position using equation (15). where feature selection in kernel space will be used to determine average, zero mean, covariance matrix, eigenvalue and eigenvector in feature space. These values are yielded by using kernel trick as nonlinear component. Nonlinear component is linear component (principal component) improvement. So, it is clear that the biggest value of these kernels is improvement of the PCA performance. The average value of Equation (16) can be expressed in the following equation

$$\phi\left(\overline{X}\right) = \frac{\sum_{k=1}^{m}\phi\left(X_{1,N}^{(k)}\right)}{m} \tag{17}$$

So, zero mean in the feature space can be found by using the following equation

$$\phi(\Phi_M) = \phi(X_M) - \phi\left(\overline{X}\right) \tag{18}$$

Where $\forall M$, $M \in 1..m$. The result of Equation (18) has $m \times m$. To obtain the eigenvalues and the eigenvectors in feature space, it is necessary to calculated the covariance matrix in feature space. It can be computed by using the following equation

$$\phi(C) = (\phi(X) - \phi(\overline{X})).\left(\phi(X) - \phi(\overline{X})\right)^{T} \tag{19}$$

Based on Equation (19), the eigenvalues and the eigenvectors in feature space can be determined by using the following equation

$$\begin{aligned}
\phi(C).\phi(\Lambda) &= \phi(\lambda).\phi(\Lambda) \\
\phi(C).\phi(\Lambda) &= \phi(\lambda).I.\phi(\Lambda) \\
\left(\phi(\lambda)I - \phi(C)\right).\phi(\Lambda) &= 0 \\
Det\left(\phi(\lambda)I - \phi(C)\right) &= 0
\end{aligned} \tag{20}$$

The eigenvalues and eigenvectors yielded by Equation (20) can be written in following matrices

$$\phi(\lambda) = \begin{pmatrix}
\phi(\lambda_{1,1}) & 0 & 0 & 0 & 0 \\
0 & \phi(\lambda_{2,2}) & 0 & 0 & 0 \\
0 & 0 & .... & 0 & 0 \\
0 & 0 & 0 & \phi(\lambda_{M-1,M-1}) & 0 \\
0 & 0 & 0 & 0 & \phi(\lambda_{M,M})
\end{pmatrix} \tag{21}$$

$$\phi(\Lambda) = \begin{pmatrix}
\phi(\Lambda_{1,1}) & \phi(\Lambda_{1,2}) & ..... & \phi(\Lambda_{1,m-1}) & \phi(\Lambda_{1,m}) \\
\phi(\Lambda_{2,1}) & \phi(\lambda_{2,2}) & ...... & \phi(\Lambda_{2,m-1}) & \phi(\Lambda_{2,m}) \\
...... & ...... & .... & ...... & ...... \\
\phi(\Lambda_{m-1,1}) & \phi(\Lambda_{m-1,2}) & ..... & \phi(\Lambda_{m-1,m-1}) & \phi(\Lambda_{m-1,m}) \\
\phi(\Lambda_{m,1}) & \phi(\Lambda_{m,2}) & ..... & \phi(\Lambda_{m,m-1}) & \phi(\lambda_{m,m})
\end{pmatrix} \tag{22}$$

To obtain the value of the most until the less dominant feature, the Equation (21) will be sorted decreasingly and followed by Equation (20) [Arif et al., 2008b, Mauridhi, 2010]. The bigger value of the eigenvalue in the feature space, the more dominant the corresponding eigenvector in feature space. The result of sorting Equation (21) can be shown in the following equation

$$\phi(\lambda) = \left[\phi(\lambda_{1,1}) \ \phi(\lambda_{2,2}) \ \phi(\lambda_{3,3}) \dots\dots\dots\dots\dots\dots \phi(\lambda_{m-1,m-1}) \ \phi(\lambda_{m,m}) \right] \qquad (23)$$

5. Experimental results of face recognition by using maximum value selection of kernel principal component analysis as feature extraction in feature space

In this chapter, the experimental results of "The Maximum Value Selection of Kernel Principal Component Analysis for Face Recognition" will be explained. We use Olivetti-Att-ORL (ORL) [Research Center of Att, 2007] and YALE face image databases [Yale Center for Computational Vision and Control, 2007] as experimental material.

5.1 Experimental results using the ORL face image database

ORL face image database consist of 40 persons, 36 of them are men and the other 4 are women. Each of them has 10 poses. The poses were taken at different time with various kinds of lighting and expressions (eyes open/close, smiling/not smiling) [Research Center of Att, 2007]. The face position is frontal with 10 up to 20% angles. The face image size is 92x112 pixels as shown in Figure 2.

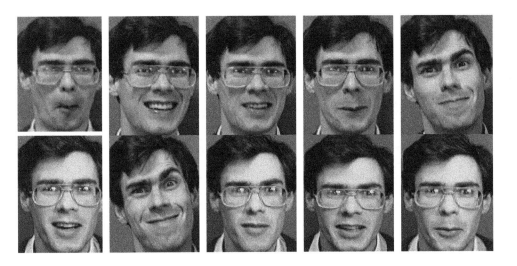

Fig. 2. Face Images of ORL Database

The experiments are employed for 5 times, and for each experiment 5, 6, 7, 8 and 9 poses for each person are used. The rest of training set, i.e. 5, 4, 3, 2 and 1, will be used as the testing [Arif et al., 2008b] as seen in Table 1

Scenario	Data Quantity			
	For Each Person		Total	
	Training	Testing	Training	Testing
1st	5	5	200	200
2nd	6	4	240	160
3rd	7	3	280	120
4th	8	2	320	80
5th	9	1	360	40

Table 1. The Scenario of ORL Face Database Experiment

In this experiment, each scenario used different dimension. The 1st, 2nd, 3rd, 4th, and 5th scenarios used 200, 240, 280, 320, and 360 dimensions respectively. The result of the 1st scenario can be seen on the Figure 3 [Arif et al., 2008b]

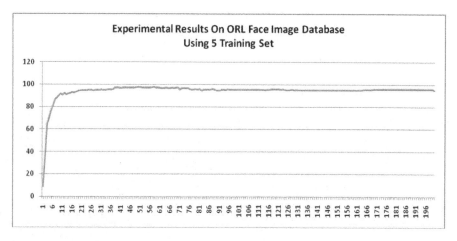

Fig. 3. Experimental Results on ORL Face Image Database Using 5 Training Set

Figure 3 shows that the more number of dimensional used, the higher recognition rate, but the recognition decreased on the certain dimension. As seen in Figure 3, recognition rate decreased into 95% when 200 dimensions were used. The first maximum recognition rate, which was 97.5%, occurred when 49 dimensions were used in this experiment [Arif et al., 2008b].

In the 2nd scenario, the maximum dimension used was 240 (240=40*6) training set. The first maximum recognition rate occurred when 46 dimensions were used, this was 99.375%. When 1 until 46 dimensions were used, recognition rate increased proportionally to the number of dimension used, but when 47 until the 240 dimensions were used, the recognition rate tended to be stable, with insignificant fluctuations as seen in Figure 4 [Arif et al., 2008b].

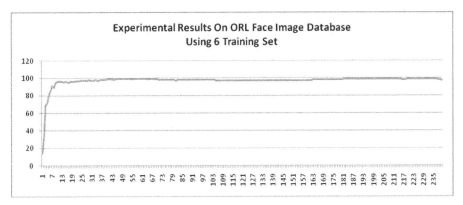

Fig. 4. Experimental Results on ORL Face Image Database Using 6 Training Set

In the 3rd scenario, training set used for each person was 7, whereas the number of dimensions used was 280. The more number of training set used, the number of dimension is increased. In this scenario, the maximum recognition rate was 100%, it occurred when 23 until 53 dimensions were used, whereas when more than 53 dimensions were used, recognition rate decreased to be 99.67% as seen in Figure 5 [Arif et al., 2008b].

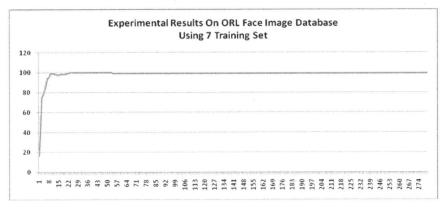

Fig. 5. Experimental Results on ORL Face Image Database Using 7 Training Set

Figure 6 is the experimental results of the 4th scenario. In this scenario, 8 training sets for each person were used, whereas the number of dimensions used was 320. Figure 6 shows that the recognition rate tended to increase significantly for experimental results using less than 23 dimensions, whereas 100% recognition rate occurred for experimental results using more than 24 dimensions [Arif et al., 2008b].

In the last scenario, 9 training sets were used, whereas the number of dimension used was 360, as seen in Figure 7. Similarly to the previous scenario, the recognition rate tended to increase when experimental used less than 6 dimensions, while using 7 dimensions resulted in 100% recognition rate, using 8 dimension resulted in 97% recognition rate, and 100% recognition rate was yielded from experimental results using more than 9 dimensions, as shown in Figure 7 [Arif et al., 2008b].

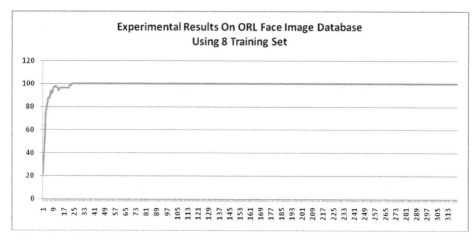

Fig. 6. Experimental Results on ORL Face Image Database Using 8 Training Set

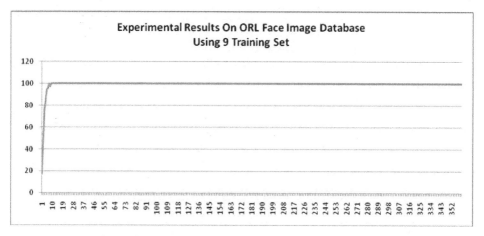

Fig. 7. Experimental Results on ORL Face Image Database Using 9 Training Set

The maximum recognition rate for all scenarios can be seen in Table 2. This table shows that the more number of training set used, the higher recognition rate achieved, whereas the first maximum recognition rate tended to occur on the lower dimension inversely proportional to the number of dimensions used [Arif et al., 2008b].

5.2 Experimental results using the YALE face image database

In this last experiment, the YALE face database was used. It contains 15 people, each of them were doing 11 poses. The poses were taken in various kinds of lighting (left lighting and center lighting), various expressions (normal, smiling, sad, sleepy, surprising, and wink) and accessories (wearing or not wearing glasses) [Yale Center for Computational Vision and Control, 2007] as shown in Figure 8.

Scenario	Number of Training Sample for Each Person	The First Maximum Recognition Rate	Dimension
1st	5	97.5	49
2nd	6	99.375	46
3rd	7	100	23
4th	8	100	24
5th	9	100	7

Table 2. The ORL Face Database Recognition Rate using Maximum Feature Value Selection Method of Nonlinear Function based on KPCA

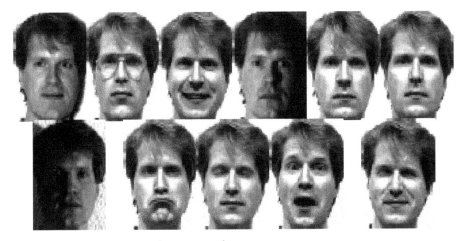

Fig. 8. Face Sample of Images of YALE Database

The experiments were conducted for 6 scenarios, for each scenario, 5, 6, 7, 8, 9, and 10 training set were used. The rest of each data sample for every experiment, i.e. 6, 5, 4, 3, 2 and 1, were used as testing set as listed in Table 3 [Arif et al., 2008b].

Scenario	Data Quantity			
	For Each Person		Total	
	Training	Testing	Training	Testing
1st	5	6	75	90
2nd	6	5	90	75
3rd	7	4	105	60
4th	8	3	120	45
5th	9	2	135	30
6th	10	1	150	15

Table 3. The Scenario of the YALE Face Database Experiment

In the first scenario, 5 training sets were used, where the rest of the YALE data experiment was used for testing. In this scenario, the number of dimensions used was 75. The completed experimental results can be seen in Figure 8. This figure shows that the number of recognition rate increased significantly when less than 9 dimensions were used, which were 16.67% until 92.22%. Whereas the maximum recognition rate occurred when 13, 14, and 15 dimensions were used, that was 94.44% [Arif et al., 2008b]. For experimental results using more than 16 dimensions, the recognition rate fluctuated insignificantly as seen in Figure 8.

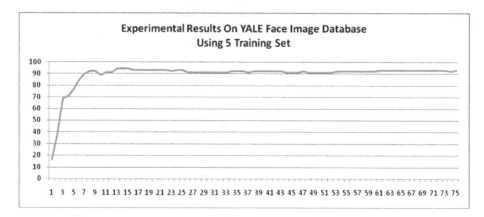

Fig. 8. Experimental Results on YALE Face Image Database Using 5 Training Set

The experimental results of the 2nd scenario were shown in Figure 9. The recognition rate increased from 22.67% until 97.33% when using less than 10 dimensions, recognition rate decreased insignificantly when using 16 dimensions, and recognition rate tended to be stable around 97.33% when experiments used more than 17 dimensions, [Arif et al., 2008b].

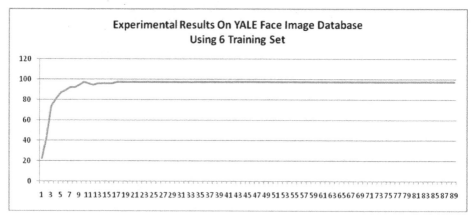

Fig. 9. Experimental Results on YALE Face Image Database Using 6 Training Set

Similarly, it occurred in the 3rd scenario. In this scenario, the recognition rate increased significantly when the number of dimensions was less than 13, though on the certain number of dimensions the recognition rate decreased. But when the number of dimensions used was more than 14, experimental results yielded its maximum rate, which is 98.33% as seen in Figure 10 [Arif et al., 2008b].

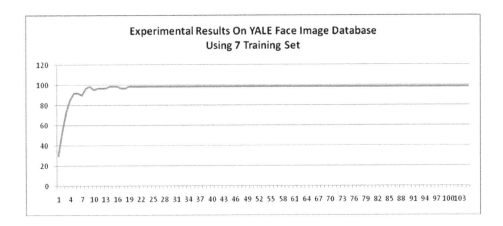

Fig. 10. Experimental Results on YALE Face Image Database Using 7 Training Set

In the last three scenarios as seen in Figure 11, 12, and 13, experimental results have shown that the recognition rate also tended to increase when the number of dimensions used was less than 7, whereas experimental results that used more than 8 dimensions achieved 100% recognition rate [Arif et al., 2008b].

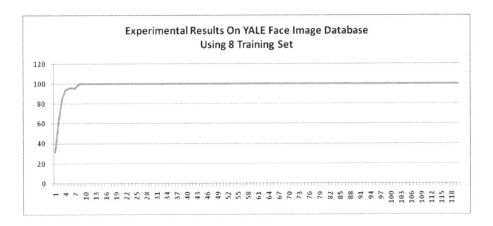

Fig. 11. Experimental Results on YALE Face Image Database Using 8 Training Set

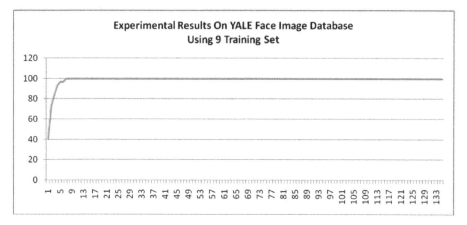

Fig. 12. Experimental Results on YALE Face Image Database Using 9 Training Set

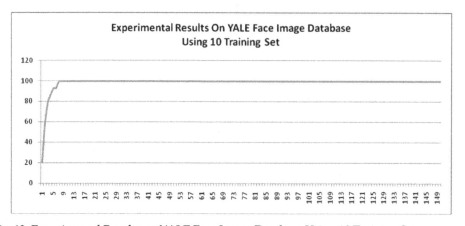

Fig. 13. Experimental Results on YALE Face Image Database Using 10 Training Set

As seen in Table 4, the maximum recognition rate for 5, 6, 7, 8, 9, and 10 training sets were 94.444%, 97.333%, 98.333%, 100%, 100% and 100% respectively. Based on Table 4, the maximum recognition rate increased proportionally to the number of training sets used. The more number of training set used, the faster maximum recognition rate is reached [Arif et al., 2008b].

The experimental results of the 1st, 2nd, and 3rd scenarios were compared to other methods, such as PCA, LDA/QR, and LPP/QR as seen in Table 5, whereas for the 4th and 5th scenarios were not compared, since they have achieved maximum result (100%). The recognition rate of 5, 6, and 7 training set, for both on the ORL and the YALE face database, "The Maximum Value Selection of Kernel Principal Component Analysis", outperformed the other methods.

Scenario	Number of Training Sample for Each Person	The First Maximum Recognition Rate	Dimension
1st	5	94.444	13
2nd	6	97.333	10
3rd	7	98.333	9
4th	8	100	8
5th	9	100	7
6th	10	100	7

Table 4. The YALE Face Database Recognition Rate using Maximum Feature Value Selection Method of Non linear Function based on KPCA

Database	Number of Training Set	The Maximum Recognition Rate (%)			
		PCA	LDA/QR	LPP/QR	The Maximum Value Selection of Kernel Principal Component Analysis
ORL	5	79.50	86.5	94.00	97.50
	6	83.13	91.25	94.37	99.38
	7	85.00	92.50	95.83	100.00
YALE	5	81.11	84.44	86.67	94.44
	6	85.33	86.67	94.67	97.33
	7	95.00	95.00	96.67	98.33

Table 5. The Comparative Results for Face Recognition Rate

6. The maximum value selection of kernel linear preserving projection as extension of kernel principal component analysis

Kernel Principal Component Analysis as appearance method in feature space yields global structure to characterized an object. Besides global structure, local structure is also important. Kernel Linear Preserving Projection as known as KLPP is method used to preserve the intrinsic geometry of the data and local structure in feature space [Cai et al., 2005; Cai et al.,, 2006; Kokiopoulou, 2004; Mauridhi et al., 2010]. The objective of LPP in feature space is written in the following equation [Mauridhi et al., 2010]

$$\min \sum_{ij} (\phi(y)_i - \phi(y)_j)^2 \cdot \phi(S_{ij}) \tag{24}$$

In this case the value of $S_{i,j}$ can be defined as

$$\phi(S_{ij}) = \begin{cases} e^{\frac{(\phi(x_i)-\phi(x_j))^2}{t}} & ||\phi(x_i)-\phi(x_j)|| < \varepsilon \\ 0 & otherwise \end{cases} \tag{25}$$

Where $\varepsilon > 0$, but it is sufficiently small compared to the local neighborhood radius. Minimizing the objective function ensures the closeness between points that is located in the same class. If neighboring points of $\phi(x_i)$ and $\phi(x_j)$ are mapped far apart in feature space and if $(\phi(y_i) - \phi(y_j))$ is large, then $\phi(S_{ij})$ incurs a heavy penalty in feature space. Suppose a set of data and a weighted graph $G = (V, E)$ is constructed from data points where the data points that are closed to linked by the edge. Suppose maps of a graph to a line is chosen to minimize the objective function of KLPP in Equation (24) on the limits (constraints) as appropriate. Suppose a represents transformation vector, whereas the ith column vector of X is symbolized by using x_i. By simple algebra formulation step, the objective function in feature space can be reduced in the following equation [Mauridhi et al., 2010]

$$
\begin{aligned}
&\frac{1}{2}\sum_{ij}(\phi(y_i) - \phi(y_j))^2 \phi(S)_{ij} \\
&= \frac{1}{2}\sum_{ij}(\phi(a^T)\phi(x_i) - \phi(a^T)\phi(x_j))^2 \phi(S_{ij}) \\
&= \frac{1}{2}\sum_{ij}((\phi(a^T)\phi(x_i))^2 - 2\phi(a^T)\phi(x_i)\phi(a^T)\phi(x_j) + (\phi(a^T)\phi(x_j))^2)\phi(S_{ij}) \\
&= \frac{1}{2}\sum_{ij}(\phi(a^T)\phi(x_i)\phi(x_i^T)\phi(a) - 2\phi(a^T)\phi(x_i)\phi(x_j^T)\phi(a) + \phi(a^T)\phi(x_j)\phi(x_j^T)\phi(a))\phi(S_{ij}) \\
&= \frac{1}{2}\sum_{ij}(2\phi(a^T)\phi(x_i)\phi(x_i^T)\phi(a) - 2\phi(a^T)\phi(x_i)\phi(x_j^T)\phi(a))\phi(S_{ij}) \\
&= \sum_{ij}\phi(a^T)\phi(x_i)\phi(S_{ij})\phi(x_i^T)\phi(a) - \sum_{ij}\phi(a^T)\phi(x_i)\phi(S_{ij})\phi(x_j^T)\phi(a) \\
&= \sum_{i}\phi(a^T)\phi(x_i)\phi(D_{ii})\phi(x_i^T)\phi(a) - \phi(a^T)\phi(X)\phi(S)\phi(X^T)\phi(a) \\
&= \phi(a^T)\phi(X)\phi(D)\phi(X^T)\phi(a) - \phi(a^T)\phi(X)\phi(S)\phi(X^T)\phi(a) \\
&= \phi(a^T)\phi(X)(\phi(D) - \phi(S))\phi(X^T)\phi(a) \\
&= \phi(a^T)\phi(X)\phi(L)\phi(X^T)\phi(a)
\end{aligned}
\tag{26}
$$

In this case, $\phi(X)=[\ \phi(x_1),\ \phi(x_2),\ \ldots\ldots\ \phi(x_M)]$, $\phi(D_{ii}) = \phi(\Sigma_j\ S_{ij})$ and $\phi(L) = \phi(D)- \phi(S)$ represent *Laplacian* matrices in feature space known as *Laplacianlips*, when these are implemented in smiling stage classification. The minimum of the objective function in feature space is given by the minimum eigenvalue solution in feature space by using the following equation

$$
\begin{aligned}
\phi(X)(\phi(D) - \phi(S))\phi(X^T)\phi(w) &= \phi(\lambda)\phi(x)\phi(D)\phi(x^T) \\
\phi(X)\phi(L)\phi(X^T)\phi(w) &= \phi(\lambda)\phi(x)\phi(D)\phi(x^T)
\end{aligned}
\tag{27}
$$

Eigenvalues and eigenvectors in feature space can be calculated by using Equation (27). The most until the less dominant features can be achieved by sorting eigenvalues decreasingly and followed by sorting corresponding eigenvectors in feature space.

7. Experimental results of smile stage classification based on the maximum value selection of kernel linear preserving projection

To evaluate the **Maximum Value Selection of Kernel Linear Preserving Projection Method**, it is necessary to conduct the experiment. In this case, 30 persons were used as experiment. Each person consists of 3 patterns, which are smiling pattern I, III and IV, while smiling pattern II is not used. The image size was 640x 640 pixels and every face image was changed the size into 50x50 pixels (Figure 14). Before feature extraction process, face image had been manually cropped against a face data at oral area to produce spatial coordinate [5.90816 34.0714 39.3877 15.1020] [Mauridhi et al., 2010]. This was conducted to simplify calculation process. In this case, cropped data were used for both training and testing set. This process caused the face data size reduction into 40x16 pixels as seen in Figure 15.

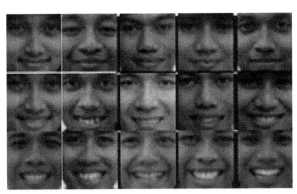

Fig. 14. Original Sample of Smiling Pattern

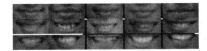

Fig. 15. Cropping Result Sample of Smiling Pattern

Experiments were applied by using 3 scenarios. In the first scenario, the first of 2/3 data (20 of 30 persons) became the training set and the rest (10 persons) were used as testing set. In the second scenario, the first of 10 persons (10 of 30) were used as testing set and the last 20 (20 of 30) persons were used as training set. In the last scenario, the first and the last of 10 persons (20 of 30) were used as training set and the middle 10 persons (10 of 30) were used for testing set. It means, data were being rotated without overlap, thus each of them has got the experience of becoming testing data. Due to smiling pattern III, the numbers of training and testing data were 20*3=60 and 10*3=30 images respectively. In this experiment, 60 dimensions were used. To measure similarity, the angular separation and Canberra were used. The Equation (17) and (18) are similarity measure for the angular separation and Canberra [Mauridhi et al., 2010]. To achieve classification rate percentage, equation (19) was used. The result of classification using the 1st, 2nd, and 3rd scenario can be seen in Figure 16, 17, and 18 respectively [Mauridhi et al., 2010].

The 1st, 2nd, and 3rd scenario had similarity trend as seen in Figure 16, 17, and 18. Recognition rate increased significantly from the 1st until 10th dimension, whereas recognition rate using more than 11 dimensions slightly fluctuated. The maximum and the average recognition rate in the 1st scenario were not different, which was 93.33%. In the 2nd scenario, the maximum recognition rate was 90%, when Canberra similarity measure was used. In the 3rd scenario, the maximum recognition rate was 100%, when angular separation was used. The maximum recognition rate was 93.33%, for both Angular Separation and Canberra Similarity Measure [Mauridhi et al., 2010] as seen in Table 6.

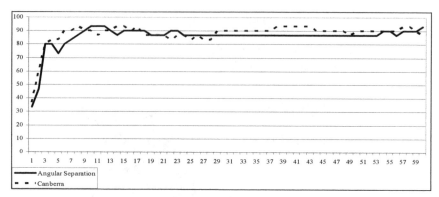

Fig. 16. Smile Stage Classification Recognition Rate Based on the Maximum Value Selection of Kernel Linear Preserving Projection Method Using 1st Scenario

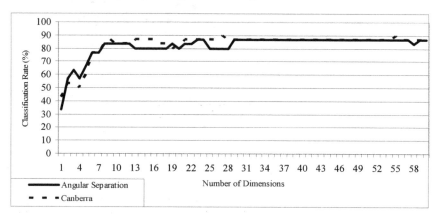

Fig. 17. Smile Stage Classification Recognition Rate Based on the Maximum Value Selection of Kernel Linear Preserving Projection Method Using 2nd Scenario

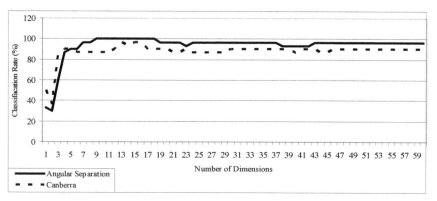

Fig. 18. Smile Stage Classification Recognition Rate Based on the Maximum Value Selection of Kernel Linear Preserving Projection Method Using 3rd Scenario

Similarity Methods	The Maximum Recognition Rate in the Scenario (%)			Average
	1st	2nd	3rd	
Angular Separation	93.33	86.67	100	93.33
Canberra	93.33	90.00	96.67	93.33
Maximum	93.33	90.00	100	94.44
Average	93.33	88.34	98.34	93.33

Table 6. The Smile Stage Classification Recognition Rate using Maximum Feature Value Selection Method of Non linear Kernel Function based on Kernel Linear Preserving Projection

The experimental results of **the Maximum Value Selection of Kernel Linear Preserving Projection Method** have been compared to "Two Dimensional Principal Component Analysis (2D-PCA) and Support Vector Machine (SVM) as its classifier" [Rima et al., 2010] and have been combined with some methods, which were Principal Component Analysis (PCA)+Linear Discriminant Analysis (LDA) and SVM as its classifier [Gunawan et al., 2009] as seen Figure 19

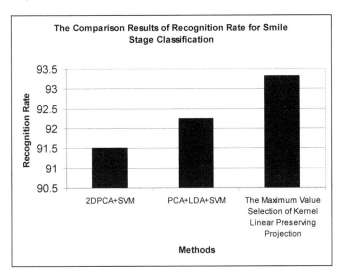

Fig. 19. The Comparison Results of Recognition Rate for Smile Stage Classification

8. Conclusion

For both the maximum non-linear feature selection of Kernel Principal Component Analysis and Kernel Linear Preserving Projection has yielded local feature structure for extraction, which is more important than global structures in feature space. It can be shown that, the maximum non-linear feature selection of Kernel Principal Component Analysis for face recognition has outperformed the PCA, LDA/QR and LPP/QR on the ORL and the YALE face databases. Whereas the maximum value selection of Kernel Linear Preserving Projection as extension of Kernel Principal Component Analysis has outperformed the 2D-PCA+SVM and the PCA+LDA+SVM for smile stage classification.

9. References

A.M. Martinez and A.C. Kak, "PCA versus LDA," IEEE Trans. Pattern Analysis Machine Intelligence, vol. 23, no. 2, pp. 228-233, Feb. 2001.

Arif Muntasa, Mochamad Hariadi, Mauridhi Hery Purnomo, "*Automatic Eigenface Selection for Face Recognition*", The 9th Seminar on Intelligent Technology and Its Applications (2008a) 29 – 34.

Arif Muntasa, Mochamad Hariadi, Mauridhi Hery Purnomo, "Maximum Feature Value Selection of Nonlinear Function Based On Kernel PCA For Face Recognition", Proceeding of The 4th Conference On Information & Communication Technology and Systems, Surabaya, Indonesia, 2008b

Cai, D., He, X., and Han, J. Using Graph Model for Face Analysis, University of Illinois at Urbana-Champaign and University of Chicago, 2005.

Cai, D., X. He, J. Han, and H.-J. Zhang. Orthogonal laplacianfaces for face recognition. *IEEE Transactions on Image Processing*, 15(11):3608–3614, 2006.

Gunawan Rudi Cahyono, Mochamad Hariadi, Mauridhi Hery Purnomo, "Smile Stages Classification Based On Aesthetic Dentistry Using Eigenfaces, Fisherfaces And Multiclass Svm", 2009

J.H.P.N. Belhumeur, D. Kriegman, "Eigenfaces vs. fisherfaces: Recognition using class specific linear projection", IEEE Trans. on PAMI, 19(7):711–720, 1997..

Jon Shlens, "*A Tutorial On Principal Component Analysis And Singular Value Decomposition*", http://mathworks.com , 2003

Kokiopoulou, E. and Saad, Y. Orthogonal Neighborhood Preserving Projections, University of Minnesota, Minneapolis, 2004.

M. Kirby and L. Sirovich, "Application of the KL Procedure for the Characterization of Human Faces," IEEE Trans. Pattern Anal. Mach. Intell., vol. 12, no. 1, pp. 103-108, 1990.

M. Turk, A. Pentland, "Eigenfaces for recognition", Journal of Cognitive Science, pages 71–86, 1991.

Mauridhi Hery Purnomo, Tri Arif, Arif Muntasa, "Smiling Stage Classification Based on Kernel Laplacianlips Using Selection of Non Linier Function Maximum Value", 2010 IEEE International Conference on Virtual Environments, Human-Computer Interfaces, and Measurement Systems (VECIMS 2010) Proceedings, pp 151-156

Mika, S., Ratsch, G.,Weston, J., Scholkopf, B. and Mller, K.R.: Fisher discriminant analysis with kernels. IEEE Workshop on Neural Networks for Signal Processing IX, (1999) 41-48

Research Center of Att, UK, *Olivetti-Att-ORL FaceDatabase*, http://www.uk.research. att.com/facedabase.html, Accessed in 2007

Rima Tri Wahyuningrum, Mauridhi Hery Purnomo, I Ketut Eddy Purnama, "Smile Stages Recognition in Orthodontic Rehabilitation Using 2D-PCA Feature Extraction", 2010

Sch"olkopf, B., Mika, S., Burges, C. J. C., Knirsch, P., Mller, K. R., Raetsch, G. and Smola, A.: Input Space vs. Feature Space in Kernel Based Methods, IEEE Trans. on NN, Vol 10. No. 5, (1999) 1000-1017

Sch"olkopf, B., Smola, A.J. and Mller, K.R.: Nonlinear Component Analysis as a Kernel Eigen-value Problem, Neural Computation, 10(5), (1998) 1299-1319

X. He, S. Yan, Y. Hu, P. Niyogi, and H.-J. Zhang. Face recognition using laplacianfaces. *IEEE Transactions on Pattern Analysis and Machine Intelligence*, 27(3):328–340, 2005.

Yale Center for Computational Vision and Control, *Yale Face Database*, http://cvc.yale. edu/projects/yalefaces/yalefaces.html, Accessed 2007

Yambor, W.S . Analysis of PCA-Based and Fisher Discriminant-Based Image Recognition Algorithms, Tesis of Master, Colorado State University, 2000

9

The Basics of Linear Principal Component Analysis

Yaya Keho

Ecole Nationale Supérieure de Statistique et d'Economie Appliquée (ENSEA), Abidjan
Côte d'Ivoire

1. Introduction

When you have obtained measures on a large number of variables, there may exist redundancy in those variables. Redundancy means that some of the variables are correlated with one another, possibly because they are measuring the same "thing". Because of this redundancy, it should be possible to reduce the observed variables into a smaller number of variables. For example, if a group of variables are strongly correlated with one another, you do not need all of them in your analysis, but only one since you can predict the evolution of all the variables from that of one. This opens the central issue of how to select or build the representative variables of each group of correlated variables. The simplest way to do this is to keep one variable and discard all others, but this is not reasonable. Another alternative is to combine the variables in some way by taking perhaps a weighted average, as in the line of the well-known Human Development Indicator published by UNDP. However, such an approach calls the basic question of how to set the appropriate weights. If one has sufficient insight into the nature and magnitude of the interrelations among the variables, one might choose weights using one's individual judgment. Obviously, this introduces a certain amount of subjectivity into the analysis and may be questioned by practitioners. To overcome this shortcoming, another method is to let the data set uncover itself the relevant weights of variables. Principal Components Analysis (PCA) is a variable reduction method that can be used to achieve this goal. Technically this method delivers a relatively small set of synthetic variables called principal components that account for most of the variance in the original dataset.

Introduced by Pearson (1901) and Hotelling (1933), Principal Components Analysis has become a popular data-processing and dimension-reduction technique, with numerous applications in engineering, biology, economy and social science. Today, PCA can be implemented through statistical software by students and professionals but it is often poorly understood. The goal of this Chapter is to dispel the magic behind this statistical tool. The Chapter presents the basic intuitions for how and why principal component analysis works, and provides guidelines regarding the interpretation of the results. The mathematics aspects will be limited. At the end of this Chapter, readers of all levels will be able to gain a better understanding of PCA as well as the when, the why and the how of applying this technique. They will be able to determine the number of meaningful components to retain from PCA, create factor scores and interpret the components. More emphasis will be placed on examples explaining in detail the steps of implementation of PCA in practice.

We think that the well understanding of this Chapter will facilitate that of the following chapters and novel extensions of PCA proposed in this book (sparse PCA, Kernel PCA, Multilinear PCA, …).

2. The basic prerequisite – Variance and correlation

PCA is useful when you have data on a large number of quantitative variables and wish to collapse them into a smaller number of artificial variables that will account for most of the variance in the data. The method is mainly concerned with identifying variances and correlations in the data. Let us focus our attention to the meaning of these concepts. Consider the dataset given in Table 1. This dataset will serve to illustrate how PCA works in practice.

ID	X1	X2	X3	X4	X5
1	24	21.5	5	2	14
2	16.7	21.4	6	2.5	17
3	16.78	23	7	2.2	15
4	17.6	22	8.7	3	20
5	22	25.7	6.4	2	14.2
6	15.3	16	8.7	2.21	15.3
7	10.2	19	4.3	2.2	15.3
8	11.9	17.1	4.5	2	14
9	14.3	19.1	6	2.2	15
10	8.7	14.3	4.1	2.24	15.5
11	6.7	10	3.8	2.23	16
12	7.1	13	2.8	2.01	12
13	10.3	16	4	2	14.5
14	7.1	13	3.9	2.4	16.4
15	7.9	13.6	4	3.1	20.2
16	3	8	3.4	2.1	14.7
17	3	9	3.3	3	20.2
18	1	7.5	3	2	14
19	0.8	7	2.8	2	15.8
20	1	4	3.1	2.2	15.3

Table 1. **Example dataset, 5 variables obtained for 20 observations.**

The variance of a given variable x is defined as the average of the squared differences from the mean:

$$\sigma_x^2 = \frac{1}{n}\sum_{i=1}^{n}\left(x_i - \overline{x}\right)^2$$ (1)

The square root of the variance is the standard deviation and is symbolized by the small Greek sigma σ_x. It is a measure of how spread out numbers are.

The variance and the standard deviation are important in data analysis because of their relationships to correlation and the normal curve. Correlation between a pair of variables measures to what extent their values co-vary. The term covariance is undoubtedly associatively prompted immediately. There are numerous models for describing the behavioral nature of a simultaneous change in values, such as linear, exponential and more. The linear correlation is used in PCA. The linear correlation coefficient for two variables x and y is given by:

$$\rho(x,y) = \frac{\frac{1}{n}\sum_{i=1}^{n}(x_i - \bar{x})(y_i - \bar{y})}{\sigma_x \sigma_y} \tag{2}$$

where σ_x and σ_y denote the standard deviation of x and y, respectively. This definition is the most widely-used type of correlation coefficient in statistics and is also called Pearson correlation or product-moment correlation. Correlation coefficients lie between -1.00 and +1.00. The value of -1.00 represents a perfect negative correlation while a value of +1.00 represents a perfect positive correlation. A value of 0.00 represents a lack of correlation. Correlation coefficients are used to assess the degree of collinearity or redundancy among variables. Notice that the value of correlation coefficient does not depend on the specific measurement units used.

When correlations among several variables are computed, they are typically summarized in the form of a correlation matrix. For the five variables in Table 1, we obtain the results reported in Table 2.

	X1	X2	X3	X4	X5
X1	1.00	0.94	0.77	-0.03	-0.08
X2		1.00	0.74	0.02	-0.04
X3			1.00	0.21	0.19
X4				1.00	0.95
X5					1.00

Table 2. **Correlations among variables**

In this Table a given row and column intersect shows the correlation between the two corresponding variables. For example, the correlation between variables X_1 and X_2 is 0.94.

As can be seen from the correlations, the five variables seem to hang together in two distinct groups. First, notice that variables X_1, X_2 and X_3 show relatively strong correlations with one another. This could be because they are measuring the same "thing". In the same way, variables X_4 and X_5 correlate strongly with each another, a possible indication that they measure the same "thing" as well. Notice that those two variables show very weak correlations with the rest of the variables.

Given that the 5 variables contain some "redundant" information, it is likely that they are not really measuring five different independent constructs, but two constructs or underlying factors. What are these factors? To what extent does each variable measure each of these factors? The purpose of PCA is to provide answers to these questions. Before presenting the mathematics of the method, let's see how PCA works with the data in Table 1.

In linear PCA each of the two artificial variables is computed as the linear combination of the original variables.

$$Z = \alpha_1 X_1 + \alpha_2 X_2 + ... + \alpha_5 X_5 \tag{3}$$

where α_j is the weight for variable j in creating the component Z. The value of Z for a subject represents the subject's score on the principal component.

Using our dataset, we have:

$$Z_1 = 0.579 X_1 + 0.577 X_2 + 0.554 X_3 + 0.126 X_4 + 0.098 X_5 \tag{4}$$

$$Z_2 = -0.172 X_1 - 0.14 X_2 + 0.046 X_3 + 0.685 X_4 + 0.693 X_5 \tag{5}$$

Notice that different coefficients were assigned to the original variables in computing subject scores on the two components. X_1, X_2 and X_3 are assigned relatively large weights that range from 0.554 to 0.579, while variables X_4 and X_5 are assigned very small weights ranging from 0.098 to 0.126. As a result, component Z_1 should account for much of the variability in the first three variables. In creating subject scores on the second component, much weight is given to X_4 and X_5, while little weight is given to X_1, X_2 and X_3. Subject scores on each component are computed by adding together weighted scores on the observed variables. For example, the value of a subject along the first component Z_1 is 0.579 times the standardized value of X_1 plus 0.577 times the standardized value of X_2 plus 0.554 times the standardized value of X_3 plus 0.126 times the standardized value of X_4 plus 0.098 times the standardized value of X_5.

At this stage of our analysis, it is reasonable to wonder how the weights from the preceding equations are determined. Are they optimal in the sense that no other set of weights could produce components that best account for variance in the dataset? How principal components are computed?

3. Heterogeneity and standardization of data

3.1 Graphs and distances among points

Our dataset in **Table 1** can be represented into two graphs: one representing the subjects, and the other the variables. In the first, we consider each subject (individual) as a vector with coordinates given by the 5 observations of the variables. Clearly, the cloud of points belongs to a R^5 space. In the second one each variable is regarded as a vector belonging to a R^{20} space.

We can calculate the centroide of the cloud of points which coordinates are the 5 means of the variables, that is $g = (\overline{X}_1, ..., \overline{X}_5)$. Again, we can compute the overall variance of the points by summing the variance of each variable:

$$I = \frac{1}{n} \sum_{i=1}^{n} \sum_{j=1}^{p} \left(X_{ij} - \overline{X}_j \right)^2 = \frac{1}{n} \sum_{i=1}^{n} d^2(s_i, g) = \sum_{j=1}^{p} \sigma_j^2 \tag{6}$$

This quantity measures how spread out the points are around the centroid. We will need this quantity when determining principal components.

We define the distance between subjects s_i and $s_{i'}$ using the Euclidian distance as follows:

$$d^2(s_i, s_{i'}) = \|s_i - s_{i'}\|^2 = \sum_{j=1}^{p=5} (X_{ij} - X_{i'j})^2 \tag{7}$$

Two subjects are close one to another when they take similar values for all variables. We can use this distance to measure the overall dispersion of the data around the centroid or to cluster the points as in classification methods.

3.2 How work when data are in different units?

There are different problems when variables are measured in different units. The first problem is the meaning of the variance: how to sum quantities with different measurement units? The second problem is that the distance between points can be greatly influenced. To illustrate this point, let us consider the distances between subjects 7, 8 and 9. Applying Eq.(7), we obtain the following results:

$$d^2(s_7, s_8) = (10.2 - 11.9)^2 + (19 - 17.1)^2 + ... + (15.3 - 14)^2 = 8.27 \tag{8}$$

$$d^2(s_7, s_9) = (10.2 - 14.3)^2 + (19 - 19.1)^2 + ... + (15.3 - 15)^2 = 19.8 \tag{9}$$

Subject 7 is closer to subject 8 than to subject 9. Multiplying the values of variable X_5 by 10 yields:

$$d^2(s_7, s_8) = (10.2 - 11.9)^2 + (19 - 17.1)^2 + ... + (153 - 140)^2 = 175.58 \tag{10}$$

$$d^2(s_7, s_9) = (10.2 - 14.3)^2 + (19 - 19.1)^2 + ... + (153 - 150)^2 = 28.71 \tag{11}$$

Now we observe that subject 7 is closer to subject 9 than to subject 8. It is hard to accept how the measurement units of the variables can change greatly the comparison results among subjects. Indeed, we could by this way render a tall man as shorter as we want!

As seen, PCA is sensitive to scale. If you multiply one variable by a scalar you get different results. In particular, the principal components are dependent on the units used to measure the original variables as well as on the range of values they assume (variance). This makes comparison very difficult. It is for these reasons we should *often* standardize the variables prior to using PCA. A common standardization method is to subtract the mean and divide by the standard deviation. This yields the following:

$$X_i^* = \frac{X_i - \bar{X}}{\sigma_x} \tag{12}$$

where $\bar{X}$ and σ_x are the mean and standard deviation of X, respectively.

Thus, the new variables all have zero mean and unit standard deviation. Therefore the total variance of the data set is the number of observed variables being analyzed.

Throughout, we assume that the data have been centered and standardized. Graphically, this implies that the centroid or center of gravity of the whole dataset is at the origin. In this case, the PCA is called normalized principal component analysis, and will be based on the correlation matrix (and not on variance-covariance matrix). The variables will lie on the unit sphere; their projection on the subspace spanned by the principal components is the "correlation circle". Standardization allows the use of variables which are not measured in the same units (e.g. temperature, weight, distance, size, etc.). Also, as we will see later, working with standardized data makes interpretation easier.

4. The mathematics of PCA: An eigenvalue problem

Now we have understood the intuitions of PCA, we present the mathematics behind the method by considering a general case. More details on technical aspects can be found in Cooley & Lohnes (1971), Stevens (1986), Lebart, Morineau & Piron (1995), Cadima & Jolliffe (1995), Hyvarinen, Karhunen & Oja (2001), and Jolliffe (2002).

Consider a dataset consisting of p variables observed on n subjects. Variables are denoted by $(x_1, x_2, ..., x_p)$. In general, data are in a table with the rows representing the subjects (individuals) and the columns the variables. The dataset can also be viewed as a $n \times p$ rectangular matrix X. Note that variables are such that their means make sense. The variables are also standardized.

We can represent these data in two graphs: on the one hand, in a subject graph where we try to find similarities or differences between subjects, on the other, in a variable graph where we try to find correlations between variables. Subjects graph belongs to an p-dimensional space, i.e. to R^p, while variables graph belongs to an n-dimensional space, i.e. to R^n. We have two clouds of points in high-dimensional spaces, too large for us to plot and see something in them. We cannot see beyond a three-dimensional space! The PCA will give us a subspace of reasonable dimension so that the projection onto this subspace retains "as much as possible" of the information present in the dataset, i.e., so that the projected clouds of points be as "dispersed" as possible. In other words, the goal of PCA is to compute another basis that best re-express the dataset. The hope is that this new basis will filter out the noise and reveal hidden structure.

$$x_i = \begin{pmatrix} x_{1i} \\ x_{2i} \\ \cdot \\ \cdot \\ \cdot \\ x_{pi} \end{pmatrix} \rightarrow \text{reduce dimensionality} \rightarrow z_i = \begin{pmatrix} z_{1i} \\ z_{2i} \\ \cdot \\ \cdot \\ z_{qi} \end{pmatrix} \text{ with } q < p \qquad (13)$$

Dimensionality reduction implies information loss. How to represent the data in a lower-dimensional form without losing too much information? Preserve as much information as possible is the objective of the mathematics behind the PCA procedure.

We first of all assume that we want to project the data points on a 1-dimensional space. The principal component corresponding to this axis is a linear combination of the original variables and can be expressed as follows:

$$z_1 = \alpha_{11}x_1 + \alpha_{12}x_2 + ... + \alpha_{1p}x_p = Xu_1 \tag{14}$$

where $u_1 = (\alpha_{11}, \alpha_{12}, ... \alpha_{1p})'$ is a column vector of weights. The principal component z_1 is determined such that the overall variance of the resulting points is as large as possible. Of course, one could make the variance of z_1 as large as possible by choosing large values for the weights $\alpha_{11}, \alpha_{12}, ..., \alpha_{1p}$. To prevent this, weights are calculated with the constraint that their sum of squares is one, that is u_1 is a unit vector subject to the constraint:

$$\alpha_{11}^2 + \alpha_{12}^2 + ... + \alpha_{1p}^2 = \|u_1\|^2 = 1 \tag{15}$$

Eq.(14) is also the projections of the n subjects on the first component. PCA finds u_1 so that

$$Var(z_1) = \frac{1}{n}\sum_{i=1}^{n} z_{1i}^2 = \frac{1}{n}\|z_1\|^2 = \frac{1}{n}u'_1 X'Xu_1 \text{ is maximal} \tag{16}$$

The matrix $C = \frac{1}{n}X'X$ is the correlation matrix of the variables. The optimization problem is:

$$\underset{\substack{u_1 \\ \|u_1\|^2 = 1}}{Max\, u'_1\, Cu_1} \tag{17}$$

This program means that we search for a unit vector u_1 so as to maximize the variance of the projection on the first component. The technique for solving such optimization problems (linearly constrained) involves a construction of a Lagrangian function.

$$\Im_1 = u'_1 Cu_1 - \lambda_1(u'_1 u_1 - 1) \tag{18}$$

Taking the partial derivative $\partial\Im_1 / \partial u_1 = Cu_1 - \lambda_1 u_1$ and solving the equation $\partial\Im_1 / \partial u_1 = 0$ yields:

$$Cu_1 = \lambda_1 u_1 \tag{19}$$

By premultiplying each side of this condition by u'_1 and using the condition $u'_1 u_1 = 1$ we get:

$$u'_1 Cu_1 = \lambda_1 u'_1 u_1 = \lambda_1 \tag{20}$$

It is known from matrix algebra that the parameters u_1 and λ_1 that satisfy conditions (19) and (20) are the maximum eigenvalue and the corresponding eigenvector of the correlation matrix C. Thus the optimum coefficients of the original variables generating the first principal component z_1 are the elements of the eigenvector corresponding to the largest eigenvalue of the correlation matrix. These elements are also known as loadings.

The second principal component is calculated in the same way, with the condition that it is uncorrelated (orthogonal) with the first principal component and that it accounts for the largest part of the remaining variance.

$$z_2 = \alpha_{21}x_1 + \alpha_{22}x_2 + ... + \alpha_{2p}x_p = Xu_2 \tag{21}$$

where $u_2 = (\alpha_{21}, \alpha_{22}, ...\alpha_{2p})'$ is the direction of the component. This axis is constrained to be orthogonal to the first one. Thus, the second component is subject to the constraints:

$$\alpha_{21}^2 + \alpha_{22}^2 + ... + \alpha_{2p}^2 = \|u_2\|^2 = 1, \quad u'_1 u_2 = 0 \tag{22}$$

The optimization problem is therefore:

$$\underset{\substack{u_2 \\ \|u_2\|^2 = 1 \\ u'_1 u_2 = 0}}{Max} \ u'_2 Cu_2 \tag{23}$$

Using the technique of Lagrangian function the following conditions:

$$Cu_2 = \lambda_2 u_2 \tag{24}$$

$$u'_2 Cu_2 = \lambda_2 \tag{25}$$

are obtained again. So once more the second vector comes to be the eignevector corresponding to the second highest eigenvalue of the correlation matrix.

Using induction, it can be proven that PCA is a procedure of eigenvalue decomposition of the correlation matrix. The coefficients generating the linear combinations that transform the original variables into uncorrelated variables are the eigenvectors of the correlation matrix. This is a good new, because finding eigenvectors is something which can be done rapidly using many statistical packages (SAS, Stata, R, SPSS, SPAD...), and because eigenvectors have many nice mathematical properties. Note that rather than maximizing variance, it might sound more plausible to look for the projection with the smallest average (mean-squared) distance between the original points and their projections on the principal components. This turns out to be equivalent to maximizing the variance (Pythagorean Theorem).

An interesting property of the principal components is that they are all uncorrelated (orthogonal) to one another. This is because matrix C is a real symmetric matrix and then linear algebra tells us that it is diagonalizable and the eigenvectors are orthogonal to one another. Again because C is a covariance matrix, it is a positive matrix in the sense that $u'Cu \geq 0$ for any vector u. This tells us that the eigenvalues of C are all non-negative.

$$var(z) = \begin{bmatrix} \lambda_1 & 0 & . & 0 \\ 0 & \lambda_2 & & \\ & . & & \\ 0 & & & \lambda_p \end{bmatrix} \tag{26}$$

The eigenvectors are the "preferential directions" of the data set. The principal components are derived in decreasing order of importance; and have a variance equal to their corresponding eigenvalue. The first principal component is the direction along which the data have the most variance. The second principal component is the direction orthogonal to the first component with the most variance. It is clear that all components explain together 100% of the variability in the data. This is why we say that PCA works like a change of basis. Analyzing the original data in the canonical space yields the same results than examining it in the components space. However, PCA allows us to obtain a linear projection of our data, originally in R^p, onto R^q, where $q < p$. The variance of the projections on to the first q principal components is the sum of the eigenvalues corresponding to these components. If the data fall near a q-dimensional subspace, then $p-q$ of the eigenvalues will be nearly zero.

Summarizing the computational steps of PCA

Suppose $x_1, x_2, ..., x_p$ are $p \times 1$ vectors collected from n subjects. The computational steps that need to be accomplished in order to obtain the results of PCA are the following:

Step 1. Compute mean: $\bar{x} = \dfrac{1}{n} \sum\limits_{i=1}^{n} x_i$

Step 2. Standardize the data: $\Phi_i = \dfrac{x_i - \bar{x}}{\sigma_x}$

Step 3. Form the matrix $A = \left[\Phi_1, \Phi_2, ..., \Phi_p \right]$ ($p \times n$ matrix), then compute:

$$C = \frac{1}{n} \sum_{i=1}^{n} \Phi'_i \, \Phi_i$$

Step 4. Compute the eigenvalues of C: $\lambda_1 > \lambda_2 > ... > \lambda_p$

Step 5. Compute the eigenvectors of C: $u_1, u_2, ..., u_p$

Step 6. Proceed to the linear tranformation $R^p \rightarrow R^q$ that performs the dimensionality reduction.

Notice that, in this analysis, we gave the same weight to each subject. We could have give more weight to some subjects, to reflect their representativity in the population.

5. Criteria for determining the number of meaningful components to retain

In principal component analysis the number of components extracted is equal to the number of variables being analyzed (under the general condition $n > p$). This means that an analysis of our 5 variables would actually result in 5 components, not two. However, since PCA aims at reducing dimensionality, only the first few components will be important enough to be retained for interpretation and used to present the data. It is therefore reasonable to wonder how many independent components are necessary to best describe the data.

Eigenvalues are thought of as quantitative assessment of how much a component represents the data. The higher the eigenvalues of a component, the more representative it is of the data. Eigenvalues are therefore used to determine the meaningfulness of components. Table 3 provides the eigenvalues from the PCA applied to our dataset. In the column headed "Eigenvalue", the eigenvalue for each component is presented. Each raw in the table presents information about one of the 5 components: the raw "1" provides information about the first component (PCA1) extracted, the raw "2" provides information about the second component (PCA2) extracted, and so forth. Eigenvalues are ranked from the highest to the lowest.

It can be seen that the eigenvalue for component 1 is 2.653, while the eigenvalue for component 2 is 1.98. This means that the first component accounts for 2.653 units of total variance while the second component accounts for 1.98 units. The third component accounts for about 0.27 unit of variance. Note that the sum of the eigenvalues is 5, which is also the number of variables. How do we determine how many components are worth interpreting?

Component	Eigenvalue	% of variance	Cumulative %
1	2.653	53.057	53.057
2	1.980	39.597	92.653
3	0.269	5.375	98.028
4	0.055	1.095	99.123
5	0.044	0.877	100.000

Table 3. **Eigenvalues from PCA**

Several criteria have been proposed for determining how many meaningful components should be retained for interpretation. This section will describe three criteria: the Kaiser eigenvalue-one criterion, the Cattell Scree test, and the cumulative percent of variance accounted for.

5.1 Kaiser method

The Kaiser (1960) method provides a handy rule of thumb that can be used to retain meaningful components. This rule suggests keeping only components with eigenvalues greater than 1. This method is also known as the eigenvalue-one criterion. The rationale for this criterion is straightforward. Each observed variable contributes one unit of variance to the total variance in the data set. Any component that displays an eigenvalue greater than 1 is accounts for a greater amount of variance than does any single variable. Such a component is therefore accounting for a meaningful amount of variance, and is worthy of being retained. On the other hand, a component with an eigenvalue of less than 1 accounts for less variance than does one variable. The purpose of principal component analysis is to reduce variables into a relatively smaller number of components; this cannot be effectively achieved if we retain components that account for less variance than do individual variables. For this reason, components with eigenvalues less than 1 are of little use and are not retained. When a covariance matrix is used, this criterion retains components whose eigenvalue is greater than the average variance of the data (Kaiser-Guttman criterion).

However, this method can lead to retaining the wrong number of components under circumstances that are often encountered in research. The thoughtless application of this rule can lead to errors of interpretation when differences in the eigenvalues of successive components are trivial. For example, if component 2 displays an eigenvalue of 1.01 and component 3 displays an eigenvalue of 0.99, then component 2 will be retained but component 3 will not; this may mislead us into believing that the third component is meaningless when, in fact, it accounts for almost exactly the same amount of variance as the second component. It is possible to use statistical tests to test for difference between successive eigenvalues. In fact, the Kaiser criterion ignores error associated with each eigenvalue due to sampling. Lambert, Wildt and Durand (1990) proposed a bootstrapped version of the Kaiser approach to determine the interpretability of eigenvalues.

Table 3 shows that the first component has an eigenvalue substantially greater than 1. It therefore explains more variance than a single variable, in fact 2.653 times as much. The second component displays an eigenvalue of 1.98, which is substantially greater than 1, and the third component displays an eigenvalue of 0.269, which is clearly lower than 1. The application of the Kaiser criterion leads us to retain unambiguously the first two principal components.

5.2 Cattell scree test

The scree test is another device for determining the appropriate number of components to retain. First, it graphs the eigenvalues against the component number. As eigenvalues are constrained to decrease monotonically from the first principal component to the last, the scree plot shows the decreasing rate at which variance is explained by additional principal components. To choose the number of meaningful components, we next look at the scree plot and stop at the point it begins to level off (Cattell, 1966; Horn, 1965). The components that appear *before* the "break" are assumed to be meaningful and are retained for interpretation; those appearing *after* the break are assumed to be unimportant and are not retained. Between the components before and after the break lies a scree.

The scree plot of eigenvalues derived from Table 3 is displayed in Figure 1. The component numbers are listed on the horizontal axis, while eigenvalues are listed on the vertical axis. The Figure shows a relatively large break appearing between components 2 and 3, meaning the each successive component is accounting for smaller and smaller amounts of the total variance. This agrees with the preceding conclusion that two principal components provide a reasonable summary of the data, accounting for about 93% of the total variance.

Sometimes a scree plot will display a pattern such that it is difficult to determine exactly where a break exists. When encountered, the use of the scree plot must be supplemented with additional criteria, such as the Kaiser method or the cumulative percent of variance accounted for criterion.

5.3 Cumulative percent of total variance accounted for

When determining the number of meaningful components, remember that the subspace of components retained must account for a reasonable amount of variance in the data. It is usually typical to express the eigenvalues as a percentage of the total. The fraction of an eigenvalue out of the sum of all eigenvalues represents the amount of variance accounted by the corresponding principal component. The cumulative percent of variance explained by the first q components is calculated with the formula:

$$r_q = \frac{\sum_{j=1}^{q} \lambda_j}{\sum_{j=1}^{p} \lambda_j} \times 100 \qquad (27)$$

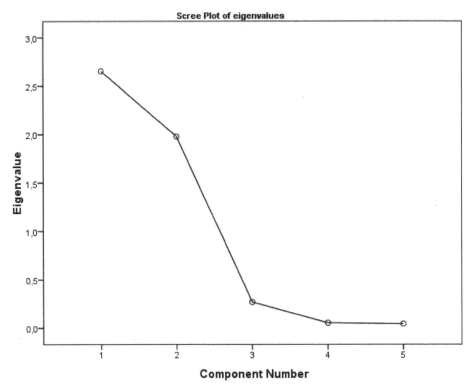

Fig. 1. **Scree plot of eigenvalues**

How many principal components we should use depends on how big an r_q we need. This criterion involves retaining all components up to a total percent variance (Lebart, Morineau & Piron, 1995; Jolliffe, 2002). It is recommended that the components retained account for at least 60% of the variance. The principal components that offer little increase in the total variance explained are ignored; those components are considered to be noise. When PCA works well, the first two eigenvalues usually account for more than 60% of the total variation in the data.

In our current example, the percentage of variance accounted for by each component and the cumulative percent variance appear in Table 3. From this Table we can see that the first component alone accounts for 53.057% of the total variance and the second component alone accounts for 39.597% of the total variance. Adding these percentages together results in a sum of 92.65%. This means that the cumulative percent of variance accounted for by the first two components is about 93%. This provides a reasonable summary of the data. Thus we can keep the first two components and "throw away" the other components.

A number of other criteria have been proposed to select the number of components in PCA and factorial analysis. Users can read Lawley (1956), Horn (1965), Humphreys and Montanelli (1975), Horn and Engstrom (1979), Zwick and Velicer (1986), Hubbard and Allen (1987) and Jackson (1993), among others.

6. Interpretation of principal components

Running a PCA has become easy with statistical software. However, interpreting the results can be a difficult task. Here are a few guidelines that should help practitioners through the analysis.

6.1 The visual approach of correlation

Once the analysis is complete, we wish to assign a name to each retained component that describes its content. To do this, we need to know what variables explain the components. Correlations of the variables with the principal components are useful tools that can help interpreting the meaning of components. The correlations between each variable and each principal component are given in Table 4.

Variables	PCA 1	PCA 2
X1	0.943	-0.241
X2	0.939	-0.196
X3	0.902	0.064
X4	0.206	0.963
X5	0.159	0.975

Notes : PCA1 and PCA2 denote the first and second principal component, respectively.

Table 4. **Correlation variable-component**

Those correlations are also known as component loadings. A coefficient greater than 0.4 in absolute value is considered as significant (see, Stevens (1986) for a discussion). We can interpret PCA1 as being highly positively correlated with variables X_1, X_2 and X_3, and weakly positively correlated to variables X_4 and X_5. So X_1, X_2 and X_3 are the most important variables in the first principal component. PCA2, on the other hand, is highly positively correlated with X_4 and X_5, and weakly negatively related to X_1 and X_2. So X_4 and X_5 are most important in explaining the second principal component. Therefore, the name of the first component comes from variables X_1, X_2 and X_3 while that of the second component comes from X_4 and X_5.

It can be shown that the coordinate of a variable on a component is the correlation coefficient between that variable and the principal component. This allows us to plot the reduced dimension representation of variables in the plane constructed from the first two components. Variables highly correlated with a component show a small angle. Figure 2 represents this graph for our dataset. For each variable we have plotted on the horizontal dimension its loading on component 1, on the vertical dimension its loading on component 2.

The graph also presents a visual aspect of correlation patterns among variables. The cosine of the angle θ between two vectors x and y is computed as:

$$< x,y >= \|x\|\|y\|\cos(x,y) \tag{28}$$

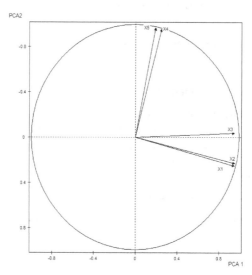

Fig. 2. **Circle of Correlation**

Replacing x and y with our transformed vectors yields:

$$\cos(x,y) = \frac{\sum\limits_{i=1}^{n}(x_i - \overline{x})(y_i - \overline{y})}{\sqrt{\left(\sum\limits_{i=1}^{n}(x_i - \overline{x})^2\right)\left(\sum\limits_{i=1}^{n}(y_i - \overline{y})\right)^2}} = \rho(x,y) \tag{29}$$

Eq.(29) shows the connection between the cosine measurement and the numerical measurement of correlation: the cosine of the angle between two variables is interpreted in terms of correlation. Variables highly positively correlated with each another show a small angle, while those are negatively correlated are directed in opposite sense, i.e. they form a flat angle. From Figure 2 we can see that the five variables hang together in two distinct groups. Variables X_1, X_2 and X_3 are positively correlated with each other, and form the first group. Variables X_4 and X_5 also correlate strongly with each other, and form the second group. Those two groups are weakly correlated. In fact, Figure 2 gives a reduced dimension representation of the correlation matrix given in Table 2.

It is extremely important, however, to notice that the angle between variables is interpreted in terms of correlation only when variables are well-represented, that is they are close to the border of the circle of correlation. Remember that the goal of PCA is to explain multiple variables by a lesser number of components, and keep in mind that graphs obtained from that reduction method are projections that optimize global criterion (i.e. the total variance). As such some relationships between variables may be greatly altered. Correlations between variables and components supply insights about variables that are not well-represented. In a subspace of components, the quality of representation of a variable is assessed by the sum-of-squared component loadings across components. This is called the communality of the variable. It measures the proportion of the variance of a variable accounted for by the components. For example, in our example, the communality of the variable X_1 is $0.943^2+0.241^2=0.948$. This means that the first two components explain about 95% of the variance of the variable X_1. This is quite substantial to enable us fully interpreting the variability in this variable as well as its relationship with the other variables. Communality can be used as a measure of goodness-of-fit of the projection. The communalities of the 5 variables of our data are displayed in Table 5. As shown by this Table, the first two components explain more than 80% of variance in each variable. This is enough to reveal the structure of correlation among the variables. Do not interpret as correlation the angle between two variables when at least one of them has a low communality. Using communality prevent potential biases that may arise by directly interpreting numerical and graphical results yielded by the PCA.

Variables	Value
X1	0.948
X2	0.920
X3	0.817
X4	0.970
X5	0.976

Table 5. **Communalities of variables**

All these interesting results show that outcomes from normalized PCA can be easily interpreted without additional complicated calculations. From a visual inspection of the graph, we can see the groups of variables that are correlated, interpret the principal components and name them.

6.2 Factor scores and their use in multivariate models

A useful by product of PCA is factor scores. Factor scores are coordinates of subjects (individuals) on each component. They indicate where a subject stands on the retained component. Factor scores are computed as weighted values on the observed variables. Results for our dataset are reported in Table 6.

Factor scores can be used to plot a reduced representation of subjects. This is displayed by Figure 3.

How do we interpret the position of points on this diagram? Recall that this graph is a projection. As such some distances could be spurious. To distinguish wrong projections from real ones and better interpret the plot, we need to use that is called "the quality of representation" of subjects. This is computed as the squared of the cosine of the angle between a subject s_i and a component z , following the formula:

$$\cos^2(s_i, z_j) = \frac{z_i^2}{\|s_i\|^2} = \frac{z_i^2}{\sum_{k=1}^{p} x_{ki}^2} \tag{30}$$

ID	PCA1	PCA2	Cos^21	Cos^22	QL12= Cos^21+ Cos^22	CTR1	CTR2
1	1.701	-1.610	0.436	0.390	0.826	5.458	6.547
2	1.701	0.575	0.869	0.099	0.969	5.455	0.837
3	1.972	-0.686	0.862	0.104	0.966	7.333	1.191
4	3.000	2.581	0.563	0.417	0.981	16.974	16.832
5	2.382	-1.556	0.687	0.293	0.980	10.700	6.116
6	1.717	-0.323	0.522	0.018	0.541	5.558	0.264
7	0.193	-0.397	0.062	0.263	0.325	0.070	0.400
8	0.084	-1.213	0.004	0.972	0.977	0.013	3.718
9	1.071	-0.558	0.765	0.208	0.974	2.162	0.787
10	-0.427	-0.110	0.822	0.054	0.877	0.344	0.030
11	-1.088	0.176	0.093	0.024	0.933	2.232	0.078
12	-1.341	-1.673	0.344	0.536	0.881	3.393	7.075
13	-0.291	-0.996	0.071	0.835	0.906	0.160	2.507
14	-0.652	0.567	0.543	0.411	0.955	0.801	0.812
15	-0.062	3.166	0.000	0.957	0.957	0.007	25.325
16	-1.830	-0.375	0.929	0.039	0.968	6.318	0.356
17	-1.181	3.182	0.119	0.868	0.988	2.630	25.572
18	-2.244	-0.751	0.877	0.098	0.976	9.493	1.424
19	-2.288	-0.150	0.933	0.004	0.937	9.874	0.057
20	-2.417	0.155	0.938	0.003	0.942	11.019	0.060

Notes: Columns PCA1 and PCA2 display the factor scores on the first and second components, respectively. Cos^21 and Cos^22 indicate the quality of representation of subjects on the first and second components, respectively. QL12= Cos^21+ Cos^22 measures the quality of representation of subjects on the plane formed by the first two components. CTR1and CTR2 are the contribution of subjects on component 1 and component 2, respectively.

Table 6. **Factor Scores of Subjects, Contributions and Quality of Representation**

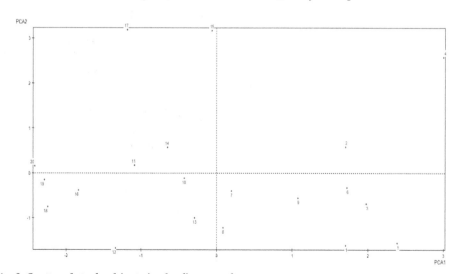

Fig. 3. **Scatterplot of subjects in the first two factors**

Cos^2 is interpreted as a measure of goodness-of-fit of the projection of a subject on a given component. Notice that in Eq. (30), $\|s_i\|^2$ is the distance of subject s_i from the origin. It measures how far the subject is from the center. So if cos^2=1 the component extracted is reproducing a great amount of the original behavior of the subject. Since the components are orthogonal, the quality of representation of a subject in a given subspace of components is the sum of the associated cos^2. This notion is similar to the concept of communality previously defined for variables.

In Table 6 we also reported these statistics. As can be seen, the two components retained explain more than 80% of the behavior of subjects, except for subjects 6 and 7. Now we are confident that almost all the subjects are well-represented, we can interpret the graph. Thus, we can tell that subjects located in the right side and having larger coordinates on the first component, i.e.1, 9, 6, 3 and 5, have values of X_1, X_2 and X_3 greater than the average. Those located in the left side and having smaller coordinates on the first axis, i.e. 20, 19, 18, 16, 12, 11 and 10, record lesser values for these variables. On the other hand, subjects 15 and 17 are characterized by highest values for variables X_4 and X_5, while subjects 8 and 13 record lowest values for these variables.

Very often a small number of subjects can determine the direction of principal components. This is because PCA uses the notions of mean, variance and correlation; and it is well known that these statistics are influenced by outliers or atypical observations in the data. To detect what are these atypical subjects we define the notion of "contribution" that measures how much a subject contributes to the variance of a component. Contributions (CTR) are computed following:

$$CTR(s_i, z_j) = \frac{z_i^2}{n\lambda_i} \times 100 \tag{31}$$

Contributions are reported in the last two columns of Table 6. Subject 4 contributes greatly to the first component with a contribution of 16.97%. This indicates that subject 4 explains alone 16.97% of the variance of the first component. Therefore, this subject takes higher values for X1, X2 and X3. This can be easily verified from the original Table 1. Regarding the second component, over 25% of the variance of the data accounted for by this component is explained by subjects 15 and 17. These subjects exhibit high values for variables X_4 and X_5.

The principal components obtained from PCA could be used in subsequent analyses (regressions, poverty analysis, classification…). For example, in linear regression models, the presence of correlated variables poses the econometric well-known problem of multicolinearity that makes instable regression coefficients. This problem is avoided when using the principal components that are orthogonal with one another. At the end of the analysis you can re-express the model with the original variables using the equations defining principal components. If there are variables that are not correlated with the other variables, you can delete them prior to the PCA, and reintroduce them in your model once the model is estimated.

7. A Case study with illustration using SPSS

We collected data on 10 socio-demographic variables for a sample of 132 countries. We use these data to illustrate how performing PCA using the SPSS software package. By following the indications provided here, user can try to reproduce himself the results obtained.

To perform a principal components analysis with SPSS, follow these steps:

1. Select **Analyze/Data Reduction/ Factor**
2. Highlight all of the quantitative variables and Click on the **Variables** button. The character variable Country is an identifier variable and should not be included in the Variables list.
3. Click on the **Descriptives** button to select **Univariate Descriptives, Initial Solution, KMO and Bartlett's test of Sphericity.**
4. Click on the **Extraction** button, and select **Method=Principal Components, Display Unrotated factor solution, Scree Plot.** Select **Extract Eigenvalue over 1** (by default).
5. Click on the **Rotation** button, and select **Display Loading Plot(s).**
6. Click on **Scores** and select **Save as variables, Method=Regression.** Select the case below.
7. Click on **Options**, and select **Exclude Cases Listwise** (option by default).

In what follows, we review and comment on the main outputs.

- **Correlation Matrix**

To discover the pattern of intercorrelations among variables, we examine the correlation matrix. That is given in Table 7:

	Life_exp	Mortality	Urban	Iliteracy	Water	Telephone	Vehicles	Fertility	Hosp_beds	Physicians
Life_exp	1.000	-0.956	0.732	-0.756	0.780	0.718	0.621	-0.870	0.514	0.702
Mortality		1.000	-0.736	0.809	-0.792	-0.706	-0.596	0.895	-0.559	-0.733
Urban			1.000	-0.648	0.692	0.697	0.599	-0.642	0.449	0.651
Iliteracy				1.000	-0.667	-0.628	-0.536	0.818	-0.603	-0.695
Water					1.000	0.702	0.633	-0.746	0.472	0.679
Telephone						1.000	0.886	-0.699	0.622	0.672
Vehicles							1.000	-0.602	0.567	0.614
Fertility								1.000	-0.636	-0.763
Hosp_beds									1.000	0.701
Physicians										1.000

Note : Figures reported in this table are correlation coefficients.

Table 7. **Correlation Matrix**

The variables can be grouped into two groups of correlated variables. We will see this later.

- **Testing for the Factorability of the Data**

Before applying PCA to the data, we need to test whether they are suitable for reduction. SPSS provides two tests to assist users:

Kaiser-Meyer-Olkin Measure of Sampling Adequacy (Kaiser, 1974): This measure varies between 0 and 1, and values closer to 1 are better. A value of 0.6 is a suggested minimum for good PCA.

Bartlett's Test of Sphericity (Bartlett, 1950): This tests the null hypothesis that the correlation matrix is an identity matrix in which all of the diagonal elements are 1 and all off diagonal elements are 0. We reject the null hypothesis when the level of significance exceeds 0.05.

The results reported in Table 8 suggest that the data may be grouped into smaller set of underlying factors.

Kaiser-Meyer-Olkin Measure of Sampling Adequacy.		.913
Bartlett's Test of Sphericity	Approx. Chi-Square	1407.151
	df	45
	Sig.	.000

Table 8. **Results of KMO and Bartlett's Test**

- **Eigenvalues and number of meaningful components**

Table 9 displays the eigenvalues, percent of variance and cumulative percent of variance from the observed data. Earlier it was stated that the number of components computed is equal to the number of variables being analyzed, necessitating that we decide how many components are truly meaningful and worthy of being retained for interpretation.

Here only component 1 demonstrates an eigenvalue greater than 1.00. So the Kaiser eigenvalue-one criterion would lead us to retain and interpret only this component. The first component provides a reasonable summary of the data, accounting for about 72% of the total variance of the 10 variables. Subsequent components each contribute less than 8%.

Component	Initial Eigenvalues			Extraction Sums of Squared Loadings		
	Total	% of Variance	Cumulative %	Total	% of Variance	Cumulative %
1	7.194	71.940	71.940	7.194	71.940	71.940
2	.780	7.801	79.741	.780	7.801	79.741
3	.667	6.675	86.416			
4	.365	3.654	90.070			
5	.302	3.022	93.092			
6	.236	2.361	95.453			
7	.216	2.162	97.615			
8	.106	1.065	98.680			
9	.095	.946	99.626			
10	.037	.374	100.000			

Table 9. **Eigenvalues**

The scree plot is displayed in Figure 4. From the second component on, we observe that the line is almost flat with a relatively large break following component 1. So the scree test would lead us to retain only the first component. The components appearing after the break (2-10) would be regarded as trivial (less than 10%).

Scree Plot

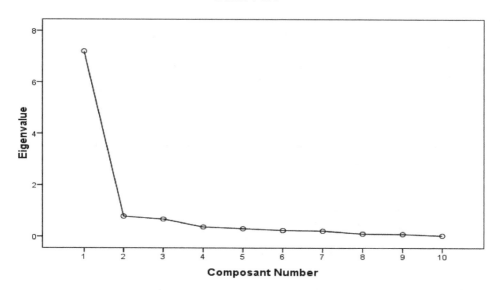

Fig. 4. **Scree Plot**

In conclusion, the dimensionality of the data could be reduced to 1. Nevertheless, we shall add the second component for representation purpose. Plot in a plane is easier to interpret than a three or 10-dimensional plot. Note that by default SPSS uses the Kaiser criterion to extract components. It belongs to the user to specify the number of components to be extracted if the Kaiser-criterion under-estimate the appropriate number. Here we specified 2 as the number of components to be extracted.

- **Component loadings**

Table 10 displays the loading matrix. The entries in this matrix are correlations between the variables and the components. As can be seen, all the variables load heavily on the first component. It is now necessary to turn to the content of the variables being analyzed in order to decide how this component should be named. What common construct do variables seem to be measuring?

In Figure 5 we observe two opposite groups of variables. The right-side variables are positively correlated one with another, and deal with social status of the countries. The left-side variables are also positively correlated one with another, and talk about another aspect of social life. It is therefore appropriate to name the first component the "social development" component.

	Component	
	1	2
Life_exp	.911	-.268
Mortality	-.926	.287
Urbanisation	.809	-.093
Iliteracy	-.848	.200
Water	.850	-.139
Telephone	.862	.355
Vehicles	.780	.483
Fertility	-.911	.183
Hosp_beds	.713	.396
Physicians	.850	.087

Table 10. **Component Matrix**

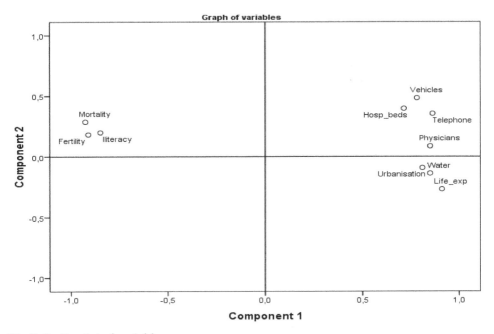

Fig. 5. **Scatterplot of variables**

- **Factor scores and Scatterplot of the Countries**

Since we have named the component, it is desirable to assign scores to each country to indicate where that country stands on the component. Here scores are indicating the level of social development of the countries. The values of the scores are to be interpreted paying attention to the signs of component loadings. From Figure 5 we say that countries with high positive scores on the first component demonstrate higher level of social development relatively to countries with negative scores. In Figure 6 we can see that countries such as Burkina Faso, Niger, Sierra Leone, Tchad, Burundi, Centrafrique and Angola belong to the under-developed group.

SPSS does not provide directly the scatterplot for subjects. Since factor scores have been created and saved as variables, we can use the Graph menu to request a scatterplot. This is an easy task on SPSS. The character variable Country is used as an identifier variable. Notice that in SPSS factor scores are standardized with a mean zero and a standard deviation of 1.

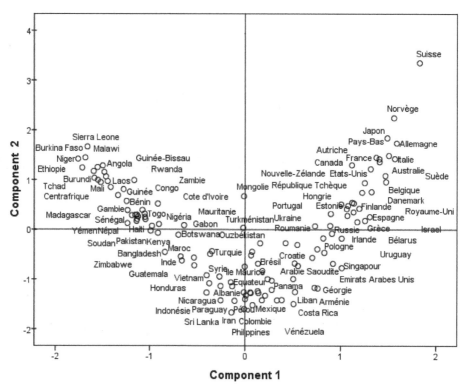

Fig. 6. Scatterplot of the Countries

A social development index is most useful to identify the groups of countries in connection with their level of development. The construction of this index assigns a social development-ranking score to each country. We rescale factor scores as follows:

$$SI_i = \frac{F_i - F_{min}}{F_{max} - F_{min}} \times 100 \tag{32}$$

where F_{min} and F_{max} are the minimum and maximum values of the factor scores F. Using the rescaled-scores, countries are sorted in ascending. Lower scores identify socially under-developed countries, whereas higher scores identify socially developed countries.

8. Conclusion

Principal components analysis (PCA) is widely used in statistical multivariate data analysis. It is extremely useful when we expect variables to be correlated to each other and want to reduce them to a lesser number of factors. However, we encounter situations where variables are non linearly related to each other. In such cases, PCA would fail to reduce the dimension of the variables. On the other hand, PCA suffers from the fact each principal component is a linear combination of all the original variables and the loadings are typically nonzero. This makes it often difficult to interpret the derived components. Rotation techniques are commonly used to help practitioners to interpret principal components, but we do not recommend them.

Recently, other new methods of data analysis have been developed to generalize linear PCA. These include Sparse Principal Components Analysis (Tibshirani, 1996; Zou, Hastie & Tibshirani, 2006), Independent Component Analysis (Vasilescu & Terzopoulos, 2007), Kernel Principal Components Analysis (Schölkopf, Smola & Müller, 1997, 1998), and Multilinear Principal Components Analysis (Haiping, Plataniotis & Venetsanopoulos, 2008).

9. Appendix

9.1 Data for the case study

Pays	Life_exp	Mortality	Urban	Iliteracy	Water	Telephone	Vehicles	Fertility	Hosp_beds	Physicians
Albanie	72.00	25.00	40.00	16.50	76.00	31.00	27.00	2.5	3.2	1.4
Algérie	71.00	35.00	59.00	35.00	90.00	53.00	25.00	3.5	2.1	0.8
Angola	47.00	124.00	33.00	41.00	32.00	6.00	18.00	6.7	1.3	0
Argentine	73.00	19.00	89.00	3.00	65.00	203.00	137.00	2.6	3.3	2.7
Arménie	74.00	15.00	69.00	2.00	99.00	157.00	0	1.3	7.6	3
Australie	79.00	5.00	85.00	3.00	99.00	512.00	488.00	1.8	8.5	2.5
Autriche	78.00	5.00	65.00	2.00	100.00	491.00	481.00	1.3	9.2	2.8
Azerbeidjan	71.00	17.00	57.00	3.00	97.00	89.00	36.00	2	9.7	3.8
Bangladesh	59.00	73.00	23.00	60.00	84.00	3.00	1.00	3.1	0.3	0.2
Bélarus	68.00	11.00	71.00	0.5	100.00	241.00	2.00	1.3	12.2	4.3
Belgique	78.00	6.00	97.00	2.00	100.00	500.00	435.00	1.6	7.2	3.4
Bénin	53.00	87.00	41.00	61.50	50.00	7.00	7.00	5.7	0.2	0.1
Bolivie	62.00	60.00	61.00	15.50	55.00	69.00	32.00	4.1	1.7	1.3
Botswana	46.00	62.00	49.00	24.50	70.00	65.00	15.00	4.2	1.6	0.2
Brésil	67.00	33.00	80.00	16.00	72.00	121.00	88.00	2.3	3.1	1.3
Bulgarie	71.00	14.00	69.00	1.50	99.00	329.00	220.00	1.1	10.6	3.5
Burkina Faso	44.00	104.00	17.00	77.50	42.00	4.00	4.00	6.7	1.4	0
Burundi	42.00	118.00	8.00	54.00	52.00	3.00	2.00	6.2	0.7	0.1
Cambodge	54.00	102.00	15.00	61.50	13.00	2.00	5.00	4.5	2.1	0.1
Cameroun	54.00	77.00	47.00	26.50	41.00	5.00	7.00	5	2.6	0.1
Canada	79.00	5.00	77.00	35.00	99.00	634.00	455.00	1.6	4.2	2.1
Centrafrique	44.00	98.00	40.00	55.50	19.00	3.00	0	4.8	0.9	0.1
Tchad	48.00	99.00	23.00	60.00	24.00	1.00	3.00	6.4	0.7	0
Chili	75.00	10.00	85.00	4.50	85.00	205.00	71.00	2.2	2.7	1.1
Chine	70.00	31.00	31.00	17.00	90.00	70.00	3.00	1.9	2.9	2
Hong Kong	79.00	3.00	100.00	7.50	100.00	558.00	56.00	1.1		1.3
Colombie	70.00	23.00	73.00	9.00	78.00	173.00	21.00	2.7	1.5	1.1
Congo Démocratique	51.00	90.00	30.00	41.00	27.00	-	9.00	6.3	1.4	0.1
Congo	48.00	90.00	61.00	21.50	47.00	8.00	14.00	6	3.4	0.3
Costa Rica	77.00	13.00	47.00	5.00	92.00	172.00	85.00	2.6	1.9	1.4
Cote d'Ivoire	46.00	88.00	45.00	55.50	72.00	12.00	18.00	5	0.8	0.1
Croatie	73.00	8.00	57.00	2.00	63.00	348.00	17.00	1.5	5.9	2
République Tchèque	75.00	5.00	75.00	3.00	97.00	364.00	358.00	1.2	9.2	2.9

Pays	Life_exp	Mortality	Urban	Iliteracy	Water	Telephone	Vehicles	Fertility	Hosp_beds	Physicians
Danemark	76.00	5.00	85.00	1.00	100.00	660.00	355.00	1.8	4.7	2.9
Equateur	70.00	32.00	63.00	9.50	70.00	78.00	41.00	2.9	1.6	1.7
Egypte	67.00	49.00	45.00	46.50	64.00	60.00	23.00	3.2	2	2.1
El Salvador	69.00	31.00	46.00	22.00	55.00	80.00	30.00	3.3	1.6	1
Erythrée	51.00	61.00	18.00	48.00	7.00	7.00	1.00	5.7		0
Estonie	70.00	9.00	69.00	2.00	100.00	343.00	312.00	1.2	7.4	3.1
Ethiopie	43.00	107.00	17.00	64.00	27.00	3.00	1.00	6.4	0.2	0
Finlande	77.00	4.00	66.00	1.00	98.00	554.00	145.00	1.8	9.2	2.8
France	78.00	5.00	75.00	1.00	100.00	570.00	442.00	1.8	8.7	2.9
Gabon	53.00	86.00	79.00	29.00	67.00	33.00	14.00	5.1	3.2	0.2
Gambie	53.00	76.00	31.00	65.50	76.00	21.00	8.00	5.6	0.6	0
Géorgie	73.00	15.00	60.00	4.00	100.00	115.00	80.00	1.3	4.8	3.8
Allemagne	77.00	5.00	87.00	1.00	100.00	567.00	506.00	1.4	9.6	3.4
Ghana	60.00	65.00	37.00	31.00	56.00	8.00	5.00	4.8	1.5	
Grèce	78.00	6.00	60.00	3.50	100.00	522.00	238.00	1.3	5	3.9
Guatemala	64.00	42.00	39.00	32.50	67.00	41.00	12.00	4.4	1	0.9
Guinée	47.00	118.00	31.00	41.00	62.00	5.00	2.00	5.4	0.6	0.2
Guinée-Bissau	44.00	128.00	23.00	63.00	53.00	7.00	6.00	5.6	1.5	0.2
Haiti	54.00	71.00	34.00	52.00	28.00	8.00	4.00	4.3	0.7	0.2
Honduras	69.00	36.00	51.00	27.00	65.00	38.00	7.00	4.2	1.1	0.8
Hongrie	71.00	10.00	64.00	1.00	99.00	336.00	233.00	1.3	9.1	3.4
Inde	63.00	70.00	28.00	45.00	81.00	22.00	5.00	3.2	0.8	0.4
Indonésie	65.00	43.00	39.00	14.50	62.00	27.00	12.00	2.7	0.7	0.2
Iran	71.00	26.00	61.00	25.50	83.00	112.00	26.00	2.7	1.6	0.9
Irlande	76.00	6.00	59.00	1.00	100.00	435.00	279.00	1.9	3.7	2.1
Israel	78.00	6.00	91.00	4.00	99.00	471.00	215.00	2.7	6	4.6
Italie	78.00	5.00	67.00	1.50	100.00	451.00	539.00	1.2	6.5	5.5
Jamaique	75.00	21.00	55.00	14.00	70.00	166.00	40.00	2.6	2.1	1.3
Japon	81.00	4.00	79.00	2.00	96.00	503.00	394.00	1.4	16.2	1.8
Jordanie	71.00	27.00	73.00	11.50	89.00	86.00	48.00	4.1	1.8	1.7
Kazakhstan	65.00	22.00	56.00	7.00	93.00	104.00	62.00	2	8.5	3.5
Kenya	51.00	76.00	31.00	19.50	53.00	9.00	11.00	4.6	1.6	0
Corée du Sud	73.00	9.00	80.00	2.50	83.00	433.00	163.00	1.6	4.6	1.1
Koweit	77.00	12.00	97.00	19.50	100.00	236.00	359.00	2.8	2.8	1.9
Laos	54.00	96.00	22.00	54.00	39.00	6.00	3.00	5.5	2.6	0.2
Liban	70.00	27.00	89.00	15.00	100.00	194.00	21.00	2.4	2.7	2.8
Lesotho	55.00	93.00	26.00	18.00	52.00	10.00	6.00	4.6		0.1
Lituanie	72.00	9.00	68.00	0.5	97.00	300.00	265.00	1.4	9.6	3.9
Madagascar	58.00	92.00	28.00	35.00	29.00	3.00	4.00	5.7	0.9	0.3
Malawi	42.00	134.00	22.00	41.50	45.00	3.00	2.00	6.4	1.3	0
Malaisie	72.00	8.00	56.00	13.50	100.00	198.00	145.00	3.1	2	0.5
Mali	50.00	117.00	29.00	61.50	37.00	3.00	3.00	6.5	0.2	0.1
Mauritanie	54.00	90.00	55.00	58.50	64.00	6.00	8.00	5.4	0.7	0.1
Ile Maurice	71.00	19.00	41.00	16.50	98.00	214.00	71.00	2	3.1	0.9
Mexique	72.00	30.00	74.00	9.00	83.00	104.00	97.00	2.8	1.2	1.2
Mongolie	66.00	50.00	62.00	38.50	45.00	37.00	16.00	2.5	11.5	2.6
Maroc	67.00	49.00	55.00	53.00	52.00	54.00	38.00	3	1	0.5
Mozambique	45.00	134.00	38.00	57.50	32.00	4.00	0	5.2	0.9	
Namibie	54.00	67.00	30.00	19.00	57.00	69.00	46.00	4.8		0.2
Népal	58.00	77.00	11.00	60.50	44.00	8.00	0	4.4	0.2	0
Pays-Bas	78.00	5.00	89.00	1.00	100.00	593.00	391.00	1.6	11.3	2.6
Nouvelle-Zélande	77.00	5.00	86.00	2.00	97.00	479.00	470.00	1.9	6.1	2.1
Nicaragua	68.00	36.00	55.00	30.50	81.00	31.00	18.00	3.7	1.5	0.8
Niger	46.00	118.00	20.00	85.50	53.00	2.00	4.00	7.3	0.1	0
Nigéria	53.00	76.00	42.00	39.00	39.00	4.00	9.00	5.3	1.7	0.2
Norvège	78.00	4.00	75.00	1.00	100.00	660.00	402.00	1.8	15	2.5
Oman	73.00	18.00	81.00	32.50	68.00	92.00	103.00	4.6	2.2	1.3
Pakistan	62.00	91.00	36.00	56.50	60.00	19.00	5.00	4.9	0.7	0.6
Panama	74.00	21.00	56.00	8.50	84.00	151.00	79.00	2.6	2.2	1.7

Pays	Life_exp	Mortality	Urban	Iliteracy	Water	Telephone	Vehicles	Fertility	Hosp_beds	Physicians
Paraguay	70.00	24.00	55.00	7.50	70.00	55.00	14.00	3.9	1.3	1.1
Pérou	69.00	40.00	72.00	11.00	80.00	67.00	26.00	3.1	1.5	0.9
Philippines	69.00	32.00	57.00	5.00	83.00	37.00	10.00	3.6	1.1	0.1
Pologne	73.00	10.00	65.00	0	98.00	228.00	230.00	1.4	5.4	2.3
Portugal	75.00	8.00	61.00	8.50	82.00	413.00	309.00	1.5	4.1	3
Roumanie	69.00	21.00	56.00	2.00	62.00	162.00	116.00	1.3	7.6	1.8
Russie	67.00	17.00	77.00	0.5	95.00	197.00	120.00	1.2	12.1	4.6
Rwanda	41.00	123.00	6.00	31.00	56.00	2.00	1.00	6.1	1.7	0
Arabie Saoudite	72.00	20.00	85.00	26.50	93.00	143.00	98.00	5.7	2.3	1.7
Sénégal	52.00	69.00	46.00	64.50	50.00	16.00	10.00	5.5	0.4	0.1
Sierra Leone	37.00	169.00	35.00	39.00	34.00	4.00	5.00	6	1.2	0.1
Singapour	77.00	4.00	100.00	8.00	100.00	562.00	108.00	1.5	3.6	1.4
Slovaquie	73.00	9.00	57.00	3.00	92.00	286.00	222.00	1.4	7.5	3
Slovénie	75.00	5.00	50.00	0	98.00	375.00	403.00	1.2	5.7	2.1
Afrique du Sud	63.00	51.00	53.00	15.50	70.00	115.00	85.00	2.8		0.6
Espagne	78.00	5.00	77.00	3.00	100.00	414.00	385.00	1.2	3.9	4.2
Sri Lanka	73.00	16.00	23.00	9.00	46.00	28.00	15.00	2.1	2.7	0.2
Soudan	55.00	69.00	34.00	44.50	50.00	6.00	9.00	4.6	1.1	0.1
Suède	79.00	4.00	83.00	1.00	100.00	674.00	428.00	1.5	5.6	3.1
Suisse	79.00	4.00	68.00	1.00	100.00	675.00	477.00	1.5	20.8	3.2
Syrie	69.00	28.00	54.00	32.50	85.00	95.00	9.00	3.9	1.5	1.4
Tadjikistan	69.00	23.00	28.00	1.00	69.00	37.00	0	3.4	8.8	2.1
Tanzanie	47.00	85.00	31.00	26.50	49.00	4.00	1.00	5.4	0.9	0
Thailande	72.00	29.00	21.00	5.00	94.20	84.00	27.00	1.9	2	0.4
Togo	49.00	78.00	32.00	45.00	63.00	7.00	19.00	5.1	1.5	0.1
Tunisie	72.00	28.00	64.00	31.50	99.00	81.00	30.00	2.2	1.7	0.7
Turquie	69.00	38.00	73.00	16.00	49.00	254.00	64.00	2.4	2.5	1.1
Turkménistan	66.00	33.00	45.00	8.00	60.00	82.00	1.00	2.9	11.5	0.2
Ouganda	42.00	101.00	14.00	35.00	34.00	3.00	2.00	6.5	0.9	0
Ukraine	67.00	14.00	68.00	0.5	55.00	191.00	0	1.3	11.8	4.5
Emirats Arabes Unis	75.00	8.00	85.00	25.00	98.00	389.00	11.00	3.4	2.6	1.8
Royaume-Uni	77.00	6.00	89.00	1.00	100.00	557.00	375.00	1.7	4.5	1.6
Etats-Unis	77.00	7.00	77.00	2.00	100.00	661.00	483.00	2	4	2.6
Uruguay	74.00	16.00	91.00	2.50	99.00	250.00	154.00	2.4	4.4	3.7
Ouzbékistan	69.00	22.00	38.00	12.00	57.00	65.00	0	2.8	8.3	3.3
Vénézuela	73.00	21.00	86.00	8.00	79.00	117.00	69.00	2.9	1.5	2.4
Vietnam	68.00	34.00	20.00	7.00	36.00	26.00	1.00	2.3	3.8	0.4
Yémen	56.00	82.00	24.00	56.00	74.00	13.00	14.00	6.3	0.7	0.2
Zambie	43.00	114.00	39.00	23.50	43.00	9.00	15.00	5.5	3.5	0.1
Zimbabwe	51.00	73.00	34.00	12.50	77.00	17.00	28.00	3.7	0.5	0.1

10. References

Bartlett M.S. (1950). Tests of Significance in Factor Analysis. *The British Journal of Psychology*, 3: 77-85.

Cadima J. & Jolliffe I. T. (1995). Loadings and Correlations in the Interpretation of Principal Components. *Journal of Applied Statistics*, 22(2):203–214.

Cattell R. B. (1966). The Scree Test for the Number of Factors. *Multivariate Behavioral Research*, 1,245-276.

Cooley W.W. & Lohnes P.R. (1971). *Multivariate Data Analysis*. John Wiley & Sons, Inc., New York.

Haiping L., Plataniotis K.N. & Venetsanopoulos A.N. (2008). MPCA: Multilinear Principal Component Analysis of Tensor Objects, *Neural Networks, IEEE Transactions on*, 19(1): 18 – 39.

Hotelling H. (1933). Analysis of a Complex of Statistical Variables into Principal Components. *Journal of Educational Psychology*, 24(6):417–441.

Horn J. L. (1965). A Rationale and Test for the Number Factors in Factor Analysis. *Psychometrika*, 30, 179-185.

Horn J. L. & Engstrom R. (1979). Cattell's Scree Test in Relation to Bartlett's Chi-Square Test and other Observations on the Number of Factors Problem. *Multivariate Behavioral Research*, 14, 283-300.

Hubbard R. & Allen S. J. (1987). An Empirical Comparison of Alternative Methods for Principal Component Extraction. *Journal of Business Research*, 15, 173-190.

Humphreys L. G. & Montanelli R. G. (1975). An Investigation of the Parallel Analysis Criterion for Determining the Number of Common Factors. *Multivariate Behavioral Research*, 10, 193-206.

Hyvarinen A., Karhunen J. & Oja E. (2001). *Independent Component Analysis*. New York: Wiley.

Jackson D.A. (1993). Stopping Rules in Principal Components Analysis: A Comparison of Heuristical and Statistical Approaches. *Ecology* 74(8): 2204-2214.

Jolliffe I. T. (2002). *Principal Component Analysis*. Second ed. New York: Springer-Verlag

Jolliffe I.T., Trendafilov N.T. & Uddin M. (2003). A Modified Principal Component Technique based on the LASSO. *Journal of Computational and Graphical Statistics*, 12(3):531–547.

Kaiser H. F. (1960). The Application of Electronic Computers to Factor Analysis. *Educational and Psychological Measurement*, 20:141-151.

Kaiser H. F. (1974). An Index of Factorial Simplicity. *Psychometrika*, 39:31-36.

Lambert Z. V., Wildt A.R. & Durand R.M. (1990). Assessing Sampling Variation Relative to number-of-factor Criteria. *Educational and Psychological Measurement*, 50:33-49.

Lawley D.N. (1956). Tests of Significance for the Latent Roots of Covariance and Correlation Matrices. *Biometrika*, 43: 128-136.

Lebart L., Morineau A. & Piron M. (1995). *Statistique Exploratoire Multidimensionnelle*. Paris : Dunod.

Pearson K. (1901). On Lines and Planes of Closest Fit to Systems of Points in Space. *Philosophical Magazine*, Series 6, 2(11), 559–572.

Stevens J. (1986). *Applied Multivariate Statistics for the Social Sciences*. Hillsdale, NJ: Lawrence Erlbaum Associates.

Tibshirani R. (1996). Regression Shrinkage and Selection via the Lasso. *Journal of the Royal Statistical Society*, series B 58(267-288).

Schölkopf B., Smola A.J. & Müller K.-R. (1997). Kernel Principal Component Analysis. *Lecture Notes in Computer Science*, Vol.1327, Artificial Neural Networks – ICANN'97, pp.583-588.

Schölkopf B., Smola A.J. & Müller K.-R. (1998). Nonlinear Component Analysis as a Kernel Eigenvalue Problem. *Neural Computation*, 10(5):1299–1319.

Vasilescu M. A. O. & Terzopoulos D. (2007). Multilinear (Tensor) ICA and Dimensionality Reduction. *Lecture Notes in Computer Science*, Vol. 4666, Independent Component Analysis and Signal Separation, pp. 818-826.

Zou H., Hastie T. & Tibshirani R. (2006). Sparse Principal Component Analysis. *Journal of Computational and Graphical Statistics*, 15(2): 265-286.

Zwick W.R. & Velicer W.F. (1986). Comparison of Five Rules for Determining the Number of Components to Retain. *Psychological Bulletin*, 99, 432-442.

10

FPGA Implementation for GHA-Based Texture Classification

Shiow-Jyu Lin[1,2], Kun-Hung Lin[1] and Wen-Jyi Hwang[1]
[1]*Department of Computer Science and Information Engineering,*
National Taiwan Normal University and
[2]*Department of Electronic Engineering,*
National Ilan University
Taiwan

1. Introduction

Principal components analysis (PCA) (Alpaydin, 2010; Jolliffe, 2002) is an effective unsupervised feature extraction algorithm for pattern recognition, classification, computer vision or data compression (Bravo et al., 2010; Zhang et al., 2006; Kim et al., 2005; Liying & Weiwei, 2009; Pavan et al., 2007; Qian & James, 2008). The goal of PCA is to obtain a compact and accurate representation of the data that reduces or eliminates statistically redundant components. Basic approaches for PCA involve the computation of the covariance matrix and the extraction of eigenvalues and eigenvectors. A drawback of the basic approaches is the high computational complexity and large memory requirement for data with high vector dimension. Therefore, these approaches may not be well suited for real time applications requiring fast feature extraction.

A number of fast algorithms (Dogaru et al., 2004; El-Bakry, 2006; Gunter et al., 2007; Sajid et al., 2008; Sharma & Paliwal, 2007) have been proposed to reduce the computation time of PCA. However, only moderate acceleration can be achieved because most of these algorithms are based on software. Although hardware implementation of PCA and its variants are possible, large storage size and complicated circuit control management are usually necessary. The PCA hardware implementation may therefore be possible only for small dimensions (Boonkumklao et al., 2001; Chen & Han, 2009).

An alternative for the PCA implementation is to use the generalized Hebbian algorithm (GHA) (Haykin, 2009; Oja, 1982; Sanger, 1989). The principal computation by the GHA is based on an effective incremental updating scheme for reducing memory utilization. Nevertheless, slow convergence of the GHA (Karhunen & Joutsensalo, 1995) is usually observed. A large number of iterations therefore is required, resulting in long computational time for many GHA-based algorithms. The hardware implementation of GHA has been found to be effective for reducing the computation time. However, since the number of multipliers in the circuit grows with the dimension, the circuits may be suitable only for PCA with small dimensions. Although analog GHA hardware architectures (Carvajal et al., 2007; 2009) can be used to lift the constraints on the vector dimensions, these architectures are difficult to be directly used for digital devices.

In light of the facts stated above, a digital GHA hardware architecture capable of performing fast PCA for large vector dimension is presented. Although large amount of arithmetic computations are required for GHA, the proposed architecture is able to achieve fast training with low area cost. The proposed architectures can be divided into three parts: the synaptic weight updating (SWU) unit, the principal components computing (PCC) unit, and memory unit. The memory unit is the on-chip memory storing synaptic weight vectors. Based on the synaptic weight vectors stored in the memory unit, the SWU and PCC units are then used to compute the principal components and update the synaptic weight vectors, respectively.

In the SWU unit, one synaptic weight vector is computed at a time. The results of precedent weight vectors will be used for the computation of subsequent weight vectors for expediting training speed. In addition, the computation of different weight vectors shares the same circuit for lowering the area cost. Moreover, in the PCC unit, the input vectors are allowed to be separated into smaller segments for the delivery over data bus with limited width. Both the SWU and PCC units can also operate concurrently to further enhance the throughput.

To demonstrate the effectiveness of the proposed architecture, a texture classification system on a system-on-programmable-chip (SOPC) platform is constructed. The system consists of the proposed architecture, a softcore NIOS II processor (Altera Corp., 2010), a DMA controller, and a SDRAM. The proposed architecture is adopted for finding the PCA transform by the GHA training, where the training vectors are stored in the SDRAM. The DMA controller is used for the DMA delivery of the training vectors. The softcore processor is only used for coordinating the SOPC system. It does not participate the GHA training process. As compared with its software counterpart running on Intel $i7$ CPU, our system has significantly lower computational time for large training set. All these facts demonstrate the effectiveness of the proposed architecture.

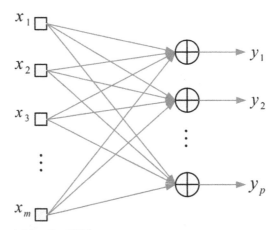

Fig. 1. The neural model for the GHA.

2. Preliminaries

Figure 1 shows the neural model for GHA, where $\mathbf{x}(n) = [x_1(n), ..., x_m(n)]^T$, and $\mathbf{y}(n) = [y_1(n), ..., y_p(n)]^T$ are the input and output vectors to the GHA model, respectively. The output vector $\mathbf{y}(n)$ is related to the input vector $\mathbf{x}(n)$ by

$$y_j(n) = \sum_{i=1}^{m} w_{ji}(n)x_i(n),\tag{1}$$

where the $w_{ji}(n)$ stands for the weight from the i-th synapse to the j-th neuron at iteration n. Each synaptic weight vector $\mathbf{w}_j(n)$ is adapted by the Hebbian learning rule:

$$w_{ji}(n+1) = w_{ji}(n) + \eta[y_j(n)x_i(n) - y_j(n)\sum_{k=1}^{j} w_{ki}(n)y_k(n)],\tag{2}$$

where η denotes the learning rate. After a large number of iterative computation and adaptation, $\mathbf{w}_j(n)$ will asymptotically approach to the eigenvector associated with the j-th principal component λ_j of the input vector, where $\lambda_1 > \lambda_2 > ... > \lambda_p$. To reduce the complexity of computing implementation, eq.(2) can be rewritten as

$$w_{ji}(n+1) = w_{ji}(n) + \eta y_j(n)[x_i(n) - \sum_{k=1}^{j} w_{ki}(n)y_k(n)].\tag{3}$$

A more detailed discussion of GHA can be found in (Haykin, 2009; Sanger, 1989)

3. The proposed GHA architecture

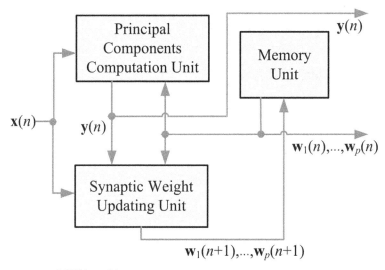

Fig. 2. The proposed GHA architecture.

As shown in Figure 2, the proposed GHA architecture consists of three functional units: the memory unit, the synaptic weight updating (SWU) unit, and the principal components computing (PCC) unit. The memory unit is used for storing the *current* synaptic weight vectors. Assume the *current* synaptic weight vectors $\mathbf{w}_j(n), j = 1,...,p$, are now stored in the memory unit. In addition, the input vector $\mathbf{x}(n)$ is available. Based on $\mathbf{x}(n)$ and $\mathbf{w}_j(n), j = 1,...,p$, the goal of PCC unit is to compute output vector $\mathbf{y}(n)$. Using $\mathbf{x}(n)$, $\mathbf{y}(n)$

and $\mathbf{w}_j(n), j = 1, ..., p$, the SWU unit produces the new synaptic weight vectors $\mathbf{w}_j(n+1), j = 1, ..., p$. It can be observed from Figure 2 that the new synaptic weight vectors will be stored back to the memory unit for subsequent training.

3.1 SWU unit

The design of SWU unit is based on eq.(3). Although the direct implementation of eq.(3) is possible, it will consume large hardware resources. To further elaborate this fact, we first see from eq.(3) that the computation of $w_{ji}(n+1)$ and $w_{ri}(n+1)$ shares the same term $\sum_{k=1}^{r} w_{ki}(n) y_k(n)$ when $r \leq j$. Consequently, independent implementation of $w_{ji}(n+1)$ and $w_{ri}(n+1)$ by hardware using eq.(3) will result in large hardware resource overhead.

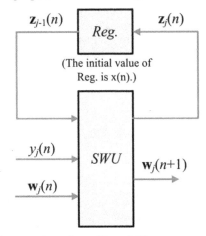

Fig. 3. The hardware implementation of eqs.(5) and (6).

To reduce the resource consumption, we first define a vector $z_{ji}(n)$ as

$$z_{ji}(n) = x_i(n) - \sum_{k=1}^{j} w_{ki}(n) y_k(n), j = 1, ..., p, \tag{4}$$

and $\mathbf{z}_j(n) = [z_{j1}(n), ..., z_{jm}(n)]^T$. Integrating eq.(3) and (4), we obtain

$$w_{ji}(n+1) = w_{ji}(n) + \eta y_j(n) z_{ji}(n), \tag{5}$$

where $z_{ji}(n)$ can be obtained from $z_{(j-1)i}(n)$ by

$$z_{ji}(n) = z_{(j-1)i}(n) - w_{ji}(n) y_j(n), j = 2, ..., p. \tag{6}$$

When $j = 1$, from eq.(4) and (6), it follows that

$$z_{0i}(n) = x_i(n). \tag{7}$$

Figure 3 depicts the hardware implementation of eqs.(5) and (6). As shown in the figure, the SWU unit produces one synaptic weight vector at a time. The computation of $\mathbf{w}_j(n+1)$, the j-th weight vector at the iteration $n+1$, requires the $\mathbf{z}_{j-1}(n)$, $\mathbf{y}(n)$ and $\mathbf{w}_j(n)$ as inputs.

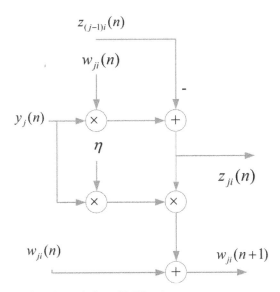

Fig. 4. The architecture of each module in SWU unit.

In addition to $\mathbf{w}_j(n+1)$, the SWU unit also produces $\mathbf{z}_j(n)$, which will then be used for the computation of $\mathbf{w}_{j+1}(n+1)$. Hardware resource consumption can then be effectively reduced.

One way to implement the SWU unit is to produce $\mathbf{w}_j(n+1)$ and $\mathbf{z}_j(n)$ in one shot. In SWU unit, m identical modules may be required because the dimension of vectors is m. Figure 4 shows the architecture of each module. The area cost of the SWU unit then will grows linearly with m. To further reduce the area cost, each of the output vectors $\mathbf{w}_j(n+1)$ and $\mathbf{z}_j(n)$ are separated into b segments, where each segment contains q elements. The SWU unit only computes one segment of $\mathbf{w}_j(n+1)$ and $\mathbf{z}_j(n)$ at a time. Therefore, it will take b clock cycles to produce complete $\mathbf{w}_j(n+1)$ and $\mathbf{z}_j(n)$.

Let

$$\hat{\mathbf{w}}_{j,k}(n) = [w_{j,(k-1)q+1}(n), ..., w_{j,(k-1)q+q}(n)]^T, k = 1, ..., b. \qquad (8)$$

and

$$\hat{\mathbf{z}}_{j,k}(n) = [z_{j,(k-1)q+1}(n), ..., z_{j,(k-1)q+q}(n)]^T, k = 1, ..., b. \qquad (9)$$

be the k-th segment of $\mathbf{w}_j(n)$ and $\mathbf{z}_j(n)$, respectively. The computation $\mathbf{w}_j(n)$ and $\mathbf{z}_j(n)$ take b clock cycles. At the k-th clock cycle, $k = 1, ..., b$, the SWU unit computes $\hat{\mathbf{w}}_{j,k}(n+1)$ and $\hat{\mathbf{z}}_{j,k}(n)$. Because each of $\hat{\mathbf{w}}_{j,k}(n+1)$ and $\hat{\mathbf{z}}_{j,k}(n)$ contains only q elements, the SWU unit consists of q identical modules. The architecture of each module is also shown in Figure 4. The SWU unit can be used for GHA with different vector dimension m. As m increases, the area cost therefore remains the same at the expense of large number of clock cycles b for the computation of $\hat{\mathbf{w}}_{j,k}(n+1)$ and $\hat{\mathbf{z}}_{j,k}(n)$.

Figures 5, 6 and 7 show the operation of the q modules. For the sake of simplicity, the computation of the first weight vector $\mathbf{w}_1(n+1)$ (i.e., $j = 1$) and the corresponding $\mathbf{z}_1(n)$ are considered in the figures. Based on eq.(7), the input vector $\mathbf{z}_{j-1}(n)$ is actually the training

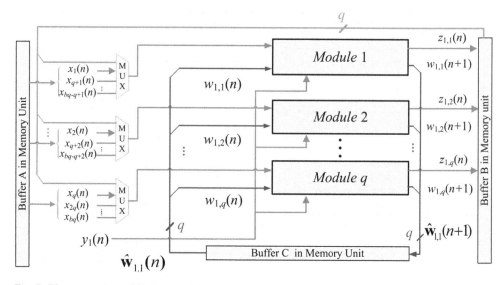

Fig. 5. The operation of SWU unit for computing the first segment of $\mathbf{w}_1(n+1)$.

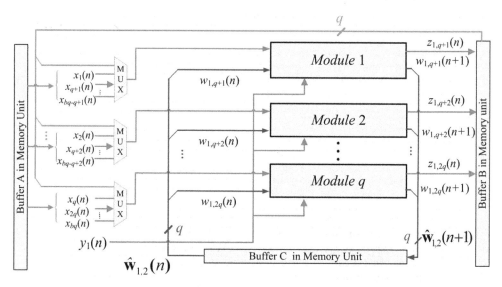

Fig. 6. The operation of SWU unit for computing the second segment of $\mathbf{w}_1(n+1)$.

vector $\mathbf{x}(n)$, which is also separated into b segments, where the k-th segment is given by

$$\hat{\mathbf{z}}_{0,k}(n) = [x_{(k-1)q+1}(n), ..., x_{(k-1)q+q}(n)]^T, k = 1, ..., b. \tag{10}$$

They are then multiplexed to the q modules. The $\hat{\mathbf{z}}_{0,1}(n)$ and $\hat{\mathbf{w}}_{1,1}(n)$ are used for the computation of $\hat{\mathbf{z}}_{1,1}(n)$ and $\hat{\mathbf{w}}_{1,1}(n+1)$ in Figure 5. Similarly, the $\hat{\mathbf{z}}_{0,k}(n)$ and $\hat{\mathbf{w}}_{1,k}(n)$ are

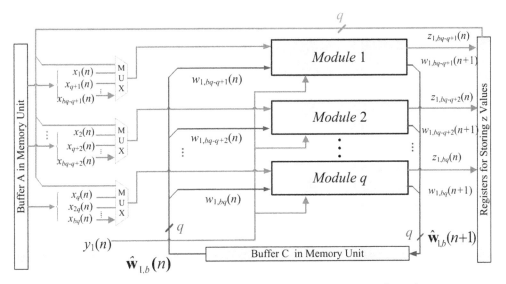

Fig. 7. The operation of SWU unit for computing the b-th segment of $\mathbf{w}_1(n+1)$.

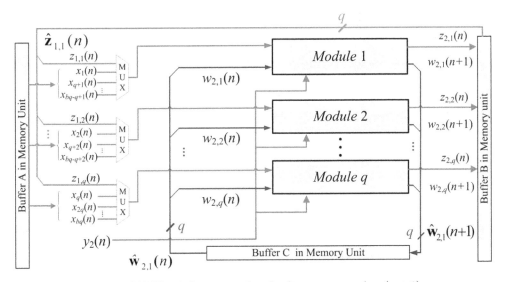

Fig. 8. The operation of SWU unit for computing the first segment of $\mathbf{w}_2(n+1)$.

used for the computation of $\hat{\mathbf{z}}_{1,k}(n)$ and $\hat{\mathbf{w}}_{1,k}(n+1)$ in Figures 6 and 7 for $k = 2$ and $k = b$, respectively.

After the computation of $\mathbf{w}_1(n+1)$ is completed, the vector $\mathbf{z}_1(n)$ is available as well. The vector $\mathbf{z}_1(n)$ is then used for the computation of $\mathbf{w}_2(n+1)$. Figure 8 shows the computation of the first segment of $\mathbf{w}_2(n+1)$ (i.e., $\hat{\mathbf{w}}_{2,1}(n+1)$) based on the first segment of $\mathbf{z}_1(n)$ (i.e., $\hat{\mathbf{z}}_{1,1}(n)$). The same process proceeds for the subsequent segments until the computation of the entire vectors $\mathbf{w}_2(n+1)$ and $\mathbf{z}_2(n)$ are completed. The vector $\mathbf{z}_2(n)$ is then used for the computation of $\mathbf{w}_3(n+1)$. The weight vector updating process at the iteration $n+1$ will be completed until the SWU unit produces the weight vector $\mathbf{w}_p(n+1)$.

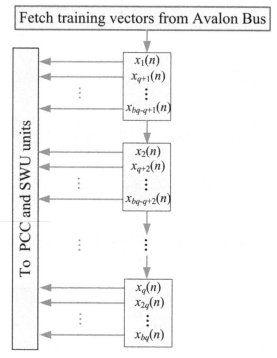

Fig. 9. The architecture of Buffer A in memory unit.

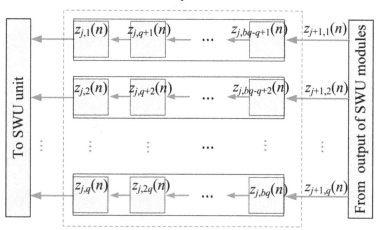

Fig. 10. The architecture of Buffer B in memory unit.

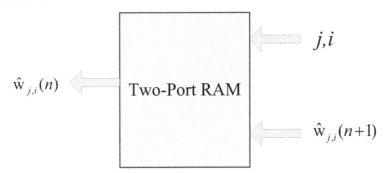

Fig. 11. The architecture of Buffer C in memory unit.

3.2 Memory unit

The memory unit contains three buffers: Buffer A, Buffer B and Buffer C. As shown in Figure 9, Buffer A stores training vector $\mathbf{x}(n)$. It consists of q sub-buffers, where each sub-buffer contains b elements. All the sub-buffers are connected to the SWU and PCC units.

The architecture of Buffer B is depicted in Figure 10, which holds the values of $\mathbf{z}_j(n)$. Each segment of $\mathbf{z}_j(n)$ computed from SWU unit is stored in Buffer B. After all the segments are produced, the Buffer B then deliver the segments of $\mathbf{z}_j(n)$ to SWU unit in the first-in-first-out (FIFO) fashion.

The Buffer C is used for storing the synaptic weight vectors $\mathbf{w}_j(n)$, $j = 1, ..., p$. It is a two-port RAM for reading and writing weight vectors, as revealed in Figure 11. The address for the RAM is expressed in terms of indices j and i for reading or writing the i-th segment of the weight vector $\mathbf{w}_j(n)$.

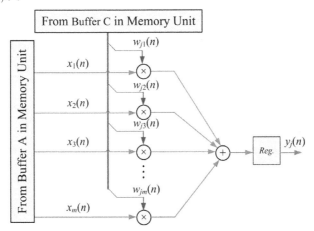

Fig. 12. The architecture of PCC unit.

3.3 PCC unit

The PCC operations are based on eq.(1). Therefore, the PCC unit of the proposed architecture contains adders and multipliers. Figure 12 shows the architecture of PCC. The training vector

$\mathbf{x}(n)$ and synaptic weight vector $\mathbf{w}_j(n)$ are obtained from the Buffer A and Buffer C of the memory unit, respectively. When both $\mathbf{x}(n)$ and $\mathbf{w}_j(n)$ are available, the proposed PCC unit then computes $y_j(n)$. Note that, after the $y_j(n)$ is obtained, the SWU unit can then compute $\mathbf{w}_j(n+1)$. Figure 13 reveals the timing diagram of the proposed architecture. It can be observed from Figure 13 that both the $y_{j+1}(n)$ and $\mathbf{w}_j(n+1)$ are computed concurrently. The throughput of the proposed architecture can then be effectively enhanced.

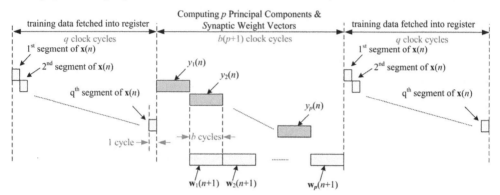

Fig. 13. The timing diagram of the proposed architecture.

3.4 SOPC-based GHA training system

The proposed architecture is used as a custom user logic in a SOPC system consisting of softcore NIOS CPU (Altera Corp., 2010), DMA controller and SDRAM, as depicted in Figure 14. All training vectors are stored in the SDRAM and then transported to the proposed circuit via the Avalon bus. The softcore NIOS CPU runs on a simple software to coordinate different components, including the proposed custom circuit in the SOPC. The proposed circuit operates as a hardware accelerator for GHA training. The resulting SOPC system is able to perform efficient on-chip training for GHA-based applications.

4. Experimental results

This section presents some experimental results of the proposed architecture applied to texture classification. The target FPGA device for all the experiments in this paper is Altera Cyclone III (Altera Corp., 2010). The design platform is Altera Quartus II with SOPC Builder and NIOS II IDE. Two sets of textures are considered in the experiments. The first set of textures, shown in Figure 15, consists of three different textures. The second set of textures is revealed in Figure 16, which contains four different textures. The size of each texture in Figures 15 and 16 is 320×320.

In the experiment, the principal component based k nearest neighbor (PC-kNN) rule is adopted for texture classification. Two steps are involved in the PC-kNN rule. In the first step, the GHA is applied to the input vectors to transform m dimensional data into p principal components. The synaptic weight vectors after the convergence of GHA training are adopted to span the linear transformation matrix. In the second step, the kNN method is applied to the principal subspace for texture classification.

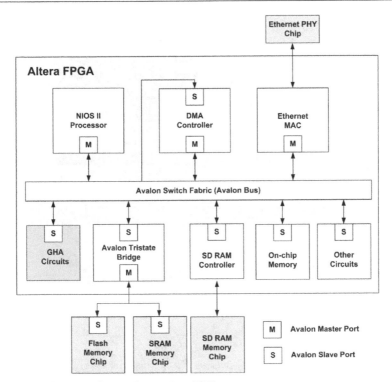

Fig. 14. The SOPC system for implementing GHA.

Fig. 15. The first set of textures for the experiments.

Figures 17 and 18 show the distribution of classification success rates (CSR) of the proposed architecture for the texture sets in Figures 15 and 16, respectively. The classification success rate is defined as the number of test vectors which are successfully classified divided by the total number of test vectors. The number of principal components is $p = 4$. The vector dimension is $m = 16 \times 16$. The distribution is based on 20 independent GHA training processes. The distribution of the architecture presented in (Lin et al., 2011) with the same p is also included for comparison purpose. The vector dimension for (Lin et al., 2011) is $m = 4 \times 4$.

Fig. 16. The second set of textures for the experiments.

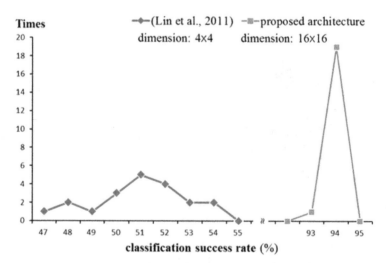

Fig. 17. The distribution of CSR of the proposed architecture for the texture set in Figure 15.

It can be observed from Figures 17 and 18 that the proposed architecture has better CSR. This is because the vector dimension of the proposed architecture is higher than that in (Lin et al., 2011). Spatial information of textures therefore is more effectively exploited for improving CSR by the proposed architecture. In fact, the vector dimension in the proposed architecture is $m = 16 \times 16$. The proposed architecture is able to implement hardware GHA for larger

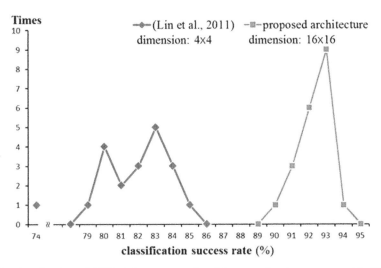

Fig. 18. The distribution of CSR of the proposed architecture for the texture set in Figure 16.

vector dimension because the area cost of the SWU unit in the architecture is independent of vector dimension. By contrast, the area cost of the SWU unit in (Lin et al., 2011) grows with the vector dimension. Therefore, only smaller vector dimension (i.e., $m = 4 \times 4$) can be implemented.

To further elaborate these facts, Tables 1 and 2 show the hardware resource consumption of the proposed architecture and the architecture in (Lin et al., 2011) for vector dimensions $m = 4 \times 4$ and $m = 16 \times 16$. Three different area costs are considered in the table: logic elements (LEs), embedded memory bits, and embedded multipliers. It can be observed from Tables 1 and 2 that given the same $m = 4 \times 4$ and the same p, the proposed architecture consumes significantly less hardware resources as compared with the architecture in (Lin et al., 2011). Although the area costs of the proposed architecture increase as m becomes 16×16, as shown in Table 1, they are only slightly higher than those of (Lin et al., 2011) in Table 2. The proposed architecture therefore is well suited for GHA training with larger vector dimension due to better spatial information exploitation.

| p | Proposed GHA with $m = 4 \times 4$ | | | Proposed GHA with $m = 16 \times 16$ | | |
	LEs	Memory Bits	Embedded Multipliers	LEs	Memory Bits	Embedded Multipliers
3	3942	1152	36	63073	1152	569
4	4097	1152	36	65291	1152	569
5	4394	1280	36	70668	1280	569
6	4686	1280	36	75258	1280	569
7	4988	1280	36	79958	1280	569

Table 1. Hardware resource consumption of the proposed GHA architecture for vector dimensions $m = 4 \times 4$ and $m = 16 \times 16$.

	GHA in (Lin et al., 2011) with $m = 4 \times 4$		
p	LEs	Memory Bits	Embedded Multipliers
3	22850	0	204
4	31028	0	272
5	38261	0	340
6	45991	0	408
7	53724	0	476

Table 2. Hardware resource consumption of the GHA architecture (Lin et al., 2011) for vector dimension $m = 4 \times 4$.

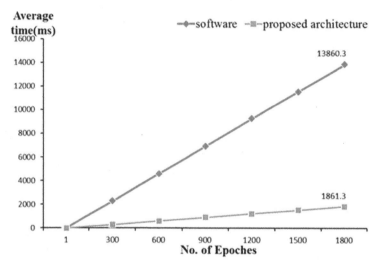

Fig. 19. The CPU time of the NIOS-based SOPC system using the proposed architecture as the hardware accelerator for various numbers of training iterations with $p = 4$.

Figure 19 shows the CPU time of the NIOS-based SOPC system using the proposed architecture as the hardware accelerator for various numbers of training iterations with $p = 4$. The clock rate of NIOS CPU in the system is 50 MHz. The CPU time of the software counterpart also is depicted in the Figure 19 for comparison purpose. The software training is based on the general purpose 2.67-GHz Intel $i7$ CPU. It can be clearly observed from Figure 16 that the proposed hardware architecture attains high speedup over its software counterpart. In particular, when the number of training iterations reaches 1800, the CPU time of the proposed SOPC system is 1861.3 ms. By contrast, the CPU time of Intel $i7$ is 13860.3 ms. The speedup of the proposed architecture over its software counterpart therefore is 7.45.

5. Concluding remarks

Experimental results reveal that the proposed GHA architecture has superior speed performance over its software counterparts. In addition, the architecture is able to attain higher classification success rate for texture classification as compared with other existing

GHA architectures. The architecture also has low area cost for PCA analysis with high vector dimension. The proposed architecture therefore is an effective alternative for on-chip learning applications requiring low area cost, high classification success rate and high speed computation.

6. References

Alpaydin, E. (2010). *Introduction to Machine Learning*, second ed., MIT Press, Massachusetts, USA.

Altera Corporation (2010). *NIOS II Processor Reference Handbook ver 10.0.*
http://www.altera.com/literature/lit-nio2.jsp

Altera Corporation (2010). *Cyclone III Device Handbook.*
http://www.altera.com/products/devices/cyclone3/cy3-index.jsp

Boonkumklao, W., Miyanaga, Y., and Dejhan, K. (2001). Flexible PCA architectura realized on FPGA. *International Symposium on Communications and Information Technologies*, pp. 590 - 593.

Bravo, I., Mazo, M., Lazaro, J.L., Gardel, A., Jimenez, P., and Pizarro, D. (2010). An Intelligent Architecture Based on FPGA Designed to Detect Moving Objects by Using Principal Component Analysis. *Sensors*. 10(10), pp. 9232 - 9251.

Carvajal, G., Valenzuela, W., Figueroa, M. (2007). Subspace-Based Face Recognition in Analog VLSI. In: *Advances in Neural Information Processing Systems*, 20, pp. 225 - 232, MIT Press, Cambridge.

Carvajal, G., Valenzuela, W., and Figueroa M. (2009). Image Recognition in Analog VLSI with On-Chip Learning. *Lecture Notes in Computer Science* (ICANN 2009), Vol.5768, pp. 428 - 438.

Chen, D., and Han, J.-Q. (2009). An FPGA-based face recognition using combined 5/3 DWT with PCA methods. *Journal of Communication and Computer*, Vol. 6, pp.1 - 8.

Chengcui, Z., Xin, C., and Wei-bang, C. (2006). A PCA-Based Vehicle Classification Framework. Proceedings of the 22nd International Conference on Data Engineering Workshops, pp. 17.

Dogaru, R., Dogaru, I., and Glesner, M. (2004). CPCA: a multiplierless neural PCA. Proceedings of IEEE International Joint Conference on Neural Networks, vol.2684, pp. 2689 - 2692.

El-Bakry, H.M. (2006). A New Implementation of PCA for Fast Face Detection. *International Journal of Intelligent Systems and Technologies*,1(2), pp. 145 - 153.

Gunter S., Schraudolph, N.N., and Vishwanathan, S.V.N. (2007). Fast Iterative Kernel Principal Component Analysis, *Journal of Machine Learning Research*, pp. 1893 - 1918.

Haykin, S. (2009). *Neural Networks and Learning Machines*, third Ed., Pearson.

Jolliffe, I.T. (2002). *Principal component Analysis*, second Ed., Springer, New York.

Karhunen, J. and J. Joutsensalo (1995). Generalizations of principal component analysis, optimization problems, and neural networks. *Neural Networks*, 8(4), pp. 549 - 562.

Kim, K., Franz, M.O., and Scholkopf, B. (2005). Iterative kernel principal component analysis for image modeling, *IEEE Trans. Pattern Analysis and Machine Intelligence*, pp. 1351 - 1366.

Lin, S.-J., Hung, Y.-T., and Hwang, W.-J. (2011). Efficient hardware architecture based on generalized Hebbian algorithm for texture classification. *Neurocomputing* 74(17), pp. 3248 - 3256.

Liying, L. and Weiwei, G. (2009). The Face Detection Algorithm Combined Skin Color Segmentation and PCA. International Conference on Information Engineering and Computer Science, pp. 1 - 3.

Oja, E. Simplified neuron model as a principal component analyzer. *Journal of Mathematical Biology* 15(3), pp. 267-273.

Pavan Kumar, A., Kamakoti, V., and Das, S. (2007). System-on-programmable-chip implementation for on-line face recognition. *Pattern Recognition Letters* 28(3), pp. 342 - 349.

Qian Du and J.E.F. (2008). Low-Complexity Principal Component Analysis for Hyperspectral Image Compression. *International Journal of High Performance Computing Applications*.

Sajid, I., Ahmed M.M., and Taj, I. (2008). Design and Implementation of a Face Recognition System Using Fast PCA. IEEE International Symposium on Computer Science and its Applications, pp. 126 - 130.

Sanger,T.D. (1989). Optimal unsupervised learning in a single-layer linear feedforward neural network. *Neural Networks*, vol. 12, pp. 459 - 473.

Sharma, A. and Paliwal, K.K. (2007). Fast principal component analysis using fixed-point algorithm, *Pattern Recognition Letters*, pp. 1151 - 1155.

11

Robust Density Comparison Using Eigenvalue Decomposition

Omar Arif[1] and Patricio A. Vela[2]
[1]*National University of Sciences and Technology*
[2]*Georgia Institute of Technology*
[1]*Pakistan*
[2]*USA*

1. Introduction

Many problems in various fields require measuring the similarity between two distributions. Often, the distributions are represented through samples and no closed form exists for the distribution, or it is unknown what the best parametrization is for the distribution . Therefore, the traditional approach of first estimating the probability distribution using the samples, then comparing the distance between the two distributions is not feasible. In this chapter, a method to compute the similarity between two distributions, which is robust to noise and outliers, is presented. The method works directly on the samples without requiring the intermediate step of density estimation, although the approach is closely related to density estimation. The method is based on mapping the distributions into a reproducing kernel Hilbert space, where eigenvalue decomposition is performed. Retention of only the top M eigenvectors minimizes the effect of noise on density comparison.

The chapter is organized in two parts. First, we explain the procedure to obtain the robust density comparison method. The relation between the method and kernel principal component analysis (KPCA) is also explained. The method is validated on synthetic examples. In the second part, we apply the method to the problem of visual tracking. In visual tracking, an initial target and target appearance is given, and must be found within future images. The target information is assumed to be characterized by a probability distribution. Thus tracking, in this scenario, is defined to be the problem of finding the distribution within each image of a sequence that most fits the given target distribution. Here, the object is tracked by minimizing the similarity measure between the model distribution and the candidate distribution where the target position is the optimization variable.

2. Mercer kernels

Let $\{u_i\}_{i=1}^n, u_i \in \mathbb{R}^d$, be a set of n observations. A Mercer kernel is a function $k : \mathbb{R}^d \times \mathbb{R}^d \rightarrow \mathbb{R}$, which satisfies:

1. k is continuous

2. $k(u_i, u_j) = k(u_j, u_i)$. Symmetric

3. The matrix K, with entries $K_{ij} = k(u_i, u_j)$ is positive definite.

$$\phi : \mathbb{R}^2 \to \mathbb{R}^3$$

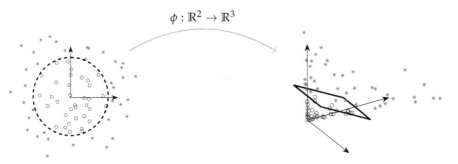

Fig. 1. Toy example: Dot product in the mapped space can be computed using the kernel in the input space.

Theorem: If k is a Mercer kernel then, there exists a high dimensional Hilbert space $\mathcal{H}$ with mapping $\phi : \mathbb{R}^d \to \mathcal{H}$ such that:

$$\phi(u_i) \cdot \phi(u_j) = k(u_i, u_j). \tag{1}$$

The Mercer kernel k implicitly maps the data to a Hilbert space $\mathcal{H}$, where the dot product is given by the kernel k.

3. Example

Figure 1 shows a simple binary classification example from Schölkopf & Smola (2001). The true decision boundary is given by the circle in the input space. The points in the input space, $u = [u_1, u_2]^T$, are mapped to $\mathbb{R}^3$ using the mapping $\phi(u) = [u_1^2, \sqrt{2}\, u_1 u_2, u_2^2]^T$. In $\mathbb{R}^3$, the decision boundary is transformed from an circle to a hyperplane, i.e. from a non-linear boundary to a linear one. There are many ways to carry out the mapping ϕ, but the above defined mapping has the important property that the dot product in the mapped space is given by the square of the dot product in the input space. This means that the dot product in the mapped space can be obtained without explicitly computing the mapping ϕ.

$$\phi(u) \cdot \phi(v) = u_1^2 v_1^2 + 2u_1 v_1 u_2 v_2 + u_2^2 v_2^2$$
$$= (u_1 v_1 + u_2 v_2)^2 = (u \cdot v)^2 = k(u, v).$$

An example Mercer kernel is Gaussian kernel:

$$k(u_i, u_j) = \frac{1}{\sqrt{|2\pi\Sigma|}} \exp\left(-\frac{1}{2}(u_i - u_j)^T \Sigma^{-1}(u_i - u_j)\right), \tag{2}$$

where Σ is $d \times d$ covariance matrix.

4. Maximum nean discrepancy

Let $\{u_i\}_{i=1}^{n_u}$, with $u_i \in \mathbb{R}^d$, be a set of n_u observations drawn from the distribution P_u. Define a mapping $\phi : \mathbb{R}^d \to \mathcal{H}$, such that $\langle \phi(u_i), \phi(u_j) \rangle = k(u_i, u_j)$, where k is a Mercer kernel function, such as the Gaussian kernel. The mean of the mapping is defined as $\mu : P_u \to \mu[P_u]$, where $\mu[P_u] = E[\phi(u_i)]$. If the finite sample of points $\{u_i\}_{i=1}^{n_u}$ are drawn from the distribution P_u, then the unbiased numerical estimate of the mean mapping $\mu[P_u]$ is $\frac{1}{n_u} \sum_{i=1}^{n_u} \phi(u_i)$. Smola et al. (2007) showed that the mean mapping can be used to compute the probability at a test point $u \in \mathbb{R}^d$ as

$$p(u) = \langle \mu[P_u], \phi(u) \rangle \approx \frac{1}{n_u} \sum_{i=1}^{n_u} k(u, u_i). \tag{3}$$

Equation (3) results in the familiar Parzen window density estimator. In terms of the Hilbert space embedding, the density function estimate results from the inner product of the mapped point $\phi(u)$ with the mean of the distribution $\mu[P_u]$. The mean map $\mu : P_u \to \mu[P_u]$ is injective, Smola et al. (2007), and allows for the definition of a similarity measure between two sampled sets P_u and P_v, sampled from the same or two different distributions. The measure is defined to be $D(P_u, P_v) := ||\mu[P_u] - \mu[P_v]||$. This similarity measure is called the maximum mean discrepancy (MMD). MMD has been used to address the two sample problem, Gretton et al. (2007). The next section introduces Robust MMD (rMMD).

5. Robust maximum mean discrepancy

In the proposed method, principal component analysis is carried out in the Hilbert space $\mathcal{H}$ and the eigenvectors corresponding to the leading eigenvalues are retained. It is assumed that the lower eigenvectors capture the noise present in the data set. Mapped points in the Hilbert space are reconstructed by projecting them onto the eigenvectors. The reconstructed points are then used to compute the robust mean map. All the computations in the Hilbert space are performed through the Mercer kernel in the input space and no explicit mapping is carried out.

5.1 Eigenvalue decomposition

Let $\{u_i\}_{i=1}^{n_u}$, with $u_i \in \mathbb{R}^d$, be a set of n_u observations. As mentioned before, if k is a Mercer kernel then there exists a high dimensional Hilbert space $\mathcal{H}$, with mapping $\phi : \mathbb{R}^d \to \mathcal{H}$. The covariance matrix $C_{\mathcal{H}}$ in the Hilbert space $\mathcal{H}$ is given by

$$C_{\mathcal{H}} = \frac{1}{n_u} \sum_{i=1}^{n_u} \phi(u_i)\phi(u_i)^T,$$

Empirical computations of $C_{\mathcal{H}}$ require one to know the mapping up front. A technique to avoid this requirement is to perform eigenvalue decomposition of the covariance matrix $C_{\mathcal{H}}$ using the inner product matrix K, called the Gram/kernel matrix, with $K_{ij} = \phi(u_i)^T \phi(u_j) = k(u_i, u_j)$. The Gram matrix allows for an eigenvalue/eigenvector decomposition of the covariance matrix without explicitly computing the mapping ϕ. The Gram kernel matrix can be computed using the Mercer kernel. If $a_i^k, i = 1, \ldots, n$, and λ^k are the k-th eigenvector components and eigenvalue of the kernel matrix K, then the k-th eigenvector of the covariance matrix $C_{\mathcal{H}}$ is given by Leventon (2002)

$$V^k = \frac{1}{\sqrt{\lambda^k}} \sum_{i=1}^{n} a_i^k \phi(u_i).$$

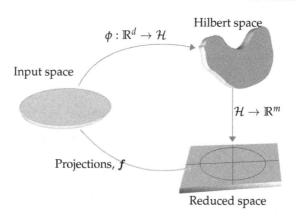

Fig. 2. Eigenvalue decomposition in the Hilbert space $\mathcal{H}$. Observations $\{u_i\}_{i=1}^{n_u}$ are mapped implicitly to the Hilbert space where eigenvalue decomposition results in an m-dimensional reduced space.

5.2 Robust density function

Let $V = [V^1, \cdots, V^m]$ be the m leading eigenvectors of the covariance matrix $C_\mathcal{H}$, where the eigenvector V^k is given by

$$V^k = \sum_{i=1}^{n_u} \alpha_i^k \phi(u_i) \qquad \text{with } \alpha_i^k = \frac{a_i^k}{\sqrt{\lambda^k}},$$

where λ^k and a_i^k are the kth eigenvalue and its associated eigenvector components, respectively, of the kernel matrix K. The reconstruction of the point $\phi(u)$ in the Hilbert space $\mathcal{H}$ using m eigenvectors V is

$$\phi_r(u) = V \cdot f(u), \tag{4}$$

where $f(u) = [f^1(u), \ldots, f^m(u)]^T$ is a vector whose components are the projections onto each of the m eigenvectors. The projections are given by

$$f^k(u) = V^k \cdot \phi(u) = \sum_{i=1}^{n_u} \alpha_i^k k(u_i, u), \tag{5}$$

This procedure is schematically described in Figure 2.

Kernel principal component analysis (**KPCA**) Scholköpf et al. (1998) is a non-linear extension of principal component analysis using a Mercer kernel k. Eigenvectors V are the principal components and the KPCA projections are given by Equation (5).

The reconstructed points, $\phi_r(u)$, are used to compute the numerical estimate of the robust mean mapping $\mu_r[P_u]$:

$$\mu_r[P_u] = \frac{1}{n_u} \sum_{i=1}^{n_u} \phi_r(u_i) = \frac{1}{n_u} \sum_{i=1}^{n_u} \sum_{k=1}^{m} V^k f^k(u_i) = \sum_{k=1}^{m} \omega^k V^k,$$

where

$$\omega^k = \frac{1}{n_u} \sum_{i=1}^{n_u} f^k(u_i) \tag{6}$$

The density at a point u is then estimated by the inner-product of the robust mean map $\mu_r[P_u]$ and the mapped point $\phi(u)$.

$$p(u) = \mu_r[P_u] \cdot \phi(u) = \sum_{k=1}^{m} \omega^k f^k(u). \tag{7}$$

Retention of only the leading eigenvectors in the procedure minimizes the effects of noise on the density estimate. Figure 3(d) shows density estimation of a multimodal Gaussian distribution in the presence of noise using the robust method. The effect of noise is less pronounced as compared to the kernel density estimation (Figure 3(b)). An alternate procedure that reaches the same result (Equation 7) from a different perspective is proposed by Girolami (2002). There, the probability density is estimated using orthogonal series of functions, which are then approximated using the KPCA eigenfunctions.

5.2.1 Example:

As mentioned before a kernel density estimate, obtained as per Equation (3), is computable using the inner product of the mapped test point and the mean mapping. The sample mean can be influenced by outliers and noise. In Kim & Scott (2008), the sample mean is replaced with a robust estimate using M-estimation Huber et al. (1981). The resulting density function is given by

$$p(u) = \langle \hat{\mu}[P_u], \phi(u) \rangle, \tag{8}$$

where the sample mean $\mu[P_u]$ is replaced with a robust mean estimator $\hat{\mu}[P_u]$. The robust mean estimator is computed using the M-estimation criterion

$$\hat{\mu}[P_u] = \arg\min_{\mu[P_u] \in \mathcal{H}} \sum_{i=1}^{n_u} \rho(||\phi(u_i) - \mu[P_u]||), \tag{9}$$

where ρ is robust loss function. The iterative re-weighted least squares (IRWLS) is used to compute the robust mean estimate. IRWLS depends only on the inner products and can be efficiently implemented using the Mercer kernel Kim & Scott (2008). The resulting density estimation function is

$$p(u) = \frac{1}{n_u} \sum_{i=1}^{n_u} \gamma_i k(u, u_i), \tag{10}$$

where $\gamma_i \geq 0, \sum \gamma_i = 1$ and γ_i are obtained through IRWLS algorithm. The γ_i values tend to be low for outlier data points.

In this section, we compare the performance of kernel density estimation (Equation (3)), robust density estimation using M-estimation (Equation (10)) and robust density estimation using eigenvalue decomposition (Equation 7).

A sample set X of 150 points is generated from a 2-dimensional multimodal Gaussian distribution

$$X \sim \frac{1}{3}\mathcal{N}_1(\mu_1, \Sigma_1) + \frac{1}{3}\mathcal{N}_2(\mu_2, \Sigma_2) + \frac{1}{3}\mathcal{N}_3(\mu_3, \Sigma_3), \tag{11}$$

where $\mu_1 = [3,3]^T$, $\mu_2 = [-3,3]^T$, $\mu_3 = [0,-3]^T$ and $\Sigma_1 = \Sigma_2 = \Sigma_3 = I$. Outliers are added from a uniform distribution over the domain $[-6,6] \times [-6,6]$. Figure 3 shows the true and the estimated density using the three methods. Data points corresponding to the outliers

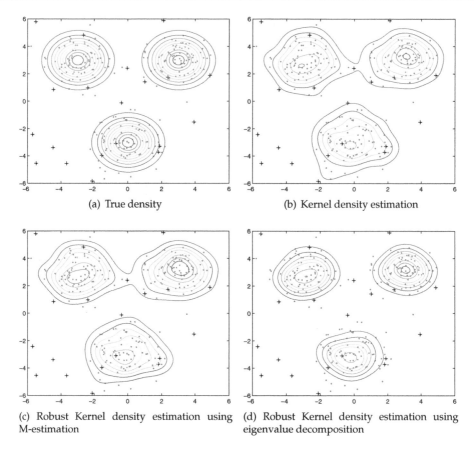

(a) True density

(b) Kernel density estimation

(c) Robust Kernel density estimation using M-estimation

(d) Robust Kernel density estimation using eigenvalue decomposition

Fig. 3. Density estimation comparisons. The effect of outliers is less pronounced in the robust density estimation using eigenvalue decomposition.

are marked as $+$. To measure the performance of the density estimates, the Bhattacharyya distance is used. The number of outliers used in the tests are $\Gamma = [20, 40, 60, 80, 100]$. At each Γ the simulations are run 50 times and the average Bhattacharyya distance is recorded. The results are shown in Figure 4. The number of eigenvectors retained for the robust density estimation were 8.

5.3 Robust maximum mean discrepancy

The robust mean map $\mu_r : P_u \to \mu_r[P_u]$, with $\mu_r[P_u] := \sum_{k=1}^{n_u} \omega^k V^k$, is used to define the similarity measure between the two distributions P_u and P_v. We call it the robust MMD (rMMD),

$$D_r(P_u, P_v) := ||\mu_r[P_u] - \mu_r[P_v]||.$$

The mean map $\mu_r[P_v]$ for the samples $\{v_i\}_{i=1}^{n_v}$ is calculated by repeating the same procedure as for P_u. This may be computationally expensive as it requires eigenvalue decomposition of the kernel matrices. Further, the two eigenspaces may be unrelated. The proposed solution is

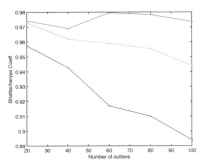

Fig. 4. Bhattacharyya distance measure between true and estimated densities. Red: Robust density estimation using eigenvalue decomposition, Green: robust density estimation using M-estimation, Blue: Kernel density estimation.

to use the eigenvectors V^k of the distribution P_u. The similarity measure between the samples is then given by

$$D_r(P_u, P_v) = ||\omega_u - \omega_v||, \tag{12}$$

where $\omega_u = [\omega_u^1, \ldots, \omega_u^m]^T$ and $\omega_v = [\omega_v^1, \ldots, \omega_v^m]^T$. Since both mean maps live in the same eigenspace, the eigenvectors V^k have been dropped from the (Equation 12).

5.4 Summary

The procedure is summarized below.

- Given samples $\{u_i\}_{i=1}^{n_u}$ and $\{v_i\}_{i=1}^{n_v}$ from two distributions P_u and P_v.
- Form kernel matrix K using the samples from the distribution P_u. Diagonalize the kernel matrix to get eigenvectors $a^k = [a_1^k, \ldots, a_{n_u}^k]$ and eigenvalues λ^k for $k = 1, \ldots, m$, where m is the total number of eigenvectors retained.
- Calculate ω_u using Equation (6), and ω_v by $\omega_v^k = \frac{1}{n_v} \sum_{i=1}^{n_v} f^k(v_i)$.
- The similarity of P_v to P_u is given by Equation (12).

5.5 Example 1

As a simple synthetic example (visual tracking examples will be given in the next section), we compute MMD and robust MMD between two distributions. The first one is a multi-modal Gaussian distribution given by

$$X \sim \frac{1}{3}\mathcal{N}_1(\mu_1, \Sigma_1) + \frac{1}{3}\mathcal{N}_2(\mu_2, \Sigma_2) + \frac{1}{3}\mathcal{N}_3(\mu_3, \Sigma_3), \tag{13}$$

where $\mu_1 = [0,0]^T$, $\mu_2 = [5,5]^T$, $\mu_3 = [5,-5]^T$ and $\Sigma_1 = \Sigma_2 = \Sigma_3 = .5 \times I$. The sample points for the other distribution are obtained from the first one by adding Gaussian noise to about 50% of the samples. Figure 5(c) shows the MMD and robust MMD measure as the standard deviation of the noise is increased. The slope of robust MMD is lower than MMD showing that it is less sensitive to noise. In Figure 6, the absolute value of the difference between the two distributions is plotted for MMD and rMMD measure. The samples from the two distributions are shown in red and blue color. The effect of noise is more pronounced in case of MMD.

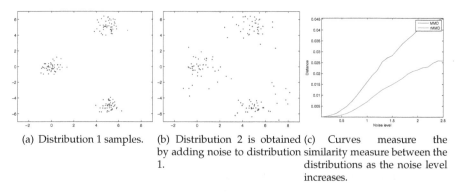

(a) Distribution 1 samples.   (b) Distribution 2 is obtained (c) Curves measure the by adding noise to distribution similarity measure between the 1. distributions as the noise level increases.

Fig. 5. MMD vs robust MMD.

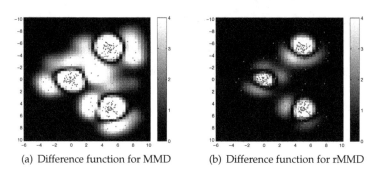

(a) Difference function for MMD   (b) Difference function for rMMD

Fig. 6. Illustration of the effect of noise on the difference between the the two distributions. The samples from the two distributions are shown in red and blue.

5.6 Example 2

Consider another example of a 2-dimensional swiss roll. The data set is generated by the following function.

$$t = \frac{3}{2} \cdot \pi \cdot (1 + 2r) \; where \; r \geq 0,$$

$$x = t \cdot cos(t),$$

$$y = t \cdot sin(t),$$

where x and y are the coordinates of the data points. 300 points are uniformly sampled and are shown in Figure 7(a). The noisy data sets are obtained by adding Gaussian noise to the original data set at standard deviations, $\sigma = [0 - 1.5]$. For example Figure 7(b) shows a noisy data at $\sigma = 1$. We measure the similarity between the two data sets using MMD and rMMD. 20 eigenvectors are retained for the rMMD computation. Figure 8 shows that the MMD and rMMD measure as the standard deviation of the noise is increased. The slope of rMMD is lower than MMD showing that it is less sensitive to noise.

As mentioned earlier, the eigenvectors corresponding to the lower eigenvalues capture noise present in the data set. The rMMD measure uses the the reconstructed points ϕ_r (Equation 4)

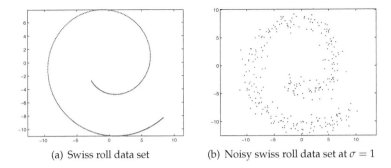

(a) Swiss roll data set   (b) Noisy swiss roll data set at $\sigma = 1$

Fig. 7. Swiss roll example

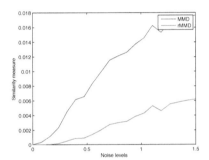

Fig. 8. Curves measure the similarity between the two data sets as the noise level is increased.

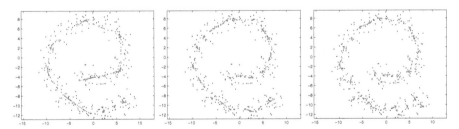

Fig. 9. Reconstruction of the noisy points using 10, 20, 30 eigenvectors Rathi, Dambreville & Tannenbaum (2006). Blue: noise data set, Red: reconstructed points.

to compute the robust mean map. The reconstructed points are obtained by using only the leading eigenvectors. Therefore, the effect of noise on the reconstructed points is reduced. We use the method descibed in Rathi, Dambreville & Tannenbaum (2006) to visualize the reconstructed points in the input space. Figure 9 shows the reconstructed points using 10, 20 and 30 eigenvectors. The blue dots are the noisy data set and the red dots are the reconstructed points. It is clear from the figure that the reconstructed data points using few leading eigenvectors match faithfully to the original data set.

6. Visual tracking through density comparison

In the first part of the chapter, we presented a technique to robustly compare two distributions represented by their samples. An application of the technique is visual target tracking. The object is tracked by finding the correspondence of the object region in consecutive images by using a template or a set of templates of the target object to define a model distribution. To track the object, each image of the sequence is searched to find the region whose candidate distribution closely matches the model distribution. A key requirement here is that the similarity measure should be robust to noise and outliers, which arise for a number of reasons such as noise in imaging procedure, background clutter, partial occlusions, etc.

One popular algorithm, the mean shift tracker Comaniciu et al. (2003), uses a histogram weighted by a spatial kernel as a probability density function of the object region. The correspondence of the target object between sequential frames is established at the region level by maximizing the Bhattacharyya coefficient between the target and the candidate distributions using mean-shift Cheng (1995). Instead of using the Bhattacharyya coefficient as a distance measure between the two distributions, Hager et al. (2004) use the Matusita distance between kernel-modulated histograms. The Matusita distance is optimized using Newton-style iterations, which provides faster convergence than the mean-shift. Histograms discard spatial information, which becomes problematic when faced with occlusions and/or the presence of target features in the background. In Birchfield & Rangarajan (2005), histograms were generalized to include spatial information, leading to spatiograms. A spatiogram augments each histogram bin with the spatial means and covariances of the pixels comprising the bin. The spatiograms captures the probability density function of the image values. The similarity between the two density functions was computed using Bhattacharyya coefficient. Elgammal Elgammal et al. (2003) employs a joint appearance-spatial density estimate and measure the similarity of the model and the candidate distributions using Kullback-Leigbler information distance.

Similarly, measuring the similarity/distance between two distributions is also required in image segmentation. For example, in some contour based segmentation algorithms (Freedman & Zhang (2004); Rathi, Malcolm & Tannenbaum (2006)), the contour is evolved either to separate the distribution of the pixels inside and outside of the contour, or to evolve the contour so that the distribution of the pixels inside matches a prior distribution of the target object. In both cases, the distance between the distributions is calculated using Bhattacharyya coefficient or Kullback-Liebler information distance.

The algorithms defined above require computing the probability density functions using the samples, which becomes computationally expensive for higher dimensions. Another problem associated with computing probability density functions is the sparseness of the observations within the d-dimensional feature space, especially when the sample set size is small. This makes the similarity measures, such as Kullback-Leibler divergence and Bhattacharyya coefficient, computationally unstable, Yang & R. Duraiswami (2005). Additionally, these techniques require sophisticated space partitioning and/or bias correction strategies Smola et al. (2007).

This section describes a method to use robust maximum mean discrepancy (rMMD) measure, described in the first part of this chapter, for visual tracking. The similarity between the two distributions can be computed directly on the samples without requiring the intermediate step of density estimation. Also, the model density function is designed to capture the appearance and spatial characteristics of the target object.

6.1 Extracting target feature vectors

The feature vector associated to a given pixel is a d-dimensional concatenation of a p-dimensional appearance vector and a 2-dimensional spatial vector $u = [\mathcal{F}(x), x]$, where $\mathcal{F}(x)$ is the p-dimensional appearance vector extracted from $\mathcal{I}$ at the spatial location x,

$$\mathcal{F}(x) = \Gamma(\mathcal{I}, x),$$

where Γ can be any mapping such as color $\mathcal{I}(x)$, image gradient, edge, texture, etc., any combination of these, or the output from a filter bank (Gabor filter, wavelet, etc.).

The feature vectors are extracted from the segmented target template image(s). The set of all feature vectors define the target input space $\mathbb{D}$,

$$\mathbb{D} = \{u_1, u_2, ..., u_n\},$$

where n is the total number of feature vectors extracted from the template image(s). The set of all pixel vectors, $\{u_i\}_{i=1}^{n_u}$, extracted from the template region R, are observations from an underlying density function P_u. To locate the object in an image, a region $\tilde{R}$ (with samples $\{v_i\}_{i=1}^{n_v}$) with density P_v is sought which minimizes the rMMD measure given by Equation (12). The kernel in this case is

$$k(u_i, u_j) = \exp\left(-\frac{1}{2}(u_i - u_j)^T \Sigma^{-1}(u_i - u_j)\right), \tag{14}$$

where Σ is a $d \times d$ diagonal matrix with bandwidths for each appearance-spatial coordinate, $\{\sigma_{F_1}, ..., \sigma_{F_p}, \sigma_{s_1}, \sigma_{s_2}\}$.

An exhaustive search can be performed to find the region or, starting from an initial guess, gradient based methods can be used to find the local minimum. For the latter approach, we provide a variational localization procedure below.

6.2 Variational target localization

Assume that the target object undergoes a geometric transformation from region R to a region $\tilde{R}$, such that $R = T(\tilde{R}, a)$, where $a = [a_1, ..., a_g]$ is a vector containing the parameters of transformation and g is the total number of transformation parameters. Let $\{u_i\}_{i=1}^{n_u}$ and $\{v_i\}_{i=1}^{n_v}$ be the samples extracted from region R and $\tilde{R}$, and let $v_i = [\mathcal{F}(\tilde{x}_i), T(\tilde{x}_i, a)]^T = [\mathcal{F}(\tilde{x}_i), x_i]^T$. The rMMD measure between the distributions of the regions R and $\tilde{R}$ is given by the Equation (12), with the L_2 norm is

$$D_r = \sum_{k=1}^{m} \left(\omega_u^k - \omega_v^k\right)^2, \tag{15}$$

where the m-dimensional robust mean maps for the two regions are $\omega_u^k = \frac{1}{n_u}\sum_{i=1}^{n_u} f^k(u_i)$ and $\omega_v^k = \frac{1}{n_v}\sum_{i=1}^{n_v} f^k(v_i)$. Gradient descent can be used to minimize the rMMD measure with respect to the transformation parameter a. The gradient of Equation (15) with respect to the transformation parameters a is

$$\nabla_a D_r = -2 \sum_{k=1}^{m} \left(\omega_u^k - \omega_v^k \right) \nabla_a \omega_{v'}^k$$

where $\nabla_a \omega_v^k = \frac{1}{n_v} \sum_{i=1}^{n_v} \nabla_a f^k(v_i)$. The gradient of $f^k(v_i)$ with respect to a is,

$$\nabla_a f^k(v_i) = \nabla_x f^k(v_i) \cdot \nabla_a T(\tilde{x}, a),$$

where $\nabla_a T(\tilde{x}, a)$ is a $g \times 2$ Jacobian matrix of T and is given by $\nabla_a T = [\frac{\partial T}{\partial a_1}, \ldots, \frac{\partial T}{\partial a_g}]^T$. The gradient $\nabla_x f^k(v_i)$ is computed as,

$$\nabla_x f^k(v_i) = \frac{1}{\sigma_s^2} \sum_{j=1}^{n_u} w_j^k k(u_j, v_i)(\pi_s(u_j) - x_i),$$

where π_s is a projection from d-dimensional pixel vector to its spatial coordinates, such that $\pi_s(u) = x$ and σ_s is the spatial bandwidth parameter used in kernel k. The transformation parameters are updated using the following equation,

$$a(t+1) = a(t) - \delta t \nabla_a D_r,$$

where δt is the time step.

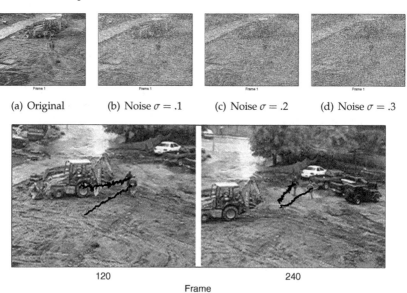

(a) Original   (b) Noise $\sigma = .1$ (c) Noise $\sigma = .2$ (d) Noise $\sigma = .3$

120   240

Frame

Fig. 10. Construction Sequence. Trajectories of the track points are shown. Red: No noise added, Green: $\sigma = .1$, Blue: $\sigma = .2$, Black: $\sigma = .3$.

Sequence	Resolution	Object size	Total Frames
Construction 1	320×240	15×15	240
Construction 2	320×240	10×15	240
Pool player	352×240	40×40	90
Fish	320×240	30×30	309
Jogging (1st row)	352×288	25×60	303
Jogging (2nd row)	352×288	30×70	111

Table 1. Tracking sequence

6.3 Results

The tracker was applied to a collection of video sequences. The pixel vectors are constructed using the color values and the spatial values. The value of σ used in the Gaussian kernel is $\sigma_F = 60$ for the color values and $\sigma_s = 4$ for the spatial domain. The number of eigenvectors, m, retained for the density estimation were chosen following Girolami (2002). In particular, given that the error associated with the eigenvector k is

$$\epsilon^k = (\omega^k)^2 = \left\{ \frac{1}{n} \sum_{i=1}^{n} f^k(u_i) \right\}^2, \tag{16}$$

the eigenvectors satisfying the following inequality were retained,

$$\left\{ \frac{1}{n} \sum_{i=1}^{n} f^k(u_i) \right\}^2 > \frac{1}{1+n} \left\{ \frac{1}{n} \sum_{i=1}^{n} (f^k(u_i))^2 \right\}. \tag{17}$$

In practice, about 25 of the top eigenvectors were kept, i.e, $M = 25$. The tracker was implemented using Matlab on an Intel Core2 1.86 GHz processor with 2GB RAM. The run time for the proposed tracker was about 0.5-1 frames/sec, depending upon the object size.

In all the experiments, we consider translation motion and the initial size and location of the target objects are chosen manually. Figure 10 shows results of tracking two people under different levels of Gaussian noise. Matlab command imnoise was used to add zero mean Gaussian noise of $\sigma = [.1, .2, .3]$. The sample frames are shown in Figure 10(b), 10(c) and 10(e). The trajectories of the track points are also shown. The tracker was able to track in all cases. The mean shift tracker (Comaniciu et al. (2003)) lost track within few frames in case of noise level $\sigma = .1$.

Figure 11 shows the result of tracking the face of a pool player. The method was able to track 100% at different noise levels. The covariance tracker Porikli et al. (2006) could detect the face correctly for 47.7% of the frames, for the case of no model update (no noise case). The mean shift tracker Comaniciu et al. (2003) lost track at noise level $\sigma = .1$.

Figure 12 shows tracking results of a fish sequence. The sequence contains noise, background clutter and fish size changes. The jogging sequence (Figure 13) was tracked in conjunction with Kalman filtering (Kalman (1960)) to successfully track through short-term total occlusions.

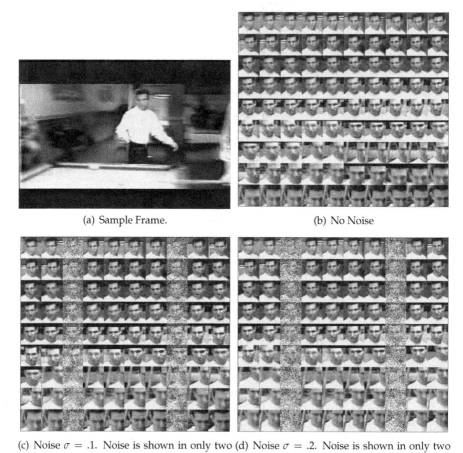

(a) Sample Frame.   (b) No Noise

(c) Noise $\sigma = .1$. Noise is shown in only two (d) Noise $\sigma = .2$. Noise is shown in only two
columns for better visualization.   columns for better visualization.

Fig. 11. Face sequence. Montages of extracted results from 90 consecutive frames for different noise levels.

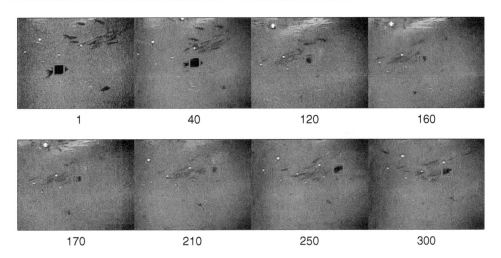

| 1 | 40 | 120 | 160 |
| 170 | 210 | 250 | 300 |

Fig. 12. Fish Sequence.

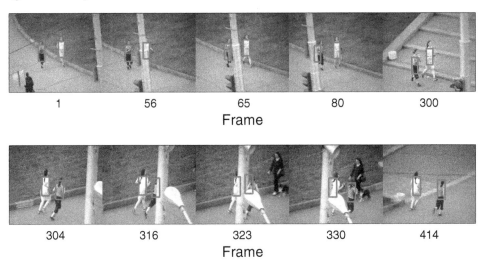

| 1 | 56 | 65 | 80 | 300 |

Frame

| 304 | 316 | 323 | 330 | 414 |

Frame

Fig. 13. Jogging sequence.

7. Conclusion

This chapter presented a novel density comparison method, given two sets of points sampled from two distributions. The method does not require explicit density estimation as an intermediate step. Instead it works directly on the data points to compute the similarity measure. The proposed similarity measure is robust to noise and outliers. Possible applications of the proposed density comparison method in computer vision are visual tracking, segmentation, image registration, and stereo registration. We used the technique for visual tracking and provided a variational localization procedure.

8. References

Birchfield, S. & Rangarajan, S. (2005). Spatiograms versus histograms for region-based tracking, *IEEE Conference on Computer Vision and Pattern Recognition*, Vol. 2, pp. 1158–1163.

Cheng, Y. (1995). Mean shift, mode seeking, and clustering, *IEEE Transactions on Pattern Analysis and Machine Intelligence* 17: 790–799.

Comaniciu, D., Meer, P. & Ramesh, V. (2003). Kernel-based object tracking, *IEEE Transactions on Pattern Analysis and Machine Intelligence* 25: 564–577.

Elgammal, A., Duraiswami, R. & Davis, L. (2003). Probabilistic tracking in joint feature-spatial spaces, *IEEE Conference on Computer Vision and Pattern Recognition*, pp. 781–788.

Freedman, D. & Zhang, T. (2004). Active contours for tracking distributions, *IEEE Transactions on Image Processing* 13(4).

Girolami, M. (2002). Orthogonal series density estimation and the kernel eigenvalue problem, *Neural Computation*. 14(3): 669–688.

Gretton, A., Borgwardt, K., Rasch, M., Schölkopf, B. & Smola, A. (2007). A kernel method for the two-sample problem, *Technical Report 157*, Max Planck Institute.

Hager, G., Dewan, M. & Stewart, C. (2004). Multiple kernel tracking with SSD, *IEEE Conference on Computer Vision and Pattern Recognition*, pp. 790–797.

Huber, P., Ronchetti, E. & MyiLibrary (1981). *Robust statistics*, Vol. 1, Wiley Online Library.

Kalman, R. (1960). A new approach to linear filtering and prediction problems, *Journal of Basic Engineering* 82(1): 35–45.

Kim, J. & Scott, C. (2008). Robust kernel density estimation, *IEEE International Conference on Acoustics, Speech and Signal Processing*, pp. 3381–3384.

Leventon, M. (2002). *Statistical models in medical image analysis*, PhD thesis, Massachusetts Institute of Technology.

Porikli, F., Tuzel, O. & Meer, P. (2006). Covariance tracking using model update based means on Riemannian manifolds, *IEEE Conference on Computer Vision and Pattern Recognition*, pp. 728–735.

Rathi, Y., Dambreville, S. & Tannenbaum, A. (2006). Statistical shape analysis using kernel PCA, *Proceedings of SPIE*, Vol. 6064, pp. 425–432.

Rathi, Y., Malcolm, J. & Tannenbaum, A. (2006). Seeing the unseen: Segmenting with distributions, *International Conference on Signal and Image Processing*.

Schölkopf, B. & Smola, A. (2001). *Learning with Kernels: Support Vector Machines, Regularization, Optimization, and Beyond*, The MIT Press.

Scholköpf, B., Smola, A. & Muller, K.-R. (1998). Nonlinear component analysis as a kernel eigenvalue problem, *Neural Computation* pp. 1299–1319.

Smola, A., Gretton, A., Song, L. & Schölkopf, B. (2007). A Hilbert space embedding for distributions, *Lecture Notes in Computer Science* .

Yang, C. & R. Duraiswami, L. D. (2005). Efficient mean-shift tracking via a new similarity measure, *IEEE Conference on Computer Vision and Pattern Recognition*, pp. 176–183.

12

Robust Principal Component Analysis for Background Subtraction: Systematic Evaluation and Comparative Analysis

Charles Guyon, Thierry Bouwmans and El-hadi Zahzah
Lab. MIA - Univ. La Rochelle
France

1. Introduction

The analysis and understanding of video sequences is currently quite an active research field. Many applications such as video surveillance, optical motion capture or those of multimedia need to first be able to detect the objects moving in a scene filmed by a static camera. This requires the basic operation that consists of separating the moving objects called "foreground" from the static information called "background". Many background subtraction methods have been developed (Bouwmans et al. (2010); Bouwmans et al. (2008)). A recent survey (Bouwmans (2009)) shows that subspace learning models are well suited for background subtraction. Principal Component Analysis (PCA) has been used to model the background by significantly reducing the data's dimension. To perform PCA, different Robust Principal Components Analysis (RPCA) models have been recently developped in the literature. The background sequence is then modeled by a low rank subspace that can gradually change over time, while the moving foreground objects constitute the correlated sparse outliers. However, authors compare their algorithm only with the PCA (Oliver et al. (1999)) or another RPCA model. Furthermore, the evaluation is not made with the datasets and the measures currently used in the field of background subtraction. Considering all of this, we propose to evaluate RPCA models in the field of video-surveillance. Contributions of this chapter can be summarized as follows:

- A survey regarding robust principal component analysis
- An evaluation and comparison on different video surveillance datasets

The rest of this paper is organized as follows: In Section 2, we firstly provide the survey on robust principal component analysis. In Section 3, we evaluate and compare robust principal component analysis in order to achieve background subtraction. Finally, the conclusion is established in Section 4.

2. Robust principal component analysis: A review

In this section, we review the original PCA and five recent RPCA models and their applications in background subtraction:

- Principal Component Analysis (PCA) (Eckart & Young (1936); Oliver et al. (1999))

- RPCA via Robust Subspace Learning (RSL) (Torre & Black (2001); Torre & Black (2003))
- RPCA via Principal Component Pursuit (PCP) (Candes et al. (2009))
- RPCA via Templates for First-Order Conic Solvers (TFOCS[1]) (Becker et al. (2011))
- RPCA via Inexact Augmented Lagrange Multiplier (IALM[2]) (Lin et al. (2009))
- RPCA via Bayesian Framework (BRPCA) (Ding et al. (2011))

2.1 Principal component analysis

Assuming that the video is composed of n frames of size $width \times height$. We arrange this training video in a rectangular matrix $A \in R^{m \times n}$ (m is the total amount of pixels), each video frame is then vectorized into column of the matrix A, and rows correspond to a specific pixel and its evolution over time. The PCA firstly consists of decomposing the matrix A in the product USV'. where $S \in \mathbb{R}^{n \times n}(diag)$ is a diagonal matrix (singular values), $U \in \mathbb{R}^{m \times n}$ and $V \in \mathbb{R}^{n \times n}$ (singular vectors). Then only the principals components are retained. To solve this decomposition, the following function is minimized (in tensor notation):

$$(S_0, U_0, V_0) = \underset{S,U,V}{\operatorname{argmin}} \sum_{r=1}^{\min(n,m)} ||A - \underset{kk}{S} \underset{ik}{U} \underset{jk}{V}||_F^2 \ , \ 1 \le k \le r \ \text{ subj } \begin{cases} \underset{ki\,kj}{UU} = \underset{ki\,kj}{VV} = 1 & \text{if } i = j \\ \underset{ij}{S} = 0 & \text{if } i \ne j \end{cases} \tag{1}$$

This imply singular values are straightly sorted and singular vectors are mutually orthogonal ($U_0'U_0 = V_0'V_0 = I_n$). The solutions S_0, U_0 and V_0 of (1) are not unique.

We can define U_1 and V_1, the set of cardinality $2^{\min(n,m)}$ of all solution;

$$U_1 = U_0 R \ , \ V_1 = R V_0 \ , \ \underset{ij}{R} = \begin{cases} \pm 1 & if \ i = j \\ 0 & elsewhere \end{cases} \ , \ m > n \tag{2}$$

We choose k (small) principal components:

$$\underset{ij}{U} = \underset{ij}{U_1} \ , \ 1 \le j \le k \tag{3}$$

The background is computed as follows:

$$Bg = UU'v \tag{4}$$

where v is the current frame. The foreground dectection is made by thresholding the difference between the current frame v and the reconstructed background image (in Iverson notation):

$$Fg = [\,|v - Bg| < T\,] \tag{5}$$

where T is a constant threshold.

Results obtained by Oliver et al. (1999) show that the PCA provides a robust model of the probability distribution function of the background, but not of the moving objects while they do not have a significant contribution to the model. As developped in Bouwmans (2009), this

---

[1] http://tfocs.stanford.edu/
[2] http://perception.csl.uiuc.edu/matrix-rank/sample_code.html

model presents several limitations. The first limitation of this model is that the size of the foreground object must be small and don't appear in the same location during a long period in the training sequence. The second limitation appears for the background maintenance. Indeed, it is computationally intensive to perform model updating using the batch mode PCA. Moreover without a mechanism of robust analysis, the outliers or foreground objects may be absorbed into the background model. The third limitation is that the application of this model is mostly limited to the gray-scale images since the integration of multi-channel data is not straightforward. It involves much higher dimensional space and causes additional difficulty to manage data in general. Another limitation is that the representation is not multimodal so various illumination changes cannot be handled correctly. In this context, several robust PCA can be used to alleviate these limitations.

2.2 RPCA via Robust Subspace Learning

Torre & Black (2003) proposed a Robust Subspace Learning (RSL) which is a batch robust PCA method that aims at recovering a good low-rank approximation that best fits the majority of the data. RSL solves a nonconvex optimization via alternative minimization based on the idea of soft-detecting andown-weighting the outliers. These reconstruction coefficients can be arbitrarily biased by an outlier. Finally, a binary outlier process is used which either completely rejects or includes a sample. Below we introduce a more general analogue outlier process that has computational advantages and provides a connection to robust M-estimation. The energy function to minimize is then:

$$(S_0, U_0, V_0) = \underset{S,U,V}{\text{argmin}} \sum_{r=1}^{\min(n,m)} \rho(A - \mu \mathbf{1_n}' - \underset{kk \; ik \; jk}{S \, U \, V}) \; , \; 1 \leq k \leq r \qquad (6)$$

where μ is the mean vector and the $\rho - function$ is the particular class of robust ρ-function (Black & Rangarajan (1996)). They use the Geman-McClure error function $\rho(x, \sigma_p) = \frac{x^2}{x^2 + \sigma_p^2}$ where σ_p is a scale parameter that controls the convexity of the robust function. Similar, the penalty term associate is $(\sqrt{L_{pi}} - 1)^2$. The robustness of De La Torre's algorithm is due to this $\rho - function$. This is confirmed by the results presented whitch show that the RSL outperforms the standard PCA on scenes with illumination change and people in various locations.

2.3 RPCA via Principal Component Pursuit

Candes et al. (2009) achieved Robust PCA by the following decomposition:

$$A = L + S \qquad (7)$$

where L is a low-rank matrix and S must be sparse matrix. The straightforward formulation is to use L_0 norm to minimize the energy function:

$$\underset{L,S}{\text{argmin}} \; Rank(L) + \lambda ||S||_0 \;\; \text{subj} \;\; A = L + S \qquad (8)$$

where λ is arbitrary balanced parameter. But this problem is NP-hard, typical solution might involve a search with combinatorial complexity. For solve this more easily, the natural way is to fix the minimization with L_1 norm that provided an approximate convex problem:

$$\underset{L,S}{\text{argmin}} ||L||_* + \lambda ||S||_1 \quad \text{subj} \quad A = L + S \tag{9}$$

where $||.||_*$ is the nuclear norm (which is the L_1 norm of singular value). Under these minimal assumptions, the PCP solution perfectly recovers the low-rank and the sparse matrices, provided that the rank of the low-rank matrix and the sparsity matrix are bounded by the follow inequality:

$$\text{rank}(L) \leq \frac{\rho_r \max(n, m)}{\mu (\log \min(n, m))^2} \quad , \quad ||S||_0 \leq \rho_s mn \tag{10}$$

where, ρ_r and ρ_s are positive numerical constants, m and n are the size of the matrix A.

For further consideration, lamda is choose as follow:

$$\lambda = \frac{1}{\sqrt{\max(m, n)}} \tag{11}$$

Results presented show that PCP outperform the RSL in case of varying illuminations and bootstraping issues.

2.4 RPCA via templates for first-order conic solvers

Becker et al. (2011) used the same idea as Candes et al. (2009) that consists of some matrix A which can be broken into two components $A = L + S$, where L is low-rank and S is sparse. The inequality constrained version of RPCA uses the same objective function, but instead of the constraints $L + S = A$, the constraints are:

$$\underset{L,S}{\text{argmin}} ||L||_* + \lambda ||S||_1 \quad \text{subj} \quad ||L + S - A||_\infty \leq \alpha \tag{12}$$

Practically, the A matrix is composed from datas generated by camera, consequently values are quantified (rounded) on 8 bits and bounded between 0 and 255. Suppose $A_0 \in \mathcal{R}^{m \times n}$ is the ideal data composed with real values, it is more exact to perform exact decomposition onto A_0. Thus, we can assert $||A_0 - A||_\infty < \frac{1}{2}$ with $A_0 = L + S$.

The result show improvements for dynamic backgrounds [3].

2.5 RPCA via inexact augmented Lagrange multiplier

Lin et al. (2009) proposed to substitute the constraint equality term by penalty function subject to a minimization under L_2 norm :

$$\underset{L,S}{\text{argmin}} \ Rank(L) + \lambda ||S||_0 + \mu \frac{1}{2} ||L + S - A||_F^2 \tag{13}$$

This algorithm solves a slightly relaxed version of the original equation. The μ constant lets balance between exact and inexact recovery. Lin et al. (2009) didn't present result on background subtraction.

---

[3] http://www.salleurl.edu/~ftorre/papers/rpca/rpca.zip

2.6 RPCA via Bayesian framework

Ding et al. (2011) proposed a hierarchical Bayesian framework that considered for decomposing a matrix (A) into low-rank (L), sparse (S) and noise matrices (E). In addition, the Bayesian framework allows exploitation of additional structure in the matrix . Markov dependency is introduced between consecutive rows in the matrix implicating an appropriate temporal dependency, because moving object are strongly correlated across consecutive frames. A spatial dependency assumption is also added and introduce the same Markov contrain as temporal utilizing the local neightborood. Indeed, it force the sparce outliers component to be spatialy and temporaly connected. Thus the decomposition is made as follows:

$$A = L + S + E = U(SB_L)V' + X \circ B_S + E \tag{14}$$

Where L is the low-rank matrix, S is the sparse matrix and E is the noise matrix. Then some assumption about components distribution are done:

- Singular vector (U and V') are drawn from normal distribution.
- Singular value and sparse matrix (S and X) value are drawn from normal-gamma distribution
- Singular sparness mask (B_L and B_S) from bernouilli-beta process.

Note that L_1 minimization is done by l_0 minimization (number of non-zero values fixed for the sparness mask), afterwards a l_2 minimization is performed on non-zero values.

The matrix A is assumed noisy, with unknown and possibly non-stationary noise statistics. The Bayesian framework infers an approximate representation for the noise statistics while simultaneously inferring the low-rank and sparse-outlier contributions: the model is robust to a broad range of noise levels, without having to change model hyperparameter settings. The properties of this Markov process are also inferred based on the observed matrix, while simultaneously denoising and recovering the low-rank and sparse components. Ding et al. (2011) applied it to background modelling and the result obtain show more robustness to noisy background, slow changing foreground and bootstrapping issue than the RPCA via convex optimization (Wright et al. (2009)).

3. Comparison

In this section, we present the evaluation of the five RPCA models (RSL, PCP, TFOCS, IALM, Bayesian) and the basic average algorithm (SUB) on three different datasets used in video-surveillance: the Wallflower dataset provided by Toyama et al. (1999), the dataset of Li et al. (2004) and dataset of Sheikh & Shah (2005). Qualitative and quantitative results are provided for each dataset.

3.1 Wallflower dataset [4]

We have chosen this particular dataset provided by Toyama et al. Toyama et al. (1999) because of how frequent its use is in this field. This frequency is due to its faithful representation of real-life situations typical of scenes susceptible to video surveillance. Moreover, it consists of seven video sequences, with each sequence presenting one of the difficulties a practical task is

---

[4] http://research.microsoft.com/en-us/um/people/jckrumm/wallflower/
testimages.htm

likely to encounter (i.e illumination changes, dynamic backgrounds). The size of the images is 160 × 120 pixels. A brief description of the Wallflower image sequences can be made as follows:

- **Moved Object (MO)**: A person enters into a room, makes a phone call, and leaves. The phone and the chair are left in a different position. This video contains 1747 images.

- **Time of Day (TOD)**: The light in a room gradually changes from dark to bright. Then, a person enters the room and sits down. This video contains 5890 images.

- **Light Switch (LS)**: A room scene begins with the lights on. Then a person enters the room and turns off the lights for a long period. Later, a person walks in the room and switches on the light. This video contains 2715 images.

- **Waving Trees (WT)**: A tree is swaying and a person walks in front of the tree. This video contains 287 images.

- **Camouflage (C)**: A person walks in front of a monitor, which has rolling interference bars on the screen. The bars include similar color to the person's clothing. This video contains 353 images.

- **Bootstrapping (B)**: The image sequence shows a busy cafeteria and each frame contains people. This video contains 3055 images.

- **Foreground Aperture (FA)**: A person with uniformly colored shirt wakes up and begins to move slowly. This video contains 2113 images.

For each sequence, the ground truth is provided for one image when the algorithm has to show its robustness to a specific change in the scene. Thus, the performance is evaluated against hand-segmented ground truth. Four terms are used in the evaluation:

- True Positive (TP) is the number of foreground pixels that are correctly marked as foreground.

- False Positive (FP) is the number of background pixels that are wrongly marked as foreground.

- True Negative (TN) is the number of background pixels that are correctly marked as background.

- False Negative (FN) is the number of foreground pixels that are wrongly marked as background.

| | | Algorithm | |
		Foreground	Background
Ground Truth	Foreground	TP	FN
	Background	FP	TN

Table 1. Measure for performance evalutation

Table 1 illustrates how to compute these different terms. Then, we computed the following metrics: the detection rate, the precision and the F-measure. Detection rate gives the percentage of corrected pixels classified as background when compared with the total number of background pixels in the ground truth:

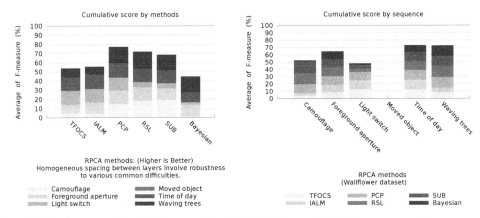

Fig. 1. Performance on the Wallflower dataset. The left (resp. right) figure concern the cumulative score by method (resp. sequence).

$$DR = \frac{TP}{TP + FN} \tag{15}$$

Precision gives the percentage of corrected pixels classified as background as compared at the total pixels classified as background by the method:

$$Precision = \frac{TP}{TP + FP} \tag{16}$$

A good performance is obtained when the detection rate is high without altering the precision. We also computed the F-measure used in (Maddalena & Petrosino (2010)) as follows:

$$F = \frac{2 \times DR \times Precision}{DR + Precision} \tag{17}$$

Table 2 shows the results obtained by the different algorithms on each sequence. For each sequence, the first column shows the original image and the corresponding ground truth.

The second part presents the sparse matrix in the first row and the optimal foreground mask in the second row. The detection rate (DR), Precision (Prec) and F-measure (F) are indicated below each foreground mask. Fig. 1 shows two cumulative histograms of F-measure: The left (resp. right) figure concern the cumulative score by method (resp. sequence). PCP gives the best result followed by RSL, IALM, TFOCS, and Bayesian RPCA. This ranking has to be taken with prrecaution because a poor performance on one sequence influences the overall F-measure and then modifies the rank for just one sequence. For example, the Bayesian obtained a bad score because of the following assumption: the background has necessarily a bigger area than the foreground. It happen in the sequences Camouflage and Light Switch. In the first case, the person hides more than half of the screen space. In the second case, when the person switch on the light all the pixels are affected and the algorithm exchanges the foreground and background. PCP seems to be robust for all critical situations.

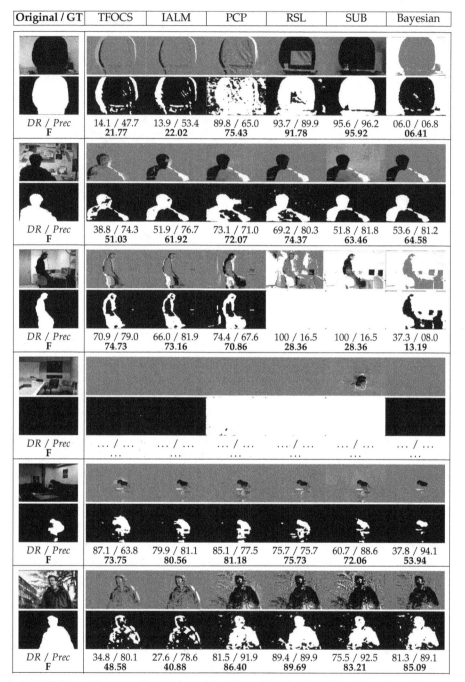

Table 2. **Wallflower dataset:** *From left to right: Ground Truth, TFOCS, IALM, PCP, RSL, SUB, Bayesian RPCA. From top to bottom: camouflage (251), foreground aperture (489), light switch (1865), moved object (985), time of day (1850), waving trees (247).*

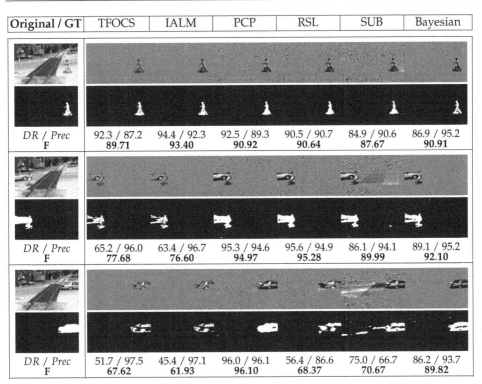

Original / GT	TFOCS	IALM	PCP	RSL	SUB	Bayesian
DR / Prec F	92.3 / 87.2 **89.71**	94.4 / 92.3 **93.40**	92.5 / 89.3 **90.92**	90.5 / 90.7 **90.64**	84.9 / 90.6 **87.67**	86.9 / 95.2 **90.91**
DR / Prec F	65.2 / 96.0 **77.68**	63.4 / 96.7 **76.60**	95.3 / 94.6 **94.97**	95.6 / 94.9 **95.28**	86.1 / 94.1 **89.99**	89.1 / 95.2 **92.10**
DR / Prec F	51.7 / 97.5 **67.62**	45.4 / 97.1 **61.93**	96.0 / 96.1 **96.10**	56.4 / 86.6 **68.37**	75.0 / 66.7 **70.67**	86.2 / 93.7 **89.82**

Table 3. **Shah dataset:** *From left to right: Ground Truth, TFOCS, IALM, PCP, RSL, SUB, Bayesian RPCA. From top to bottom: level crossing (309), level crossing (395), level crossing (462)*

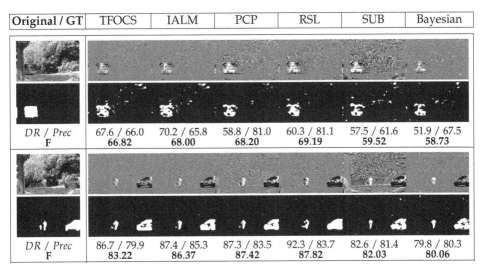

Original / GT	TFOCS	IALM	PCP	RSL	SUB	Bayesian
DR / Prec F	67.6 / 66.0 **66.82**	70.2 / 65.8 **68.00**	58.8 / 81.0 **68.20**	60.3 / 81.1 **69.19**	57.5 / 61.6 **59.52**	51.9 / 67.5 **58.73**
DR / Prec F	86.7 / 79.9 **83.22**	87.4 / 85.3 **86.37**	87.3 / 83.5 **87.42**	92.3 / 83.7 **87.82**	82.6 / 81.4 **82.03**	79.8 / 80.3 **80.06**

Table 4. **Li dataset:** *From left to right: Ground Truth, TFOCS, IALM, PCP, RSL, SUB, Bayesian RPCA. From top to bottom: campus (1650), campus (1812)*

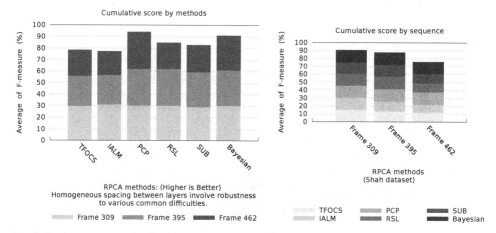

Fig. 2. Performance on the Shah dataset. The left (resp. right) figure concern the cumulative score by method (resp. sequence).

3.2 Shah's dataset [5]

This sequence involved a camera mounted on a tall tripod and comes from Sheikh & Shah (2005). It contains 500 images and the corresponding GT. The wind caused the tripod to sway back and forth causing nominal motion in the scene. Table 3 shows the results obtained by the different algorithms on three images of the sequence: Frame 309 that contains a walking person, frame 395 when a car arrived the scene and frame 462 when the same car left the scene. For each frame, the first column shows the original image and the corresponding ground truth. The second part presents the sparse matrix in the first row and the optimal foreground mask in the second row. The detection rate (DR), Precision (Prec) and F-measure (F) are indicated below each foreground mask. Fig. 2 shows two cumulative histograms of F-measure: as in previous performance evaluation. PCP gives the best result followed by Bayesian RPCA, RSL, TFOCS and IALM. We can notice that the Bayesian give better performance on this dataset because none of moving object are bigger than the background area.

3.3 Li's dataset [6]

This dataset provided by Li et al. (2004) consists of nine video sequences, which each sequence presenting dynamic backgrounds or illumination changes. The size of the images is 176*144 pixels. Among this dataset, we have chosen seven sequences that are the following ones:

- **Campus**: Persons walk and vehicles pass on a road in front of waving trees. This sequence contains 1439 images.

- **Water Surface**: A person arrives in front of the sea. There are many waves. This sequence contains 633 images.

---

[5] http://www.cs.cmu.edu/~yaser/new_backgroundsubtraction.htm
[6] http://perception.i2r.a-star.edu.sg/bk_model/bk_index.html

- **Curtain**: A person presents a course in a meeting room with a moving curtain. This sequence contains 23893 images.

- **Escalator**: This image sequence shows a busy hall where an escalator is used by people. This sequence contains 4787 images.

- **Airport**: This image sequence shows a busy hall of an airport and each frame contains people. This sequence contains 3584 images.

- **Shopping Mall**: This image sequence shows a busy shopping center and each frame contains people. This sequence contains 1286 images.

- **Restaurant**: This sequence comes from the wallflower dataset and shows a busy cafeteria. This video contains 3055 images.

The sequences Campus, Water Surface and Curtain present dynamic backgrounds whereas the sequences Restaurant, Airport, Shopping Mall show bootstrapping issues. For each sequence, the ground truth is provided for twenty images when algorithms have to show their robustness. Table 4 shows the results obtained by the different algorithms on the sequence campus. Table 5 presents the results on the dynamic background on the sequences Water Surface, Curtain, Escalator, whereas table 6 presents the result on bootstrapping issues on the sequences Airport, Shopping mall, Restaurant. For each table, the first column shows the original image and the corresponding ground truth. The second part presents the sparse matrix in the first row and the optimal foreground mask in the second row. The detection rate (DR), Precision (Prec) and F-measure (F) are indicated below each foreground mask.

Fig. 3 and 4 shows the two cumulative histograms of F-measure respectively for dynamic background and bootstrapping issues. In each case, Bayesian RPCA gives best results followed by PCP, TFOCS, IALM and RSL.

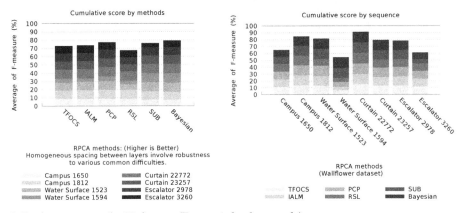

Fig. 3. Performance on the Li dataset (Dynamic backgrounds).

3.4 Implementation and time issues

Regarding the code, we have used the following implementation in *MATLAB*: RSL provided by F. De La Torre[7], PCP provided by C. Qiu[8], IALM provided by M. Chen and A. Ganesh[9],

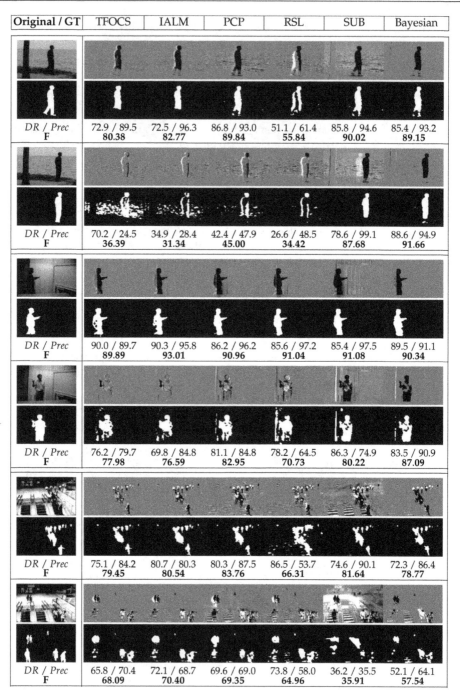

Table 5. **Li dataset:** *From left to right: Ground Truth, TFOCS, IALM, PCP, RSL, SUB, Bayesian RPCA. From top to bottom: water surface (1523), water surface (1594), curtain (22772), curtain (23257), escalator (2978), escalator (3260)*

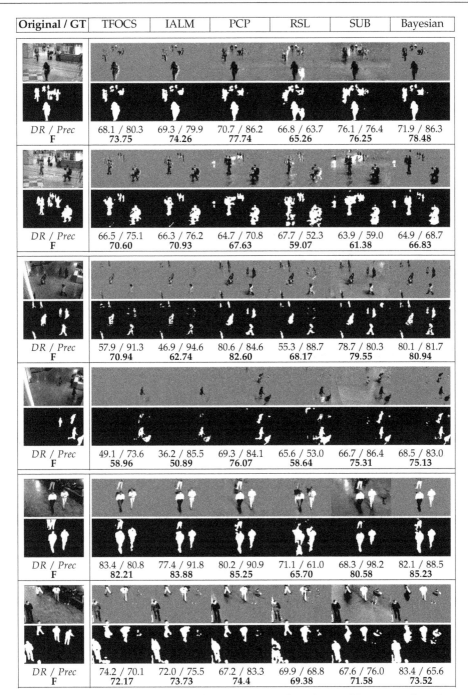

Table 6. **(Perception) Li dataset:** *From left to right: Ground Truth, TFOCS, IALM, PCP, RSL, SUB, Bayesian RPCA. From top to bottom: Airport (2926), Airport (3409), shopping mall (1862), shopping mall (1980) Restaurant (1842), Restaurant (2832).*

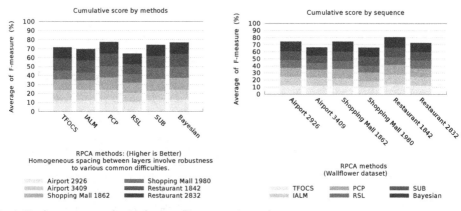

Fig. 4. Performance on the Li dataset (Bootstrap issues).

TFOCS provided by S. Becker[10] and Bayesian provided by X. Ding[11]. Additionally, a 5×5 median filter is postprocessed in order to suppress peak noise. The thresholding value is automatically choose for maximize the F-measure.

For time issues, the current implementations are faraway to achieve real-time. Indeed, the computing of the backgrounds take few hours for a training sequence with 200 frames for each algorithm. This time can be reduced by *C/Cuda* implementation as suggested in (Mu et al. (2011);Anderson et al. (2011)).

4. Conclusion

In this chapter, we started off by providing a survey on five Robust Principal Component Analysis models recently developed: Robust Subspace Learning, Principal Component Pursuit, Templates for First-Order Conic Solvers, Recursive Projected Compressive Sensing, Bayesian RPCA. We then presented a systematic evalutation and comparative analysis on different datasets used in video-surveillance. PCP demonstrates more robutness on all datasets by providing best global score. The Bayesian RPCA offers also good performance but presents a drawback related to the assumption : the background has necessarily a bigger area than the foreground. For the IALM, its performance is still acceptable.

Futur research directions may concern the evalutation of accelerate hardware implementation of robust PCA (Mu et al. (2011);Anderson et al. (2011)) and robust Independent Components Analysis (Yamazaki et al. (2006)), Incremental Non-negative Matrix Factorization (Bucak et al. (2007)) and Incremental Rank Tensor (Li et al. (2008)).

---

[7] http://www.salleurl.edu/~ftorre/papers/rpca/rpca.zip

[8] http://home.engineering.iastate.edu/~chenlu/ReProCS/ReProCS_code.zip

[9] http://perception.csl.uiuc.edu/matrix-rank/Files/inexact_alm_rpca.zip

[10] http://tfocs.stanford.edu/code

[11] http://www.ece.duke.edu/~lihan/brpca_code/BRPCA.zip

5. References

Anderson, M., Ballard, G., Demme, J. & Keutzer, K. (2011).   Communication-avoiding qr decomposition for gpu, *IEEE International Parallel and Distributed Processing Symposium, IPDPS 2011* .

Becker, S., Candes, E. & Grant, M. (2011).   Tfocs: Flexible first-order methods for rank minimization, *Low-rank Matrix Optimization Symposium, SIAM Conf. on Optimization* .

Black, M. & Rangarajan, A. (1996).   On the unification of line processes, outlier rejection, and robust statistics with applications in early vision., *International Journal of ComputerVision* pp. 57–92.

Bouwmans, T. (2009). Subspace learning for background modeling: A survey, *Recent Patents on Computer Science* 2(3): 223–234.

Bouwmans, T., Baf, F. E. & Vachon, B. (2008).   Background modeling using mixture of gaussians for foreground detection - a survey, *Recent Patents on Computer Science* 1(3): 219–237.

Bouwmans, T., Baf, F. E. & Vachon, B. (2010). Statistical background modeling for foreground detection: A survey, *Handbook of Pattern Recognition and Computer Vision, World Scientific Publishing* 4(2): 181–189.

Bucak, S., Gunsel, B. & Gursoy, O. (2007).   Incremental non-negative matrix factorization for dynamic background modeling, *International Workshop on Pattern Recognition in Information Systems* .

Candes, E., Li, X., Ma, Y. & Wright, J. (2009). Robust principal component analysis?, *Preprint* .

Ding, X., He, L. & Carin, L. (2011).   Bayesian robust principal component analysis, *IEEE Transaction on Image Processing* .

Eckart, C. & Young, G. (1936).   The approximation of one matrix by another of lower rank, *Psychometrika* 1: 211–218.

Li, L., Huang, W., Gu, I. & Tian, Q. (2004).   Statistical modeling of complex backgrounds for foreground object detection, *IEEE Transaction on Image Processing* pp. 1459–1472.

Li, X., Hu, W., Zhang, Z. & Zhang, X. (2008). Robust foreground segmentation based on two effective background models, *Multimedia Information Retrieval (MIR)* pp. 223–228.

Lin, Z., Chen, M., Wu, L. & Ma, Y. (2009). The augmented lagrange multiplier method for exact recovery of corrupted low-rank matrices, *UIUC Technical Report* .

Maddalena, L. & Petrosino, A. (2010).   A fuzzy spatial coherence-based approach to background foreground separation for moving object detection, *Neural Computing and Applications, NCA* pp. 1–8.

Mu, Y., Dong, J., Yuan, X. & Yan, S. (2011). Accelerated low-rank visual recovery by random projection, *International Conference on Computer Vision, CVPR 2011* .

Oliver, N., Rosario, B. & Pentland, A. (1999). A bayesian computer vision system for modeling human interactions, *International Conference on Vision Systems, ICVS 1998* .

Sheikh, Y. & Shah, M. (2005). Bayesian modeling of dynamic scenes for object detection, *IEEE Transactions on Pattern Analysis and Machine Intelligence* 27: 1778–1792.

Torre, F. D. L. & Black, M. (2001). Robust principal component analysis for computer vision, *International Conference on Computer Vision, ICCV 2001* .

Torre, F. D. L. & Black, M. (2003). A framework for robust subspace learning, *International Journal on Computer Vision* pp. 117–142.

Toyama, K., Krumm, J., Brumitt, B. & Meyers, B. (1999). Wallflower: Principles and practice of background maintenance, *International Conference on Computer Vision* pp. 255–261.

Wright, J., Peng, Y., Ma, Y., Ganesh, A. & Rao, S. (2009). Robust principal component analysis: Exact recovery of corrupted low-rank matrices by convex optimization, *Neural Information Processing Systems, NIPS 2009* .

Yamazaki, M., Xu, G. & Chen, Y. (2006). Detection of moving objects by independent component analysis, *Asian Conference on Computer Vision (ACCV) 2006* pp. 467–478.

13

Computing and Updating Principal Components of Discrete and Continuous Point Sets

Darko Dimitrov
Freie Universität Berlin
Germany

1. Introduction

Efficient computation and updating of the principal components are crucial in many applications including data compression, exploratory data analysis, visualization, image processing, pattern and image recognition, time series prediction, detecting perfect and reflective symmetry, and dimension detection. The thorough overview over PCA's applications can be found for example in the textbooks by Duda et al. (2001) and Jolliffe (2002).

Dynamic versions of the PCA applications, i.e., when the point set (population) changes, are of big importance and interest. Efficient solutions of those problems depend heavily on an efficient dynamic computation of the principal components (eigenvectors of the covariance matrix). Dynamic updates of variances in different settings have been studied since the sixties by Chan et al. (1979), Knuth (1998), Pébay (2008), Welford (1962) and West (1979). Pébay (2008) also investigated the dynamic maintenance of covariance matrices.

The principal components of discrete point sets can be strongly influenced by point clusters (Dimitrov, Knauer, Kriegel & Rote (2009)). To avoid the influence of the distribution of the point set, often continuous sets, especially the convex hull of a point set is considered, which lead to so-called continuous PCA. Computing PCA bounding boxes (Gottschalk et al. (1996), Dimitrov, Holst, Knauer & Kriegel (2009)), or retrieval of 3D-objects (Vranić et al. (2001)), are typical applications where continuous PCA are of interest.

The organization and the main results presented in this chapter are as follows: In Section 2, we present a standard approach of computing principal components of discrete point set in $\mathbb{R}^d$. In Section 3, we present closed-form solutions for efficiently updating the principal components of a set of n points, when m points are added or deleted from the point set. For both operations performed on a discrete point set in $\mathbb{R}^d$, we can compute the new principal components in $O(m)$ time for fixed d. This is a significant improvement over the commonly used approach of recomputing the principal components from scratch, which takes $O(n + m)$ time. In Section 4 we consider the computation of the principal components of a dynamic continuous point set. We give closed form-solutions when the point set is a convex polytope $\mathbb{R}^3$. Solutions for the cases when the point set is the boundary of a convex polytope in $\mathbb{R}^2$ or $\mathbb{R}^3$, or a convex polygon in $\mathbb{R}^2$, are presented in the appendix. Conclusion and open problems are presented in Section 5.

For implementation and verification of some the theoretical results presented here, we refer interested readers to Dimitrov, Holst, Knauer & Kriegel (2009) and Dimitrov et al. (2011).

2. Basics. Computing principal components - discrete case in $\mathbb{R}^d$

The central idea and motivation of PCA is to reduce the dimensionality of a point set by identifying *the most significant directions (principal components)*. Let $P = \{\vec{p}_1, \vec{p}_2, \ldots, \vec{p}_n\}$ be a set of vectors (points) in $\mathbb{R}^d$, and $\vec{\mu} = (\mu_1, \mu_2, \ldots, \mu_d) \in \mathbb{R}^d$ be the center of gravity of P. For $1 \leq k \leq d$, we use $p_{i,k}$ to denote the k-th coordinate of the vector p_i. Given two vectors $\vec{u}$ and $\vec{v}$, we use $\langle \vec{u}, \vec{v} \rangle$ to denote their inner product. For any unit vector $\vec{v} \in \mathbb{R}^d$, the *variance of P in direction* $\vec{v}$ is

$$\mathrm{var}(P, \vec{v}) = \frac{1}{n} \sum_{i=1}^{n} \langle p_i - \vec{\mu}, \vec{v} \rangle^2. \tag{1}$$

The most significant direction corresponds to the unit vector $\vec{v}_1$ such that $\mathrm{var}(P, \vec{v}_1)$ is maximum. In general, after identifying the j most significant directions $\vec{v}_1, \ldots, \vec{v}_j$, the $(j+1)$-th most significant direction corresponds to the unit vector $\vec{v}_{j+1}$ such that $\mathrm{var}(P, \vec{v}_{j+1})$ is maximum among all unit vectors perpendicular to $\vec{v}_1, \vec{v}_2, \ldots, \vec{v}_j$.

It can be verified that for any unit vector $\vec{v} \in \mathbb{R}^d$,

$$\mathrm{var}(P, \vec{v}) = \langle \Sigma \vec{v}, \vec{v} \rangle, \tag{2}$$

where Σ is the *covariance matrix* of P. Σ is a symmetric $d \times d$ matrix where the (i, j)-th component, $\sigma_{ij}, 1 \leq i, j \leq d$, is defined as

$$\sigma_{ij} = \frac{1}{n} \sum_{k=1}^{n} (p_{i,k} - \mu_i)(p_{j,k} - \mu_j). \tag{3}$$

The procedure of finding the most significant directions, in the sense mentioned above, can be formulated as an eigenvalue problem. If $\lambda_1 \geq \lambda_2 \geq \cdots \geq \lambda_d$ are the eigenvalues of Σ, then the unit eigenvector $\vec{v}_j$ for λ_j is the j-th most significant direction. Since the matrix Σ is symmetric positive semidefinite, its eigenvectors are orthogonal, all λ_js are non-negative and $\lambda_j = \mathrm{var}(X, \vec{v}_j)$.

Computation of the eigenvalues, when d is not very large, can be done in $O(d^3)$ time, for example with the *Jacobi* or the *QR method* (Press et al. (1995)). Thus, the time complexity of computing principal components of n points in $\mathbb{R}^d$ is $O(n + d^3)$. The additive factor of $O(d^3)$ throughout the paper will be omitted, since we will assume that d is fixed. For very large d, the problem of computing eigenvalues is non-trivial. In practice, the above mentioned methods for computing eigenvalues converge rapidly. In theory, it is unclear how to bound the running time combinatorially and how to compute the eigenvalues in decreasing order. In Cheng & Y. Wang (2008) a modification of the *Power method* (Parlett (1998)) is presented, which can give a guaranteed approximation of the eigenvalues with high probability.

3. Updating the principal components efficiently - discrete case in $\mathbb{R}^d$

In this section, we consider the problem of updating the covariance matrix Σ of a discrete point set $P = \{\vec{p}_1, \vec{p}_2, \ldots, \vec{p}_n\}$ in $\mathbb{R}^d$, when m points are added or deleted from P. We give closed-form solutions for computing the components of the new covariance matrix Σ'. Those closed-form solutions are based on the already computed components of Σ. The recent result of Pébay (2008) implies the same solution for additions that will be presented in the sequel. The main result of this section is given in the following theorem.

Theorem 3.1. *Let P be a set of n points in $\mathbb{R}^d$ with known covariance matrix Σ. Let P' be a point set in $\mathbb{R}^d$, obtained by adding or deleting m points from P. The principal components of P' can be computed in $O(m)$ time for fixed d.*

Proof. **Adding points**

Let $P_m = \{\vec{p}_{n+1}, \vec{p}_{n+2}, \ldots, \vec{p}_{n+m}\}$ be a point set with center of gravity $\vec{\mu}^m = (\mu_1^m, \mu_2^m, \ldots, \mu_d^m)$. We add P_m to P obtaining new point set P'. The j-th component, μ_j', $1 \leq j \leq d$, of the center of gravity $\vec{\mu}' = (\mu_1', \mu_2', \ldots, \mu_d')$ of P' is

$$\mu_j' = \frac{1}{n+m} \sum_{k=1}^{n+m} p_{k,j} = \frac{1}{n+m} \left(\sum_{k=1}^{n} p_{k,j} + \sum_{k=n+1}^{n+m} p_{k,j} \right) = \frac{n}{n+m} \mu_j + \frac{m}{n+m} \mu_j^m.$$

The (i,j)-th component, σ_{ij}', $1 \leq i,j \leq d$, of the covariance matrix Σ' of P' is

$$\sigma_{ij}' = \frac{1}{n+m} \sum_{k=1}^{n+m} (p_{k,i} - \mu_i')(p_{k,j} - \mu_j')$$

$$= \frac{1}{n+m} \sum_{=1}^{n} (p_{k,i} - \mu_i')(p_{k,j} - \mu_j') + \frac{1}{n+m} \sum_{k=n+1}^{n+m} (p_{k,i} - \mu_i')(p_{k,j} - \mu_j').$$

Let

$$\sigma_{ij}' = \sigma_{ij,1}' + \sigma_{ij,2}',$$

where,

$$\sigma_{ij,1}' = \frac{1}{n+m} \sum_{k=1}^{n} (p_{k,i} - \mu_i')(p_{k,j} - \mu_j'), \tag{4}$$

and

$$\sigma_{ij,2}' = \frac{1}{n+m} \sum_{k=n+1}^{n+m} (p_{k,i} - \mu_i')(p_{k,j} - \mu_j'). \tag{5}$$

Plugging-in the values of μ_i' and μ_j' in (4), we obtain:

$$\sigma_{ij,1}' = \frac{1}{n+m} \sum_{k=1}^{n} (p_{k,i} - \frac{n}{n+m}\mu_i - \frac{m}{n+m}\mu_i^m)(p_{k,j} - \frac{n}{n+m}\mu_j - \frac{m}{n+m}\mu_j^m)$$

$$= \frac{1}{n+m} \sum_{k=1}^{m} (p_{k,i} - \mu_i + \frac{m}{n+m}\mu_i - \frac{m}{n+m}\mu_i^m)(p_{k,j} - \mu_j + \frac{m}{n+m}\mu_j - \frac{m}{n+m}\mu_j^m)$$

$$= \frac{1}{n+m} \sum_{k=1}^{n} (p_{k,i} - \mu_i)(p_{k,j} - \mu_j) + \frac{1}{n+m} \sum_{k=1}^{n} (p_{k,i} - \mu_i)(\frac{m}{n+m}\mu_j - \frac{m}{n+m}\mu_j^m) +$$

$$\frac{1}{n+m} \sum_{k=1}^{n} (\frac{m}{n+m}\mu_i - \frac{m}{n+m}\mu_i^m)(p_{k,j} - \mu_j) +$$

$$\frac{1}{n+m} \sum_{k=1}^{n} (\frac{m}{n+m}\mu_i - \frac{m}{n+m}\mu_i^m)(\frac{m}{n+m}\mu_j - \frac{m}{n+m}\mu_j^m).$$

Since $\sum_{k=1}^{n}(p_{k,i}-\mu_i)=0, 1\le i\le d$, we have

$$\sigma'_{ij,1}=\frac{n}{n+m}\sigma_{ij}+\frac{nm^2}{(n+m)^3}(\mu_i-\mu_i^m)(\mu_j-\mu_j^m). \tag{6}$$

Plugging-in the values of μ'_i and μ'_j in (5), we obtain:

$$
\begin{aligned}
\sigma'_{ij,2}&=\frac{1}{n+m}\sum_{k=n+1}^{n+m}(p_{k,i}-\frac{n}{n+m}\mu_i-\frac{m}{n+m}\mu_i^m)(p_{k,j}-\frac{n}{n+m}\mu_j-\frac{m}{n+m}\mu_j^m)\\
&=\frac{1}{n+m}\sum_{k=n+1}^{n+m}(p_{k,i}-\mu_i^m+\frac{n}{n+m}\mu_i^m-\frac{n}{n+m}\mu_i)(p_{k,j}-\mu_j^m+\frac{n}{n+m}\mu_j^m-\frac{n}{n+m}\mu_j)\\
&=\frac{1}{n+m}\sum_{k=n+1}^{n+m}(p_{k,i}-\mu_i^m)(p_{k,j}-\mu_j^m)+\frac{1}{n+m}\sum_{k=n+1}^{n+m}(p_{k,i}-\mu_i^m)\frac{n}{n+m}(\mu_j^m-\mu_j)+\\
&\quad\frac{1}{n+m}\sum_{k=n+1}^{n+m}\frac{n}{n+m}(\mu_i^m-\mu_i)(p_{k,j}-\mu_j^m)+\\
&\quad\frac{1}{n+m}\sum_{k=n+1}^{n+m}\frac{n}{n+m}(\mu_i^m-\mu_i)\frac{n}{n+m}(\mu_j^m-\mu_j).
\end{aligned}
$$

Since $\sum_{k=n+1}^{n+m}(p_{k,i}-\mu_i^m)=0, 1\le i\le d$, we have

$$\sigma'_{ij,2}=\frac{m}{n+m}\sigma_{ij}^m+\frac{n^2m}{(n+m)^3}(\mu_i-\mu_i^m)(\mu_j-\mu_j^m), \tag{7}$$

where

$$\sigma_{ij}^m=\frac{1}{m}\sum_{k=n+1}^{n+m}(p_{k,i}-\mu_i^m)(p_{k,j}-\mu_j^m)), \quad 1\le i,j\le d,$$

is the i,j-th element of the covariance matrix Σ_m of the point set P_m. Finally, we have

$$\sigma'_{ij}=\sigma'_{ij,1}+\sigma'_{ij,2}=\frac{1}{n+m}(n\sigma_{ij}+m\sigma_{ij}^m)+\frac{nm}{(n+m)^2}(\mu_i-\mu_i^m)(\mu_j-\mu_j^m). \tag{8}$$

Note that σ_{ij}^m, and therefore σ'_{ij}, can be computed in $O(m)$ time. Thus, for a fixed dimension d, the covariance matrix Σ also can be computed in $O(m)$ time.

The above derivation of the new principal components is summarized in Algorithm 3.1.

Deleting points

Let $P_m=\{\vec{p}_{n-m+1},\vec{p}_{n-m},\ldots,\vec{p}_n\}$ be a subset of the point set P, and let $\vec{\mu}^m=(\mu_1^m,\mu_2^m,\ldots,\mu_d^m)$ be the center of gravity of P_m. We subtract P_m from P, obtaining a new point set P'. The j-th component, $\mu'_j, 1\le j\le d$, of the center of gravity $\vec{\mu}'=(\mu'_1,\mu'_2,\ldots,\mu'_d)$ of P' is

$$\mu'_j=\frac{1}{n-m}\sum_{k=1}^{n-m}p_{k,j}=\frac{1}{n-m}\left(\sum_{k=1}^{n}p_{k,j}-\sum_{k=n-m+1}^{n}p_{k,j}\right)=\frac{n}{n-m}\mu_j-\frac{m}{n-m}\mu_j^m.$$

Algorithm 3.1 : ADDINGPOINTS(P, μ, Σ, P_m)

Input: point set P, the center of gravity μ of P, the covariance matrix Σ of P,
       point set P_m added to P

Output: principal componenets of $P \cup P_m$

1: compute the center of gravity $\bar{\mu}^m$ of P_m
2: compute the covariance matrix Σ^m of P_m
3: compute the center of gravity $\bar{\mu}'$ of $P \cup P_m$:
4:        $\mu_j' = \frac{n}{n+m}\mu_j + \frac{m}{n+m}\mu_j^m$, for $1 \leq j \leq d$
5: compute the covariance Σ' of $P \cup P_m$:
6:        $\sigma_{ij}' = \frac{1}{n+m}(n\sigma_{ij} + m\sigma_{ij}^m) + \frac{nm}{(n+m)^2}(\mu_i - \mu_i^m)(\mu_j - \mu_j^m)$, for $1 \leq i,j \leq d$
7: **return** the eigenvectors of Σ'

The (i,j)-th component, $\sigma_{ij}', 1 \leq i,j \leq d$, of the covariance matrix Σ' of P' is

$$\sigma_{ij}' = \frac{1}{n-m} \sum_{k=1}^{n-m} (p_{k,i} - \mu_i')(p_{k,j} - \mu_j')$$

$$= \frac{1}{n-m} \sum_{k=1}^{n} (p_{k,i} - \mu_i')(p_{k,j} - \mu_j') - \frac{1}{n-m} \sum_{k=n-m+1}^{n} (p_{k,i} - \mu_i')(p_{k,j} - \mu_j').$$

Let

$$\sigma_{ij}' = \sigma_{ij,1}' - \sigma_{ij,2}',$$

where

$$\sigma_{ij,1}' = \frac{1}{n-m} \sum_{k=1}^{n} (p_{k,i} - \mu_i')(p_{k,j} - \mu_j'), \tag{9}$$

and

$$\sigma_{ij,2}' = \frac{1}{n-m} \sum_{k=n-m+1}^{n} (p_{k,i} - \mu_i')(p_{k,j} - \mu_j'). \tag{10}$$

Plugging-in the values of μ_i' and μ_j' in (9), we obtain:

$$\sigma_{ij,1}' = \frac{1}{n-m} \sum_{k=1}^{n} (p_{k,i} - \frac{n}{n-m}\mu_i + \frac{m}{n-m}\mu_i^m)(p_{k,j} - \frac{n}{n-m}\mu_j + \frac{m}{n-m}\mu_j^m)$$

$$= \frac{1}{n-m} \sum_{k=1}^{m} (p_{k,i} - \mu_i + \frac{m}{n-m}\mu_i - \frac{m}{n-m}\mu_i^m)(p_{k,j} - \mu_j + \frac{m}{n-m}\mu_j - \frac{m}{n-m}\mu_j^m)$$

$$= \frac{1}{n-m} \sum_{k=1}^{n} (p_{k,i} - \mu_i)(p_{k,j} - \mu_j) + \frac{1}{n-m} \sum_{k=1}^{n} (p_{k,i} - \mu_i)(\frac{m}{n-m}\mu_j - \frac{m}{n-m}\mu_j^m) +$$

$$\frac{1}{n-m} \sum_{k=1}^{n} (\frac{m}{n-m}\mu_i - \frac{m}{n-m}\mu_i^m)(p_{k,j} - \mu_j) +$$

$$\frac{1}{n-m} \sum_{k=1}^{n} (\frac{m}{n-m}\mu_i - \frac{m}{n-m}\mu_i^m)(\frac{m}{n-m}\mu_j - \frac{m}{n-m}\mu_i^m).$$

Since $\sum_{k=1}^{n}(p_{k,i} - \mu_i) = 0, 1 \le i \le d$, we have

$$\sigma'_{ij,1} = \frac{n}{n-m}\sigma_{ij} + \frac{nm^2}{(n-m)^3}(\mu_i - \mu_i^m)(\mu_j - \mu_j^m). \tag{11}$$

Plugging-in the values of μ_i' and μ_j' in (10), we obtain:

$$\sigma'_{ij,2} = \frac{1}{n-m}\sum_{k=n-m+1}^{n}(p_{k,i} - \frac{n}{n-m}\mu_i + \frac{m}{n-m}\mu_i^m)(p_{k,j} - \frac{n}{n-m}\mu_j + \frac{m}{n-m}\mu_j^m)$$

$$= \frac{1}{n-m}\sum_{k=n-m+1}^{n}(p_{k,i} - \mu_i^m + \frac{n}{n-m}\mu_i^m - \frac{n}{n-m}\mu_i)(p_{k,j} - \mu_j^m + \frac{n}{n-m}\mu_j^m - \frac{n}{n-m}\mu_j)$$

$$= \frac{1}{n-m}\sum_{k=n-m+1}^{n}(p_{k,i} - \mu_i^m)(p_{k,j} - \mu_j^m) + \frac{1}{n-m}\sum_{k=n-m+1}^{n-m}(p_{k,i} - \mu_i^m)\frac{n}{n-m}(\mu_j^m - \mu_j) +$$

$$\frac{1}{n-m}\sum_{k=n-m+1}^{n}\frac{n}{n-m}(\mu_i^m - \mu_i)(p_{k,j} - \mu_j^m) +$$

$$\frac{1}{n-m}\sum_{k=n-m+1}^{n}\frac{n}{n-m}(\mu_i^m - \mu_i)\frac{n}{n-m}(\mu_j^m - \mu_j).$$

Since $\sum_{k=n-m+1}^{n}(p_{k,i} - \mu_i^m) = 0, 1 \le i \le d$, we have

$$\sigma'_{ij,2} = \frac{n}{n-m}\sigma_{ij}^m + \frac{n^2 m}{(n-m)^3}(\mu_i - \mu_i^m)(\mu_j - \mu_j^m), \tag{12}$$

where

$$\sigma_{ij}^m = \frac{1}{m}\sum_{k=n-m+1}^{n}(p_{k,i} - \mu_i^m)(p_{k,j} - \mu_j^m)), \quad 1 \le i,j \le d,$$

is the i, j-th element of the covariance matrix Σ_m of the point set P_m.

Finally, we have

$$\sigma'_{ij} = \sigma'_{ij,1} + \sigma'_{ij,2} = \frac{1}{n-m}(n\sigma_{ij} - m\sigma_{ij}^m) - \frac{nm}{(n-m)^2}(\mu_i - \mu_i^m)(\mu_j - \mu_j^m). \tag{13}$$

Note that σ_{ij}^m, and therefore σ'_{ij}, can be computed in $O(m)$ time. Thus, for a fixed dimension d, the covariance matrix Σ also can be computed in $O(m)$ time. $\qquad\square$

As a corollary of (13), in the case when only one point, $\vec{p}_e$, is deleted from a point set P, the elements of the new covariance matrix are given by

$$\sigma'_{ij} = \sigma'_{ij,1} - \sigma'_{ij,2} = \frac{m}{m-1}\sigma_{ij} - \frac{m}{(m-1)^2}(p_{e,i} - \mu_i)(p_{e,j} - \mu_j), \tag{14}$$

and also can be computed in $O(1)$ time.

Similar argument holds in the case when only one point is added to a point set P, and then the new covariance matrix also can be computer in constant time.

Algorithm 3.2 : DELETINGPOINTS(P, μ, Σ, P_m)

Input: point set P, the center of gravity μ of P, the covariance matrix Σ of P,
     point set P_m deleted from P

Output: principal components of $P \setminus P_m$

1: compute the center of gravity $\vec{\mu}^m$ of P_m
2: compute the covariance matrix Σ^m of P_m
3: compute the center of gravity $\vec{\mu}'$ of $P \setminus P_m$:
4:      $\mu_j = \frac{n}{n-m}\mu_j - \frac{m}{n-m}\mu_j^m$, for $1 \le j \le d$
5: compute the covariance Σ' of $P \setminus P_m$:
6:      $\sigma'_{ij} = \sigma'_{ij,1} + \sigma'_{ij,2} = \frac{1}{n-m}\left(n\sigma_{ij} - m\sigma_{ij}^m\right) - \frac{nm}{(n-m)^2}(\mu_i - \mu_i^m)(\mu_j - \mu_j^m)$, for $1 \le i, j \le d$
7: **return** the eigenvectors of Σ'

The above derivation of the new principal components is summarized in Algorithm 3.2.

Notice, that once having closed-form solutions, one can obtain similar algorithms, as presented in this chapter, when the continuous point sets are considered. Therefore, we will omit them in the next section and in the appendix.

4. Computing and updating the principal components efficiently - mboxcontinuous case $\mathbb{R}^3$

Here, we consider the computation of the principal components of a dynamic continuous point set. We present a closed form-solutions when the point set is a convex polytope or the boundary of a convex polytope in $\mathbb{R}^2$ or $\mathbb{R}^3$. When the point set is the boundary of a convex polytope, we can update the new principal components in $O(k)$ time, for both deletion and addition, under the assumption that we know the k facets in which the polytope changes. Under the same assumption, when the point set is a convex polytope in $\mathbb{R}^2$ or $\mathbb{R}^3$, we can update the principal components in $O(k)$ time after adding points. But, to update the principal components after deleting points from a convex polytope in $\mathbb{R}^2$ or $\mathbb{R}^3$ we need $O(n)$ time. This is due to the fact that, after a deletion the center of gravity of the old convex hull (polyhedron) could lie outside the new convex hull, and therefore, a retetrahedralization is needed (see Subsection 4.1 and Subsection 6.2 for details). Due to better readability and compactness of the chapter, we present in this section only the closed-form solutions for a convex polytope in $\mathbb{R}^3$, and leave the rest of the results for the appendix.

4.1 Continuous PCA over a convex polyhedron in $\mathbb{R}^3$

Let P be a point set in $\mathbb{R}^3$, and let X be its convex hull. We assume that the boundary of X is triangulated (if it is not, we can triangulate it in a preprocessing step). We choose an arbitrary point $\vec{o}$ in the interior of X, for example, we can choose $\vec{o}$ to be the center of gravity of the boundary of X. Each triangle from the boundary together with $\vec{o}$ forms a tetrahedron. Let the number of such formed tetrahedra be n. The k-th tetrahedron, with vertices $\vec{x}_{1,k}, \vec{x}_{2,k}, \vec{x}_{3,k}, \vec{x}_{4,k} = \vec{o}$, can be represented in a parametric form by $\vec{Q}_i(s, t, u) = \vec{x}_{4,i} + s(\vec{x}_{1,i} - \vec{x}_{4,i}) + t(\vec{x}_{2,i} - \vec{x}_{4,i}) + u(\vec{x}_{3,i} - \vec{x}_{4,i})$, for $0 \le s, t, u \le 1$, and $s + t + u \le 1$. For $1 \le i \le 3$, we use $x_{i,j,k}$ to denote the i-th coordinate of the vertex $\vec{x}_j$ of the tetrahedron $\vec{Q}_k$.

The center of gravity of the k-th tetrahedron is

$$\vec{\mu}_k = \frac{\int_0^1 \int_0^{1-s} \int_0^{1-s-t} \rho(\vec{Q}_k(s,t))\vec{Q}_i(s,t)\,du\,dt\,ds}{\int_0^1 \int_0^{1-s} \int_0^{1-s-t} \rho(\vec{Q}_k(s,t))\,du\,dt\,ds},$$

where $\rho(\vec{Q}_k(s,t))$ is a mass density at a point $\vec{Q}_k(s,t)$. Since we can assume $\rho(\vec{Q}_k(s,t)) = 1$, we have

$$\vec{\mu}_k = \frac{\int_0^1 \int_0^{1-s} \int_0^{1-s-t} \vec{Q}_k(s,t)\,du\,dt\,ds}{\int_0^1 \int_0^{1-s} \int_0^{1-s-t} du\,dt\,ds} = \frac{\vec{x}_{1,k} + \vec{x}_{2,k} + \vec{x}_{3,k} + \vec{x}_{4,k}}{4}.$$

The contribution of each tetrahedron to the center of gravity of X is proportional to its volume. If M_k is the 3×3 matrix whose l-th row is $\vec{x}_{l,k} - \vec{x}_{4,k}$, for $l = 1\ldots 3$, then the volume of the k-th tetrahedron is

$$v_k = \text{volume}(Q_k) = \frac{|det(M_k)|}{3!}.$$

We introduce a weight to each tetrahedron that is proportional to its volume, define as

$$w_k = \frac{v_k}{\sum_{k=1}^n v_k} = \frac{v_k}{v},$$

where v is the volume of X. Then, the center of gravity of X is

$$\vec{\mu} = \sum_{k=1}^n w_k \vec{\mu}_k.$$

The covariance matrix of the k-th tetrahedron is

$$\Sigma_k = \frac{\int_0^1 \int_0^{1-s} \int_0^{1-s-t} (\vec{Q}_k(s,t,u) - \vec{\mu})(\vec{Q}_k(s,t,u) - \vec{\mu})^T\,du\,dt\,ds}{\int_0^1 \int_0^{1-s} \int_0^{1-s-t} du\,dt\,ds}$$

$$= \frac{1}{20}\Big(\sum_{j=1}^4 \sum_{h=1}^4 (\vec{x}_{j,k} - \vec{\mu})(\vec{x}_{h,k} - \vec{\mu})^T + \sum_{j=1}^4 (\vec{x}_{j,k} - \vec{\mu})(\vec{x}_{j,k} - \vec{\mu})^T\Big).$$

The (i,j)-th element of Σ_k, $i,j \in \{1,2,3\}$, is

$$\sigma_{ij,k} = \frac{1}{20}\Big(\sum_{l=1}^4 \sum_{h=1}^4 (x_{i,l,k} - \mu_i)(x_{j,h,k} - \mu_j) + \sum_{l=1}^4 (x_{i,l,k} - \mu_i)(x_{j,l,k} - \mu_j)\Big),$$

with $\vec{\mu} = (\mu_1, \mu_2, \mu_3)$. Finally, the covariance matrix of X is

$$\Sigma = \sum_{i=1}^n w_i \Sigma_i,$$

with (i,j)-th element

$$\sigma_{ij} = \frac{1}{20}\Big(\sum_{k=1}^n \sum_{l=1}^4 \sum_{h=1}^4 w_i(x_{i,l,k} - \mu_i)(x_{j,h,k} - \mu_j) + \sum_{k=1}^n \sum_{l=1}^4 w_i(x_{i,l,k} - \mu_i)(x_{j,l,k} - \mu_j)\Big).$$

We would like to note that the above expressions hold also for any non-convex polyhedron that can be tetrahedralized. A star-shaped object, where $\vec{o}$ is the kernel of the object, is such example.

Adding points

We add points to P, obtaining a new point set P'. Let X' be the convex hull of P'. We consider that X' is obtained from X by deleting n_d, and adding n_a tetrahedra. Let

$$v' = \sum_{k=1}^{n} v_k + \sum_{k=n+1}^{n+n_a} v_k - \sum_{k=n+n_a+1}^{n+n_a+n_d} v_k = v + \sum_{k=n+1}^{n+n_a} v_k - \sum_{k=n+n_a+1}^{n+n_a+n_d} v_k.$$

The center of gravity of X' is

$$\begin{aligned}
\vec{\mu}' &= \sum_{k=1}^{n} w'_k \vec{\mu}_k + \sum_{k=n+1}^{n+n_a} w'_k \vec{\mu}_k - \sum_{k=n+n_a+1}^{n+n_a+n_d} w'_k \vec{\mu}_k \\
&= \frac{1}{v'} \left(\sum_{k=1}^{n} v_k \vec{\mu}_k + \sum_{k=n+1}^{n+n_a} v_k \vec{\mu}_k - \sum_{k=n+n_a+1}^{n+n_a+n_d} v_k \vec{\mu}_k \right) \\
&= \frac{1}{v'} \left(v\vec{\mu} + \sum_{k=n+1}^{n+n_a} v_k \vec{\mu}_k - \sum_{k=n+n_a+1}^{n+n_a+n_d} v_k \vec{\mu}_k \right).
\end{aligned} \tag{15}$$

Let

$$\vec{\mu}_a = \frac{1}{v'} \sum_{k=n+1}^{n+n_a} v_k \vec{\mu}_k, \quad \text{and} \quad \vec{\mu}_d = \frac{1}{v'} \sum_{k=n+n_a+1}^{n+n_a+n_d} v_k \vec{\mu}_k.$$

Then, we can rewrite (15) as

$$\vec{\mu}' = \frac{v}{v'} \vec{\mu} + \vec{\mu}_a - \vec{\mu}_d. \tag{16}$$

The i-th component of $\vec{\mu}_a$ and $\vec{\mu}_d$, $1 \le i \le 3$, is denoted by $\mu_{i,a}$ and $\mu_{i,d}$, respectively. The (i,j)-th component, σ'_{ij}, $1 \le i,j \le 3$, of the covariance matrix Σ' of X' is

$$\begin{aligned}
\sigma'_{ij} &= \frac{1}{20} \left(\sum_{k=1}^{n} \sum_{l=1}^{4} \sum_{h=1}^{4} w'_k (x_{i,l,k} - \mu'_i)(x_{j,h,k} - \mu'_j) + \sum_{k=1}^{n} \sum_{l=1}^{4} w'_k (x_{i,l,k} - \mu'_i)(x_{j,l,k} - \mu'_j) \right) + \\
&\quad \frac{1}{20} \left(\sum_{k=n+1}^{n+n_a} \sum_{l=1}^{4} \sum_{h=1}^{4} w'_k (x_{i,l,k} - \mu'_i)(x_{j,h,k} - \mu'_j) + \sum_{k=n+1}^{n+n_a} \sum_{l=1}^{4} w'_k (x_{i,l,k} - \mu'_i)(x_{j,l,k} - \mu'_j) - \right. \\
&\quad \sum_{k=n+n_a+1}^{n+n_a+n_d} \sum_{l=1}^{4} \sum_{h=1}^{4} w'_k (x_{i,l,k} - \mu'_i)(x_{j,h,k} - \mu'_j) - \\
&\quad \left. \sum_{k=n+n_a+1}^{n+n_a+n_d} \sum_{l=1}^{4} w'_k (x_{i,l,k} - \mu'_i)(x_{j,l,k} - \mu'_j) \right).
\end{aligned}$$

Let

$$\sigma'_{ij} = \frac{1}{20} (\sigma'_{ij,11} + \sigma'_{ij,12} + \sigma'_{ij,21} + \sigma'_{ij,22} - \sigma'_{ij,31} - \sigma'_{ij,32}),$$

where

$$\sigma'_{ij,11} = \sum_{k=1}^{n} \sum_{l=1}^{4} \sum_{h=1}^{4} w'_k (x_{i,l,k} - \mu'_i)(x_{j,h,k} - \mu'_j), \tag{17}$$

$$\sigma'_{ij,12} = \sum_{k=1}^{n}\sum_{l=1}^{4} w'_k(x_{i,l,k} - \mu'_i)(x_{j,l,k} - \mu'_j), \tag{18}$$

$$\sigma'_{ij,21} = \sum_{k=n+1}^{n+n_a}\sum_{l=1}^{4}\sum_{h=1}^{4} w'_k(x_{i,l,k} - \mu'_i)(x_{j,h,k} - \mu'_j), \tag{19}$$

$$\sigma'_{ij,22} = \sum_{k=n+1}^{n+n_a}\sum_{l=1}^{4} w'_k(x_{i,l,k} - \mu'_i)(x_{j,l,k} - \mu'_j), \tag{20}$$

$$\sigma'_{ij,31} = \sum_{k=n+n_a+1}^{n+n_a+n_d}\sum_{l=1}^{4}\sum_{h=1}^{4} w'_k(x_{i,l,k} - \mu'_i)(x_{j,h,k} - \mu'_j), \tag{21}$$

$$\sigma'_{ij,32} = \sum_{k=n+n_a+1}^{n+n_a+n_d}\sum_{l=1}^{4} w'_k(x_{i,l,k} - \mu'_i)(x_{j,l,k} - \mu'_j). \tag{22}$$

Plugging-in the values of μ'_i and μ'_j in (17), we obtain:

$$
\begin{aligned}
\sigma'_{ij,11} &= \sum_{k=1}^{n}\sum_{l=1}^{4}\sum_{h=1}^{4} w'_k\left(x_{i,l,k} - \frac{v}{v'}\mu_i - \mu_{i,a} + \mu_{i,d}\right)\left(x_{j,h,k} - \frac{v}{v'}\mu_j - \mu_{j,a} + \mu_{j,d}\right) \\
&= \sum_{k=1}^{n}\sum_{l=1}^{4}\sum_{h=1}^{4} w'_k\left(x_{i,l,k} - \mu_i + \mu_i(1 - \frac{v}{v'}) - \mu_{i,a} + \mu_{i,d}\right) \cdot \\
&\qquad\qquad\qquad\qquad \left(x_{j,h,k} - \mu_j + \mu_j(1 - \frac{v}{v'}) - \mu_{j,a} + \mu_{j,d}\right) \\
&= \sum_{k=1}^{n}\sum_{l=1}^{4}\sum_{h=1}^{4} w'_k(x_{i,l,k} - \mu_i)(x_{j,h,k} - \mu_j) + \\
&\quad \sum_{k=1}^{n}\sum_{l=1}^{4}\sum_{h=1}^{4} w'_k(x_{i,l,k} - \mu_i)\left(\mu_j(1 - \frac{v}{v'}) - \mu_{j,a} + \mu_{j,d}\right) + \\
&\quad \sum_{k=1}^{n}\sum_{l=1}^{4}\sum_{h=1}^{4} w'_k\left(\mu_i(1 - \frac{v}{v'}) - \mu_{i,a} + \mu_{i,d}\right)(x_{j,h,k} - \mu_j) + \\
&\quad \sum_{k=1}^{n}\sum_{l=1}^{4}\sum_{h=1}^{4} w'_k\left(\mu_i(1 - \frac{v}{v'}) - \mu_{i,a} + \mu_{i,d}\right)\left(\mu_j(1 - \frac{v}{v'}) - \mu_{j,a} + \mu_{j,d}\right). \tag{23}
\end{aligned}
$$

Since $\sum_{k=1}^{n}\sum_{l=1}^{4} w'_k(x_{i,l,k} - \mu_i) = 0, 1 \le i \le 3$, we have

$$
\begin{aligned}
\sigma'_{ij,11} &= \frac{1}{v'}\sum_{k=1}^{n}\sum_{l=1}^{4}\sum_{h=1}^{4} v_k(x_{i,l,k} - \mu_i)(x_{j,h,k} - \mu_j) + \\
&\quad \frac{1}{v'}\sum_{k=1}^{n}\sum_{l=1}^{4}\sum_{h=1}^{4} v_k\left(\mu_i(1 - \frac{v}{v'}) - \mu_{i,a} + \mu_{i,d}\right)\left(\mu_j(1 - \frac{v}{v'}) - \mu_{j,a} + \mu_{j,d}\right) \\
&= \frac{1}{v'}\sum_{k=1}^{n}\sum_{l=1}^{4}\sum_{h=1}^{4} v_k(x_{i,l,k} - \mu_i)(x_{j,h,k} - \mu_j) + \\
&\quad 16\frac{v}{v'}\left(\mu_i(1 - \frac{v}{v'}) - \mu_{i,a} + \mu_{i,d}\right)\left(\mu_j(1 - \frac{v}{v'}) - \mu_{j,a} + \mu_{j,d}\right). \tag{24}
\end{aligned}
$$

Plugging-in the values of μ_i' and μ_j' in (18), we obtain:

$$\sigma_{ij,12}' = \sum_{k=1}^{n}\sum_{l=1}^{4} w_k'(x_{i,l,k} - \frac{v}{v'}\mu_i - \mu_{i,a} + \mu_{i,d})(x_{j,h,k} - \frac{v}{v'}\mu_j - \mu_{j,a} + \mu_{j,d})$$

$$= \sum_{k=1}^{n}\sum_{l=1}^{4} w_k'(x_{i,l,k} - \mu_i + \mu_i(1 - \frac{v}{v'}) - \mu_{i,a} + \mu_{i,d}) \cdot$$

$$(x_{j,h,k} - \mu_j + \mu_j(1 - \frac{v}{v'}) - \mu_{j,a} + \mu_{j,d})$$

$$= \sum_{k=1}^{n}\sum_{l=1}^{4} w_k'(x_{i,l,k} - \mu_i)(x_{j,h,k} - \mu_j) +$$

$$\sum_{k=1}^{n}\sum_{l=1}^{4} w_k'(x_{i,l,k} - \mu_i)(\mu_j(1 - \frac{v}{v'}) - \mu_{j,a} + \mu_{j,d}) +$$

$$\sum_{k=1}^{n}\sum_{l=1}^{4} w_k'(\mu_i(1 - \frac{v}{v'}) - \mu_{i,a} + \mu_{i,d})(x_{j,h,k} - \mu_j) +$$

$$\sum_{k=1}^{n}\sum_{l=1}^{4} w_k'(\mu_i(1 - \frac{v}{v'}) - \mu_{i,a} + \mu_{i,d})(\mu_j(1 - \frac{v}{v'}) - \mu_{j,a} + \mu_{j,d}). \tag{25}$$

Since $\sum_{k=1}^{n}\sum_{l=1}^{4} w_k'(x_{i,l,k} - \mu_i) = 0, 1 \le i \le 3$, we have

$$\sigma_{ij,12}' = \frac{1}{v'}\sum_{k=1}^{n}\sum_{l=1}^{4} v_k(x_{i,l,k} - \mu_i)(x_{j,h,k} - \mu_j) +$$

$$\frac{1}{v'}\sum_{k=1}^{n}\sum_{l=1}^{4} v_k(\mu_i(1 - \frac{v}{v'}) - \mu_{i,a} + \mu_{i,d})(\mu_j(1 - \frac{v}{v'}) - \mu_{j,a} + \mu_{j,d})$$

$$= \frac{1}{v'}\sum_{k=1}^{n}\sum_{l=1}^{4} v_k(x_{i,l,k} - \mu_i)(x_{j,h,k} - \mu_j) +$$

$$4\frac{v}{v'}(\mu_i(1 - \frac{v}{v'}) - \mu_{i,a} + \mu_{i,d})(\mu_j(1 - \frac{v}{v'}) - \mu_{j,a} + \mu_{j,d}). \tag{26}$$

From (25) and (26), we obtain

$$\sigma_{ij,1}' = \sigma_{ij,11}' + \sigma_{ij,12}' = \sigma_{ij} + 20\frac{v}{v'}(\mu_i(1 - \frac{v}{v'}) - \mu_{i,a} + \mu_{i,d})(\mu_j(1 - \frac{v}{v'}) - \mu_{j,a} + \mu_{j,d}). \tag{27}$$

Note that $\sigma_{ij,1}'$ can be computed in $O(1)$ time. The components $\sigma_{ij,21}'$ and $\sigma_{ij,22}'$ can be computed in $O(n_a)$ time, while $O(n_d)$ time is needed to compute $\sigma_{ij,31}'$ and $\sigma_{ij,32}'$. Thus, $\vec{\mu}'$ and

$$\sigma_{ij}' = \frac{1}{20}(\sigma_{ij,11}' + \sigma_{ij,12}' + \sigma_{ij,21}' + \sigma_{ij,22}' + \sigma_{ij,31}' + \sigma_{ij,32}')$$

$$= \frac{1}{20}(\sigma_{ij} + \sigma_{ij,21}' + \sigma_{ij,22}' + \sigma_{ij,31}' + \sigma_{ij,32}') +$$

$$\frac{v}{v'}(\mu_i(1 - \frac{v}{v'}) - \mu_{i,a} + \mu_{i,d})(\mu_j(1 - \frac{v}{v'}) - \mu_{j,a} + \mu_{j,d}) \tag{28}$$

can be computed in $O(n_a + n_d)$ time.

Deleting points

Let the new convex hull be obtained by deleting n_d tetrahedra from and added n_a tetrahedra to the old convex hull. If the interior point $\vec{o}$ (needed for a tetrahedronization of a convex polytope), after several deletions, lies inside the new convex hull, then the same formulas and time complexity, as by adding points, follow. If $\vec{o}$ lie outside the new convex hull, then, we need to choose a new interior point $\vec{o}'$, and recompute the new tetrahedra associated with it. Thus, we need in total $O(n)$ time to update the principal components.

Under certain assumptions, we can recompute the new principal components faster:

- If we know that a certain point of the polyhedron will never be deleted, we can choose $\vec{o}$ to be that point. In that case, we also have the same closed-formed solution as for adding a point.

- Let the facets of the convex polyhedron have similar (uniformly distributed) area. We choose $\vec{o}$ to be the center of gravity of the polyhedron. Then, we can expect that after deleting a point, $\vec{o}$ will remain in the new convex hull. However, after several deletions, $\vec{o}$ could lie outside the convex hull, and then we need to recompute it and the tetrahedra associated with it.

Note that in the case when we consider boundary of a convex polyhedron (Subsection 6.1 and Subsection 6.3), we do not need an interior point $\vec{o}$ and the same time complexity holds for both adding and deleting points.

5. Conclusion

In this chapter, we have presented closed-form solutions for computing and updating the principal components of (dynamic) discrete and continuous point sets. The new principal components can be computed in constant time, when a constant number of points are added or deleted from the point set. This is a significant improvement of the commonly used approach, when the new principal components are computed from scratch, which takes linear time. The advantages of some of the theoretical results were verified and presented in the context of computing dynamic PCA bounding boxes in Dimitrov, Holst, Knauer & Kriegel (2009); Dimitrov et al. (2011).

An interesting open problem is to find a closed-form solutions for dynamical point sets different from convex polyhedra, for example, implicit surfaces or B-splines. An implementation of computing principal components in a dynamic, continuous setting could be a useful practical extension of the results presented here regarding continuous point sets. Applications of the results presented here in other fields, like computer vision or visualization, are of high interest.

6. Appendix: Computing and updating the principal components efficiently - continuous case

6.1 Continuous PCA over the boundary of a polyhedron in $\mathbb{R}^3$.

Let X be a polyhedron in $\mathbb{R}^3$. We assume that the boundary of X is triangulated (if it is not, we can triangulate it in a preprocessing), containing n triangles. The k-th triangle, with vertices $\vec{x}_{1,k}, \vec{x}_{2,k}, \vec{x}_{3,k}$, can be represented in a parametric form by $\vec{T}_k(s,t) = \vec{x}_{1,k} + s(\vec{x}_{2,k} - \vec{x}_{1,k}) +$

$t\,(\vec{x}_{3,k} - \vec{x}_{1,k})$, for $0 \leq s,t \leq 1$, and $s + t \leq 1$. For $1 \leq i \leq 3$, we denote by $x_{i,j,k}$ the i-th coordinate of the vertex $\vec{x}_j$ of the triangle $\vec{T}_k$. The center of gravity of the k-th triangle is

$$\vec{\mu}_k = \frac{\int_0^1 \int_0^{1-s} \vec{T}_i(s,t)\, dt\, ds}{\int_0^1 \int_0^{1-s} dt\, ds} = \frac{\vec{x}_{1,k} + \vec{x}_{2,k} + \vec{x}_{3,k}}{3}.$$

The contribution of each triangle to the center of gravity of the triangulated surface is proportional to its area. The area of the k-th triangle is

$$a_k = \text{area}(T_k) = \frac{|(\vec{x}_{2,k} - \vec{x}_{1,k})| \times |(\vec{x}_{3,k} - \vec{x}_{1,k})|}{2}.$$

We introduce a weight to each triangle that is proportional to its area, define as

$$w_k = \frac{a_k}{\sum_{i=1}^n a_k} = \frac{a_k}{a},$$

where a is the area of X. Then, the center of gravity of the boundary of X is

$$\vec{\mu} = \sum_{k=1}^n w_k \vec{\mu}_k.$$

The covariance matrix of the k-th triangle is

$$\Sigma_k = \frac{\int_0^1 \int_0^{1-s} (\vec{T}_k(s,t) - \vec{\mu})\,(\vec{T}_k(s,t) - \vec{\mu})^T\, dt\, ds}{\int_0^1 \int_0^{1-s} dt\, ds}$$

$$= \frac{1}{12} \left(\sum_{j=1}^3 \sum_{h=1}^3 (\vec{x}_{j,k} - \vec{\mu})(\vec{x}_{h,k} - \vec{\mu})^T + \sum_{j=1}^3 (\vec{x}_{j,k} - \vec{\mu})(\vec{x}_{j,k} - \vec{\mu})^T \right).$$

The (i,j)-th element of Σ_k, $i,j \in \{1,2,3\}$, is

$$\sigma_{ij,k} = \frac{1}{12} \left(\sum_{l=1}^3 \sum_{h=1}^3 (x_{i,l,k} - \mu_i)(x_{j,h,k} - \mu_j) + \sum_{l=1}^3 (x_{i,l,k} - \mu_i)(x_{j,l,k} - \mu_j) \right),$$

with $\vec{\mu} = (\mu_1, \mu_2, \mu_3)$. Finally, the covariance matrix of the boundary of X is

$$\Sigma = \sum_{k=1}^n w_k \Sigma_k.$$

Adding points

We add points to X. Let X' be the new convex hull. We assume that X' is obtained from X by deleting n_d, and adding n_a tetrahedra. Then the sum of the areas of all triangles is

$$a' = \sum_{k=1}^n a_k + \sum_{k=n+1}^{n+n_a} a_k - \sum_{k=n+n_a+1}^{n+n_a+n_d} a_k = a + \sum_{k=n+1}^{n+n_a} a_k - \sum_{k=n+n_a+1}^{n+n_a+n_d} a_k.$$

The center of gravity of X' is

$$\vec{\mu}' = \sum_{k=1}^{n} w_k' \vec{\mu}_k + \sum_{k=n+1}^{n+n_a} w_k' \vec{\mu}_k - \sum_{k=n+n_a+1}^{n+n_a+n_d} w_k' \vec{\mu}_k = \frac{1}{a'} \left(\sum_{k=1}^{n} a_k \vec{\mu}_k + \sum_{k=n+1}^{n+n_a} a_k \vec{\mu}_k - \sum_{k=n+n_a+1}^{n+n_a+n_d} a_k \vec{\mu}_k \right)$$

$$= \frac{1}{a'} \left(a\vec{\mu} + \sum_{k=n+1}^{n+n_a} a_k \vec{\mu}_k - \sum_{k=n+n_a+1}^{n+n_a+n_d} a_k \vec{\mu}_k \right). \tag{29}$$

Let

$$\vec{\mu}_a = \frac{1}{a'} \sum_{k=n+1}^{n+n_a} a_k \vec{\mu}_k, \quad \text{and} \quad \vec{\mu}_d = \frac{1}{a'} \sum_{k=n+n_a+1}^{n+n_a+n_d} a_k \vec{\mu}_k.$$

Then, we can rewrite (29) as

$$\vec{\mu}' = \frac{a}{a'} \vec{\mu} + \vec{\mu}_a - \vec{\mu}_d. \tag{30}$$

The i-th component of $\vec{\mu}_a$ and $\vec{\mu}_d$, $1 \leq i \leq 3$, is denoted by $\mu_{i,a}$ and $\mu_{i,d}$, respectively. The (i,j)-th component, σ'_{ij}, $1 \leq i,j \leq 3$, of the covariance matrix Σ' of X' is

$$\sigma'_{ij} = \frac{1}{12} \left(\sum_{k=1}^{n} \sum_{l=1}^{3} \sum_{h=1}^{3} w_k'(x_{i,l,k} - \mu_i')(x_{j,h,k} - \mu_j') + \sum_{k=1}^{n} \sum_{l=1}^{3} w_k'(x_{i,l,k} - \mu_i')(x_{j,l,k} - \mu_j') \right) +$$

$$\frac{1}{12} \left(\sum_{k=n+1}^{n+n_a} \sum_{l=1}^{3} \sum_{h=1}^{3} w_k'(x_{i,l,k} - \mu_i')(x_{j,h,k} - \mu_j') + \sum_{k=n+1}^{n+n_a} \sum_{l=1}^{3} w_k'(x_{i,l,k} - \mu_i')(x_{j,l,k} - \mu_j') - \right.$$

$$\sum_{k=n+n_a+1}^{n+n_a+n_d} \sum_{l=1}^{3} \sum_{h=1}^{3} w_k'(x_{i,l,k} - \mu_i')(x_{j,h,k} - \mu_j') -$$

$$\left. \sum_{k=n+n_a+1}^{n+n_a+n_d} \sum_{l=1}^{3} w_k'(x_{i,l,k} - \mu_i')(x_{j,l,k} - \mu_j') \right).$$

Let

$$\sigma'_{ij} = \frac{1}{12} (\sigma'_{ij,11} + \sigma'_{ij,12} + \sigma'_{ij,21} + \sigma'_{ij,22} - \sigma'_{ij,31} - \sigma'_{ij,32}),$$

where

$$\sigma'_{ij,11} = \sum_{k=1}^{n} \sum_{l=1}^{3} \sum_{h=1}^{3} w_k'(x_{i,l,k} - \mu_i')(x_{j,h,k} - \mu_j'), \tag{31}$$

$$\sigma'_{ij,12} = \sum_{k=1}^{n} \sum_{l=1}^{3} w_k'(x_{i,l,k} - \mu_i')(x_{j,l,k} - \mu_j'), \tag{32}$$

$$1.5mm\sigma'_{ij,21} = \sum_{k=n+1}^{n+n_a} \sum_{l=1}^{3} \sum_{h=1}^{3} w_k'(x_{i,l,k} - \mu_i')(x_{j,h,k} - \mu_j'), \tag{33}$$

$$1.5mm\sigma'_{ij,22} = \sum_{k=n+1}^{n+n_a} \sum_{l=1}^{3} w_k'(x_{i,l,k} - \mu_i')(x_{j,l,k} - \mu_j'), \tag{34}$$

$$\sigma'_{ij,31} = \sum_{k=n+n_a+1}^{n+n_a+n_d} \sum_{l=1}^{3} \sum_{h=1}^{3} w'_k (x_{i,l,k} - \mu'_i)(x_{j,h,k} - \mu'_j), \tag{35}$$

$$\sigma'_{ij,32} = \sum_{k=n+n_a+1}^{n+n_a+n_d} \sum_{l=1}^{3} w'_k (x_{i,l,k} - \mu'_i)(x_{j,l,k} - \mu'_j). \tag{36}$$

Plugging-in the values of μ'_i and μ'_j in (31), we obtain:

$$\sigma'_{ij,11} = \sum_{k=1}^{n} \sum_{l=1}^{3} \sum_{h=1}^{3} w'_k \left(x_{i,l,k} - \frac{a}{a'}\mu_i - \mu_{i,a} + \mu_{i,d}\right)\left(x_{j,h,k} - \frac{a}{a'}\mu_j - \mu_{j,a} + \mu_{j,d}\right)$$

$$= \sum_{k=1}^{n} \sum_{l=1}^{3} \sum_{h=1}^{3} w'_k \left(x_{i,l,k} - \mu_i + \mu_i\left(1 - \frac{a}{a'}\right) - \mu_{i,a} + \mu_{i,d}\right) \cdot$$

$$\left(x_{j,h,k} - \mu_j + \mu_j\left(1 - \frac{a}{a'}\right) - \mu_{j,a} + \mu_{j,d}\right)$$

$$= \sum_{k=1}^{n} \sum_{l=1}^{3} \sum_{h=1}^{3} w'_k (x_{i,l,k} - \mu_i)(x_{j,h,k} - \mu_j) +$$

$$\sum_{k=1}^{n} \sum_{l=1}^{3} \sum_{h=1}^{3} w'_k (x_{i,l,k} - \mu_i)\left(\mu_j\left(1 - \frac{a}{a'}\right) - \mu_{j,a} + \mu_{j,d}\right) +$$

$$\sum_{k=1}^{n} \sum_{l=1}^{3} \sum_{h=1}^{3} w'_k \left(\mu_i\left(1 - \frac{a}{a'}\right) - \mu_{i,a} + \mu_{i,d}\right)(x_{j,h,k} - \mu_j) +$$

$$\sum_{k=1}^{n} \sum_{l=1}^{3} \sum_{h=1}^{3} w'_k \left(\mu_i\left(1 - \frac{a}{a'}\right) - \mu_{i,a} + \mu_{i,d}\right)\left(\mu_j\left(1 - \frac{a}{a'}\right) - \mu_{j,a} + \mu_{j,d}\right). \tag{37}$$

Since $\sum_{k=1}^{n} \sum_{l=1}^{3} w'_k (x_{i,l,k} - \mu_i) = 0, 1 \le i \le 3$, we have

$$\sigma'_{ij,11} = \frac{1}{a'} \sum_{k=1}^{n} \sum_{l=1}^{3} \sum_{h=1}^{3} a_k (x_{i,l,k} - \mu_i)(x_{j,h,k} - \mu_j) +$$

$$\frac{1}{a'} \sum_{k=1}^{n} \sum_{l=1}^{3} \sum_{h=1}^{3} a_k \left(\mu_i\left(1 - \frac{a}{a'}\right) - \mu_{i,a} + \mu_{i,d}\right)\left(\mu_j\left(1 - \frac{a}{a'}\right) - \mu_{j,a} + \mu_{j,d}\right)$$

$$= \frac{1}{a'} \sum_{k=1}^{n} \sum_{l=1}^{3} \sum_{h=1}^{3} a_k (x_{i,l,k} - \mu_i)(x_{j,h,k} - \mu_j) +$$

$$9\frac{a}{a'}\left(\mu_i\left(1 - \frac{a}{a'}\right) - \mu_{i,a} + \mu_{i,d}\right)\left(\mu_j\left(1 - \frac{a}{a'}\right) - \mu_{j,a} + \mu_{j,d}\right). \tag{38}$$

Plugging-in the values of μ'_i and μ'_j in (32), we obtain:

$$\sigma'_{ij,12} = \sum_{k=1}^{n}\sum_{l=1}^{3} w'_k(x_{i,l,k} - \frac{a}{a'}\mu_i - \mu_{i,a} + \mu_{i,d})(x_{j,h,k} - \frac{a}{a'}\mu_j - \mu_{j,a} + \mu_{j,d})$$

$$= \sum_{k=1}^{n}\sum_{l=1}^{3} w'_k(x_{i,l,k} - \mu_i + \mu_i(1 - \frac{a}{a'}) - \mu_{i,a} + \mu_{i,d}) \cdot$$

$$(x_{j,h,k} - \mu_j + \mu_j(1 - \frac{a}{a'}) - \mu_{j,a} + \mu_{j,d})$$

$$= \sum_{k=1}^{n}\sum_{l=1}^{3} w'_k(x_{i,l,k} - \mu_i)(x_{j,h,k} - \mu_j) +$$

$$\sum_{k=1}^{n}\sum_{l=1}^{3} w'_k(x_{i,l,k} - \mu_i)(\mu_j(1 - \frac{a}{a'}) - \mu_{j,a} + \mu_{j,d}) +$$

$$\sum_{k=1}^{n}\sum_{l=1}^{3} w'_k(\mu_i(1 - \frac{a}{a'}) - \mu_{i,a} + \mu_{i,d})(x_{j,h,k} - \mu_j) +$$

$$\sum_{k=1}^{n}\sum_{l=1}^{3} w'_k(\mu_i(1 - \frac{a}{a'}) - \mu_{i,a} + \mu_{i,d})(\mu_j(1 - \frac{a}{a'}) - \mu_{j,a} + \mu_{j,d}). \qquad (39)$$

Since $\sum_{k=1}^{n}\sum_{l=1}^{3} w'_k(x_{i,l,k} - \mu_i) = 0, 1 \leq i \leq 3$, we have

$$\sigma'_{ij,12} = \frac{1}{a'}\sum_{k=1}^{n}\sum_{l=1}^{3} a_k(x_{i,l,k} - \mu_i)(x_{j,h,k} - \mu_j) +$$

$$\frac{1}{a'}\sum_{k=1}^{n}\sum_{l=1}^{3} a_k(\mu_i(1 - \frac{a}{a'}) - \mu_{i,a} + \mu_{i,d})(\mu_j(1 - \frac{a}{a'}) - \mu_{j,a} + \mu_{j,d})$$

$$= \frac{1}{a'}\sum_{k=1}^{n}\sum_{l=1}^{3} a_k(x_{i,l,k} - \mu_i)(x_{j,h,k} - \mu_j) +$$

$$3\frac{a}{a'}(\mu_i(1 - \frac{a}{a'}) - \mu_{i,a} + \mu_{i,d})(\mu_j(1 - \frac{a}{a'}) - \mu_{j,a} + \mu_{j,d}). \qquad (40)$$

From (39) and (40), we obtain

$$\sigma'_{ij,1} = \sigma'_{ij,11} + \sigma'_{ij,12}$$

$$= \sigma_{ij} + 12\frac{a}{a'}(\mu_i(1 - \frac{a}{a'}) - \mu_{i,a} + \mu_{i,d})(\mu_j(1 - \frac{a}{a'}) - \mu_{j,a} + \mu_{j,d}). \qquad (41)$$

Note that $\sigma'_{ij,1}$ can be computed in $O(1)$ time. The components $\sigma'_{ij,21}$ and $\sigma'_{ij,22}$ can be computed in $O(n_a)$ time, while $O(n_d)$ time is needed to compute $\sigma'_{ij,31}$ and $\sigma'_{ij,32}$. Thus, $\vec{\mu}'$ and

$$\sigma'_{ij} = \frac{1}{12}(\sigma'_{ij,11} + \sigma'_{ij,12} + \sigma'_{ij,21} + \sigma'_{ij,22} + \sigma'_{ij,31} + \sigma'_{ij,32}) = \frac{1}{12}(\sigma_{ij} + \sigma'_{ij,21} + \sigma'_{ij,22} + \sigma'_{ij,31} + \sigma_{ij,32}) +$$

$$\frac{a}{a'}(\mu_i(1 - \frac{a}{a'}) - \mu_{i,a} + \mu_{i,d})(\mu_j(1 - \frac{a}{a'}) - \mu_{j,a} + \mu_{j,d}). \qquad (42)$$

can be computed in $O(n_a + n_d)$ time.

Deleting points

Let the new convex hull be obtained by deleting n_d tetrahedra from and added n_a tetrahedra to the old convex hull. Consequently, the same formulas and time complexity, as by adding points, follow.

6.2 Continuous PCA over a polygon in $\mathbb{R}^2$

We assume that the polygon X is triangulated (if it is not, we can triangulate it in a preprocessing), and the number of triangles is n. The k-th triangle, with vertices $\vec{x}_{1,k}, \vec{x}_{2,k}, \vec{x}_{3,k} = \vec{0}$, can be represented in a parametric form by $\vec{T}_i(s,t) = \vec{x}_{3,k} + s(\vec{x}_{1,k} - \vec{x}_{3,k}) + t(\vec{x}_{2,k} - \vec{x}_{3,k})$, for $0 \le s, t \le 1$, and $s + t \le 1$. The center of gravity of the k-th triangle is

$$\vec{\mu}_i = \frac{\int_0^1 \int_0^{1-s} \vec{T}_i(s,t)\, dt\, ds}{\int_0^1 \int_0^{1-s} dt\, ds} = \frac{\vec{x}_{1,k} + \vec{x}_{2,k} + \vec{x}_{3,k}}{3}.$$

The contribution of each triangle to the center of gravity of X is proportional to its area. The area of the i-th triangle is

$$a_k = \text{area}(T_k) = \frac{|(\vec{x}_{2,k} - \vec{x}_{1,k})| \times |(\vec{x}_{3,k} - \vec{x}_{1,k})|}{2},$$

where $\times$ denotes the vector product. We introduce a weight to each triangle that is proportional to its area, define as

$$w_k = \frac{a_k}{\sum_{k=1}^n a_k} = \frac{a_k}{a},$$

where a is the area of X. Then, the center of gravity of X is

$$\vec{\mu} = \sum_{k=1}^n w_k \vec{\mu}_k.$$

The covariance matrix of the k-th triangle is

$$\Sigma_k = \frac{\int_0^1 \int_0^{1-s} (\vec{T}_k(s,t) - \vec{\mu})(\vec{T}_k(s,t) - \vec{\mu})^T\, dt\, ds}{\int_0^1 \int_0^{1-s} dt\, ds}$$

$$= \frac{1}{12}\left(\sum_{j=1}^3 \sum_{h=1}^3 (\vec{x}_{j,k} - \vec{\mu})(\vec{x}_{h,k} - \vec{\mu})^T + \sum_{j=1}^3 (\vec{x}_{j,k} - \vec{\mu})(\vec{x}_{j,k} - \vec{\mu})^T \right).$$

The (i,j)-th element of Σ_k, $i,j \in \{1,2\}$, is

$$\sigma_{ij,k} = \frac{1}{12}\left(\sum_{l=1}^3 \sum_{h=1}^3 (x_{i,l,k} - \mu_i)(x_{j,h,k} - \mu_j) + \sum_{l=1}^3 (x_{i,l,k} - \mu_i)(x_{j,l,k} - \mu_j) \right),$$

with $\vec{\mu} = (\mu_1, \mu_2)$. The covariance matrix of X is

$$\Sigma = \sum_{k=1}^n w_k \Sigma_k.$$

Adding points

We add points to X. Let X' be the new convex hull. We assume that X' is obtained from X by deleting n_d, and adding n_a triangles. Then the sum of the areas of all triangles is

$$a' = \sum_{k=1}^{n} a_k + \sum_{k=n+1}^{n+n_a} a_k - \sum_{k=n+n_a+1}^{n+n_a+n_d} a_k = a + \sum_{k=n+1}^{n+n_a} a_k - \sum_{k=n+n_a+1}^{n+n_a+n_d} a_k.$$

The center of gravity of X' is

$$\vec{\mu}' = \sum_{k=1}^{n} w_k' \vec{\mu}_k + \sum_{k=n+1}^{n+n_a} w_k' \vec{\mu}_k - \sum_{k=n+n_a+1}^{n+n_a+n_d} w_k' \vec{\mu}_k = \frac{1}{a'} \left(\sum_{k=1}^{n} a_k \vec{\mu}_k + \sum_{k=n+1}^{n+n_a} a_k \vec{\mu}_k - \sum_{k=n+n_a+1}^{n+n_a+n_d} a_k \vec{\mu}_k \right)$$

$$= \frac{1}{a'} \left(a\vec{\mu} + \sum_{k=n+1}^{n+n_a} a_k \vec{\mu}_k - \sum_{k=n+n_a+1}^{n+n_a+n_d} a_k \vec{\mu}_k \right). \tag{43}$$

Let

$$\vec{\mu}_a = \frac{1}{a'} \sum_{k=n+1}^{n+n_a} a_k \vec{\mu}_k, \quad \text{and} \quad \vec{\mu}_d = \frac{1}{a'} \sum_{k=n+n_a+1}^{n+n_a+n_d} a_k \vec{\mu}_k.$$

Then, we can rewrite (43) as

$$\vec{\mu}' = \frac{a}{a'} \vec{\mu} + \vec{\mu}_a - \vec{\mu}_d. \tag{44}$$

The i-th component of $\vec{\mu}_a$ and $\vec{\mu}_d$, $1 \le i \le 2$, is denoted by $\mu_{i,a}$ and $\mu_{i,d}$, respectively. The (i,j)-th component, σ_{ij}', $1 \le i,j \le 2$, of the covariance matrix Σ' of X' is

$$\sigma_{ij}' = \frac{1}{12} \left(\sum_{k=1}^{n} \sum_{l=1}^{3} \sum_{h=1}^{3} w_k'(x_{i,l,k} - \mu_i')(x_{j,h,k} - \mu_j') + \sum_{k=1}^{n} \sum_{l=1}^{3} w_k'(x_{i,l,k} - \mu_i')(x_{j,l,k} - \mu_j') \right) +$$

$$\frac{1}{12} \left(\sum_{k=n+1}^{n+n_a} \sum_{l=1}^{3} \sum_{h=1}^{3} w_k'(x_{i,l,k} - \mu_i')(x_{j,h,k} - \mu_j') + \sum_{k=n+1}^{n+n_a} \sum_{l=1}^{3} w_k'(x_{i,l,k} - \mu_i')(x_{j,l,k} - \mu_j') - \right.$$

$$\sum_{k=n+n_a+1}^{n+n_a+n_d} \sum_{l=1}^{3} \sum_{h=1}^{3} w_k'(x_{i,l,k} - \mu_i')(x_{j,h,k} - \mu_j') -$$

$$\left. \sum_{k=n+n_a+1}^{n+n_a+n_d} \sum_{l=1}^{3} w_k'(x_{i,l,k} - \mu_i')(x_{j,l,k} - \mu_j') \right).$$

Let

$$\sigma_{ij}' = \frac{1}{12} (\sigma_{ij,11}' + \sigma_{ij,12}' + \sigma_{ij,21}' + \sigma_{ij,22}' - \sigma_{ij,31}' - \sigma_{ij,32}'),$$

where

$$\sigma_{ij,11}' = \sum_{k=1}^{n} \sum_{l=1}^{3} \sum_{h=1}^{3} w_k'(x_{i,l,k} - \mu_i')(x_{j,h,k} - \mu_j'), \tag{45}$$

$$\sigma_{ij,12}' = \sum_{k=1}^{n} \sum_{l=1}^{3} w_k'(x_{i,l,k} - \mu_i')(x_{j,l,k} - \mu_j'), \tag{46}$$

$$\sigma_{ij,21}' = \sum_{k=n+1}^{n+n_a} \sum_{l=1}^{3} \sum_{h=1}^{3} w_k'(x_{i,l,k} - \mu_i')(x_{j,h,k} - \mu_j'), \tag{47}$$

$$\sigma'_{ij,22} = \sum_{k=n+1}^{n+n_a} \sum_{l=1}^{3} w'_k (x_{i,l,k} - \mu'_i)(x_{j,l,k} - \mu'_j), \tag{48}$$

$$\sigma'_{ij,31} = \sum_{k=n+n_a+1}^{n+n_a+n_d} \sum_{l=1}^{3} \sum_{h=1}^{3} w'_k (x_{i,l,k} - \mu'_i)(x_{j,h,k} - \mu'_j), \tag{49}$$

$$\sigma'_{ij,32} = \sum_{k=n+n_a+1}^{n+n_a+n_d} \sum_{l=1}^{3} w'_k (x_{i,l,k} - \mu'_i)(x_{j,l,k} - \mu'_j). \tag{50}$$

Plugging-in the values of μ'_i and μ'_j in (45), we obtain:

$$\sigma'_{ij,11} = \sum_{k=1}^{n} \sum_{l=1}^{3} \sum_{h=1}^{3} w'_k \left(x_{i,l,k} - \frac{a}{a'}\mu_i - \mu_{i,a} + \mu_{i,d}\right)\left(x_{j,h,k} - \frac{a}{a'}\mu_j - \mu_{j,a} + \mu_{j,d}\right)$$

$$= \sum_{k=1}^{n} \sum_{l=1}^{3} \sum_{h=1}^{3} w'_k \left(x_{i,l,k} - \mu_i + \mu_i(1 - \frac{a}{a'}) - \mu_{i,a} + \mu_{i,d}\right) \cdot$$

$$\left(x_{j,h,k} - \mu_j + \mu_j(1 - \frac{a}{a'}) - \mu_{j,a} + \mu_{j,d}\right)$$

$$= \sum_{k=1}^{n} \sum_{l=1}^{3} \sum_{h=1}^{3} w'_k (x_{i,l,k} - \mu_i)(x_{j,h,k} - \mu_j) +$$

$$\sum_{k=1}^{n} \sum_{l=1}^{3} \sum_{h=1}^{3} w'_k (x_{i,l,k} - \mu_i)(\mu_j(1 - \frac{a}{a'}) - \mu_{j,a} + \mu_{j,d}) +$$

$$\sum_{k=1}^{n} \sum_{l=1}^{3} \sum_{h=1}^{3} w'_k (\mu_i(1 - \frac{a}{a'}) - \mu_{i,a} + \mu_{i,d})(x_{j,h,k} - \mu_j) +$$

$$\sum_{k=1}^{n} \sum_{l=1}^{3} \sum_{h=1}^{3} w'_k (\mu_i(1 - \frac{a}{a'}) - \mu_{i,a} + \mu_{i,d})(\mu_j(1 - \frac{a}{a'}) - \mu_{j,a} + \mu_{j,d}). \tag{51}$$

Since $\sum_{k=1}^{n} \sum_{l=1}^{3} w'_k (x_{i,l,k} - \mu_i) = 0, 1 \leq i \leq 2$, we have

$$\sigma'_{ij,11} = \frac{1}{a'} \sum_{k=1}^{n} \sum_{l=1}^{3} \sum_{h=1}^{3} a_k (x_{i,l,k} - \mu_i)(x_{j,h,k} - \mu_j) +$$

$$\frac{1}{a'} \sum_{k=1}^{n} \sum_{l=1}^{3} \sum_{h=1}^{3} a_k (\mu_i(1 - \frac{a}{a'}) - \mu_{i,a} + \mu_{i,d})(\mu_j(1 - \frac{a}{a'}) - \mu_{j,a} + \mu_{j,d})$$

$$= \frac{1}{a'} \sum_{k=1}^{n} \sum_{l=1}^{3} \sum_{h=1}^{3} a_k (x_{i,l,k} - \mu_i)(x_{j,h,k} - \mu_j) +$$

$$9\frac{a}{a'} (\mu_i(1 - \frac{a}{a'}) - \mu_{i,a} + \mu_{i,d})(\mu_j(1 - \frac{a}{a'}) - \mu_{j,a} + \mu_{j,d}). \tag{52}$$

Plugging-in the values of μ'_i and μ'_j in (46), we obtain:

$$
\begin{aligned}
\sigma'_{ij,12} &= \sum_{k=1}^{n}\sum_{l=1}^{3} w'_k\left(x_{i,l,k} - \frac{a}{a'}\mu_i - \mu_{i,a} + \mu_{i,d}\right)\left(x_{j,h,k} - \frac{a}{a'}\mu_j - \mu_{j,a} + \mu_{j,d}\right) \\
&= \sum_{k=1}^{n}\sum_{l=1}^{3} w'_k\left(x_{i,l,k} - \mu_i + \mu_i\left(1 - \frac{a}{a'}\right) - \mu_{i,a} + \mu_{i,d}\right) \cdot \\
&\qquad\qquad \left(x_{j,h,k} - \mu_j + \mu_j\left(1 - \frac{a}{a'}\right) - \mu_{j,a} + \mu_{j,d}\right) \\
&= \sum_{k=1}^{n}\sum_{l=1}^{3} w'_k\left(x_{i,l,k} - \mu_i\right)\left(x_{j,h,k} - \mu_j\right) + \\
&\quad \sum_{k=1}^{n}\sum_{l=1}^{3} w'_k\left(x_{i,l,k} - \mu_i\right)\left(\mu_j\left(1 - \frac{a}{a'}\right) - \mu_{j,a} + \mu_{j,d}\right) + \\
&\quad \sum_{k=1}^{n}\sum_{l=1}^{3} w'_k\left(\mu_i\left(1 - \frac{a}{a'}\right) - \mu_{i,a} + \mu_{i,d}\right)\left(x_{j,h,k} - \mu_j\right) + \\
&\quad \sum_{k=1}^{n}\sum_{l=1}^{3} w'_k\left(\mu_i\left(1 - \frac{a}{a'}\right) - \mu_{i,a} + \mu_{i,d}\right)\left(\mu_j\left(1 - \frac{a}{a'}\right) - \mu_{j,a} + \mu_{j,d}\right).
\end{aligned}
\tag{53}
$$

Since $\sum_{k=1}^{n}\sum_{l=1}^{3} w'_k(x_{i,l,k} - \mu_i) = 0$, $1 \leq i \leq 2$, we have

$$
\begin{aligned}
\sigma'_{ij,12} &= \frac{1}{a'}\sum_{k=1}^{n}\sum_{l=1}^{3} a_k\left(x_{i,l,k} - \mu_i\right)\left(x_{j,h,k} - \mu_j\right) + \\
&\quad \frac{1}{a'}\sum_{k=1}^{n}\sum_{l=1}^{3} a_k\left(\mu_i\left(1 - \frac{a}{a'}\right) - \mu_{i,a} + \mu_{i,d}\right)\left(\mu_j\left(1 - \frac{a}{a'}\right) - \mu_{j,a} + \mu_{j,d}\right) \\
&= \frac{1}{a'}\sum_{k=1}^{n}\sum_{l=1}^{3} a_k\left(x_{i,l,k} - \mu_i\right)\left(x_{j,h,k} - \mu_j\right) + \\
&\quad 3\frac{a}{a'}\left(\mu_i\left(1 - \frac{a}{a'}\right) - \mu_{i,a} + \mu_{i,d}\right)\left(\mu_j\left(1 - \frac{a}{a'}\right) - \mu_{j,a} + \mu_{j,d}\right).
\end{aligned}
\tag{54}
$$

From (53) and (54), we obtain

$$
\sigma'_{ij,1} = \sigma'_{ij,11} + \sigma'_{ij,12} = \sigma_{ij} + 12\frac{a}{a'}\left(\mu_i\left(1 - \frac{a}{a'}\right) - \mu_{i,a} + \mu_{i,d}\right)\left(\mu_j\left(1 - \frac{a}{a'}\right) - \mu_{j,a} + \mu_{j,d}\right).
\tag{55}
$$

Note that $\sigma'_{ij,1}$ can be computed in $O(1)$ time. The components $\sigma'_{ij,21}$ and $\sigma'_{ij,22}$ can be computed in $O(n_a)$ time, while $O(n_d)$ time is needed to compute $\sigma'_{ij,31}$ and $\sigma'_{ij,32}$. Thus, $\vec{\mu}'$ and

$$
\begin{aligned}
\sigma'_{ij} &= \frac{1}{12}\left(\sigma'_{ij,11} + \sigma'_{ij,12} + \sigma'_{ij,21} + \sigma'_{ij,22} + \sigma'_{ij,31} + \sigma'_{ij,32}\right) \\
&= \frac{1}{12}\left(\sigma_{ij} + \sigma'_{ij,21} + \sigma'_{ij,22} + \sigma'_{ij,31} + \sigma_{ij,32}\right) + \\
&\quad \frac{a}{a'}\left(\mu_i\left(1 - \frac{a}{a'}\right) - \mu_{i,a} + \mu_{i,d}\right)\left(\mu_j\left(1 - \frac{a}{a'}\right) - \mu_{j,a} + \mu_{j,d}\right).
\end{aligned}
\tag{56}
$$

can be computed in $O(n_a + n_d)$ time.

Deleting points

Let the new convex hull be obtained by deleting n_d tetrahedra from and added n_a tetrahedra to the old convex hull. If the interior point $\vec{o}$, after deleting points, lies inside the new convex hull, then the same formulas and time complexity, as by adding points, follow. However, $\vec{o}$ could lie outside the new convex hull. Then, we need to choose a new interior point $\vec{o}'$, and recompute the new tetrahedra associated with it. Thus, we need in total $O(n)$ time to update the principal components.

6.3 Continuous PCA over the boundary of a polygon $\mathbb{R}^2$

Let X be a polygon in $\mathbb{R}^2$. We assume that the boundary of X is comprised of n line segments. The k-th line segment, with vertices $\vec{x}_{1,k}, \vec{x}_{2,k}$, can be represented in a parametric form by

$$\vec{S}_k(t) = \vec{x}_{1,k} + t\,(\vec{x}_{2,k} - \vec{x}_{1,k}).$$

Since we assume that the mass density is constant, the center of gravity of the k-th line segment is

$$\vec{\mu}_k = \frac{\int_0^1 \vec{S}_k(t)\,dt}{\int_0^1 dt} = \frac{\vec{x}_{1,k} + \vec{x}_{2,k}}{2}.$$

The contribution of each line segment to the center of gravity of the boundary of a polygon is proportional to the length of the line segment. The length of the k-th line segment is

$$s_k = \text{length}(S_k) = ||\vec{x}_{2,k} - \vec{x}_{1,k}||.$$

We introduce a weight to each line segment that is proportional to its length, define as

$$w_k = \frac{s_k}{\sum_{k=1}^n s_k} = \frac{s_k}{s},$$

where s is the perimeter of X. Then, the center of gravity of the boundary of X is

$$\vec{\mu} = \sum_{k=1}^n w_k \vec{\mu}_k.$$

The covariance matrix of the k-th line segment is

$$
\Sigma_k = \frac{\int_0^1 (\vec{S}_k(t) - \vec{\mu})\,(\vec{S}_k(t) - \vec{\mu})^T\,dt}{\int_0^1 dt}
$$

$$
= \frac{1}{6}\left(\sum_{j=1}^2 \sum_{h=1}^2 (\vec{x}_{j,k} - \vec{\mu})(\vec{x}_{h,k} - \vec{\mu})^T + \sum_{j=1}^2 (\vec{x}_{j,k} - \vec{\mu})(\vec{x}_{j,k} - \vec{\mu})^T \right).
$$

The (i,j)-th element of Σ_k, $i,j \in \{1,2\}$, is

$$
\sigma_{ij,k} = \frac{1}{6}\left(\sum_{l=1}^2 \sum_{h=1}^2 (x_{i,l,k} - \mu_i)(x_{j,h,k} - \mu_j) + \sum_{l=1}^2 (x_{i,l,k} - \mu_i)(x_{j,l,k} - \mu_j) \right),
$$

with $\vec{\mu} = (\mu_1, \mu_2)$.

The covariance matrix of the boundary of X is

$$\Sigma = \sum_{k=1}^{n} w_k \Sigma_k.$$

Adding points

We add points to X. Let X' be the new convex hull. We assume that X' is obtained from X by deleting n_d, and adding n_a line segments. Then the sum of the lengths of all line segments is

$$s' = \sum_{k=1}^{n} l_k + \sum_{k=n+1}^{n+n_a} s_k - \sum_{k=n+n_a+1}^{n+n_a+n_d} s_k = s + \sum_{k=n+1}^{n+n_a} s_k - \sum_{k=n+n_a+1}^{n+n_a+n_d} s_k.$$

The center of gravity of X' is

$$\vec{\mu}' = \sum_{k=1}^{n} w'_k \vec{\mu}_k + \sum_{k=n+1}^{n+n_a} w'_k \vec{\mu}_k - \sum_{k=n+n_a+1}^{n+n_a+n_d} w'_k \vec{\mu}_k$$

$$= \frac{1}{s'} \left(\sum_{k=1}^{n} s_k \vec{\mu}_k + \sum_{k=n+1}^{n+n_a} s_k \vec{\mu}_k - \sum_{k=n+n_a+1}^{n+n_a+n_d} s_k \vec{\mu}_k \right)$$

$$= \frac{1}{s'} \left(s\vec{\mu} + \sum_{k=n+1}^{n+n_a} s_k \vec{\mu}_k - \sum_{k=n+n_a+1}^{n+n_a+n_d} s_k \vec{\mu}_k \right). \tag{57}$$

Let

$$\vec{\mu}_a = \frac{1}{s'} \sum_{k=n+1}^{n+n_a} s_k \vec{\mu}_k, \quad \text{and} \quad \vec{\mu}_d = \frac{1}{s'} \sum_{k=n+n_a+1}^{n+n_a+n_d} s_k \vec{\mu}_k.$$

Then, we can rewrite (57) as

$$\vec{\mu}' = \frac{s}{s'} \vec{\mu} + \vec{\mu}_a - \vec{\mu}_d. \tag{58}$$

The i-th component of $\vec{\mu}_a$ and $\vec{\mu}_d$, $1 \leq i \leq 2$, is denoted by $\mu_{i,a}$ and $\mu_{i,d}$, respectively. The (i,j)-th component, σ'_{ij}, $1 \leq i,j \leq 2$, of the covariance matrix Σ' of X' is

$$\sigma'_{ij} = \frac{1}{6} \left(\sum_{k=1}^{n} \sum_{l=1}^{2} \sum_{h=1}^{2} w'_k (x_{i,l,k} - \mu'_i)(x_{j,h,k} - \mu'_j) + \sum_{k=1}^{n} \sum_{l=1}^{2} w'_k (x_{i,l,k} - \mu'_i)(x_{j,l,k} - \mu'_j) \right) +$$

$$\frac{1}{6} \left(\sum_{k=n+1}^{n+n_a} \sum_{l=1}^{2} \sum_{h=1}^{2} w'_k (x_{i,l,k} - \mu'_i)(x_{j,h,k} - \mu'_j) + \sum_{k=n+1}^{n+n_a} \sum_{l=1}^{2} w'_k (x_{i,l,k} - \mu'_i)(x_{j,l,k} - \mu'_j) - \right.$$

$$\sum_{k=n+n_a+1}^{n+n_a+n_d} \sum_{l=1}^{2} \sum_{h=1}^{2} w'_k (x_{i,l,k} - \mu'_i)(x_{j,h,k} - \mu'_j) -$$

$$\left. \sum_{k=n+n_a+1}^{n+n_a+n_d} \sum_{l=1}^{2} w'_k (x_{i,l,k} - \mu'_i)(x_{j,l,k} - \mu'_j) \right).$$

Let

$$\sigma'_{ij} = \frac{1}{6}(\sigma'_{ij,11} + \sigma'_{ij,12} + \sigma'_{ij,21} + \sigma'_{ij,22} - \sigma'_{ij,31} - \sigma'_{ij,32}),$$

where

$$\sigma'_{ij,11} = \sum_{k=1}^{n}\sum_{l=1}^{2}\sum_{h=1}^{2} w'_k(x_{i,l,k} - \mu'_i)(x_{j,h,k} - \mu'_j), \tag{59}$$

$$\sigma'_{ij,12} = \sum_{k=1}^{n}\sum_{l=1}^{2} w'_k(x_{i,l,k} - \mu'_i)(x_{j,l,k} - \mu'_j), \tag{60}$$

$$\sigma'_{ij,21} = \sum_{k=n+1}^{n+n_a}\sum_{l=1}^{2}\sum_{h=1}^{2} w'_k(x_{i,l,k} - \mu'_i)(x_{j,h,k} - \mu'_j), \tag{61}$$

$$\sigma'_{ij,22} = \sum_{k=n+1}^{n+n_a}\sum_{l=1}^{2} w'_k(x_{i,l,k} - \mu'_i)(x_{j,l,k} - \mu'_j), \tag{62}$$

$$\sigma'_{ij,31} = \sum_{k=n+n_a+1}^{n+n_a+n_d}\sum_{l=1}^{2}\sum_{h=1}^{2} w'_k(x_{i,l,k} - \mu'_i)(x_{j,h,k} - \mu'_j), \tag{63}$$

$$\sigma'_{ij,32} = \sum_{k=n+n_a+1}^{n+n_a+n_d}\sum_{l=1}^{3} w'_k(x_{i,l,k} - \mu'_i)(x_{j,l,k} - \mu'_j). \tag{64}$$

Plugging-in the values of μ'_i and μ'_j in (59), we obtain:

$$\sigma'_{ij,11} = \sum_{k=1}^{n}\sum_{l=1}^{2}\sum_{h=1}^{2} w'_k(x_{i,l,k} - \frac{s}{s'}\mu_i - \mu_{i,a} + \mu_{i,d})(x_{j,h,k} - \frac{s}{s'}\mu_j - \mu_{j,a} + \mu_{j,d})$$

$$= \sum_{k=1}^{n}\sum_{l=1}^{2}\sum_{h=1}^{2} w'_k(x_{i,l,k} - \mu_i + \mu_i(1 - \frac{s}{s'}) - \mu_{i,a} + \mu_{i,d}) \cdot$$

$$(x_{j,h,k} - \mu_j + \mu_j(1 - \frac{s}{s'}) - \mu_{j,a} + \mu_{j,d})$$

$$= \sum_{k=1}^{n}\sum_{l=1}^{2}\sum_{h=1}^{2} w'_k(x_{i,l,k} - \mu_i)(x_{j,h,k} - \mu_j) +$$

$$\sum_{k=1}^{n}\sum_{l=1}^{2}\sum_{h=1}^{2} w'_k(x_{i,l,k} - \mu_i)(\mu_j(1 - \frac{s}{s'}) - \mu_{j,a} + \mu_{j,d}) +$$

$$\sum_{k=1}^{n}\sum_{l=1}^{2}\sum_{h=1}^{2} w'_k(\mu_i(1 - \frac{s}{s'}) - \mu_{i,a} + \mu_{i,d})(x_{j,h,k} - \mu_j) +$$

$$\sum_{k=1}^{n}\sum_{l=1}^{2}\sum_{h=1}^{2} w'_k(\mu_i(1 - \frac{s}{s'}) - \mu_{i,a} + \mu_{i,d})(\mu_j(1 - \frac{s}{s'}) - \mu_{j,a} + \mu_{j,d}). \tag{65}$$

Since $\sum_{k=1}^{n}\sum_{l=1}^{2} w_k'(x_{i,l,k} - \mu_i) = 0, 1 \leq i \leq 2$, we have

$$\sigma_{ij,11}' = \frac{1}{s'}\sum_{k=1}^{n}\sum_{l=1}^{2}\sum_{h=1}^{2} s_k(x_{i,l,k} - \mu_i)(x_{j,h,k} - \mu_j) +$$

$$\frac{1}{s'}\sum_{k=1}^{n}\sum_{l=1}^{2}\sum_{h=1}^{2} s_k(\mu_i(1 - \frac{s}{s'}) - \mu_{i,a} + \mu_{i,d})(\mu_j(1 - \frac{s}{s'}) - \mu_{j,a} + \mu_{j,d})$$

$$= \frac{1}{s'}\sum_{k=1}^{n}\sum_{l=1}^{2}\sum_{h=1}^{2} s_k(x_{i,l,k} - \mu_i)(x_{j,h,k} - \mu_j) +$$

$$4\frac{s}{s'}(\mu_i(1 - \frac{s}{s'}) - \mu_{i,a} + \mu_{i,d})(\mu_j(1 - \frac{s}{s'}) - \mu_{j,a} + \mu_{j,d}). \tag{66}$$

Plugging-in the values of μ_i' and μ_j' in (60), we obtain:

$$\sigma_{ij,12}' = \sum_{k=1}^{n}\sum_{l=1}^{2} w_k'(x_{i,l,k} - \frac{s}{s'}\mu_i - \mu_{i,a} + \mu_{i,d})(x_{j,h,k} - \frac{s}{s'}\mu_j - \mu_{j,a} + \mu_{j,d})$$

$$= \sum_{k=1}^{n}\sum_{l=1}^{2} w_k'(x_{i,l,k} - \mu_i + \mu_i(1 - \frac{s}{s'}) - \mu_{i,a} + \mu_{i,d}) \cdot$$

$$(x_{j,h,k} - \mu_j + \mu_j(1 - \frac{s}{s'}) - \mu_{j,a} + \mu_{j,d})$$

$$= \sum_{k=1}^{n}\sum_{l=1}^{2} w_k'(x_{i,l,k} - \mu_i)(x_{j,h,k} - \mu_j) +$$

$$\sum_{k=1}^{n}\sum_{l=1}^{2} w_k'(x_{i,l,k} - \mu_i)(\mu_j(1 - \frac{s}{s'}) - \mu_{j,a} + \mu_{j,d}) +$$

$$\sum_{k=1}^{n}\sum_{l=1}^{2} w_k'(\mu_i(1 - \frac{s}{s'}) - \mu_{i,a} + \mu_{i,d})(x_{j,h,k} - \mu_j) +$$

$$\sum_{k=1}^{n}\sum_{l=1}^{2} w_k'(\mu_i(1 - \frac{s}{s'}) - \mu_{i,a} + \mu_{i,d})(\mu_j(1 - \frac{s}{s'}) - \mu_{j,a} + \mu_{j,d}). \tag{67}$$

Since $\sum_{k=1}^{n}\sum_{l=1}^{2} w_k'(x_{i,l,k} - \mu_i) = 0, 1 \leq i \leq 2$, we have

$$\sigma_{ij,12}' = \frac{1}{s'}\sum_{k=1}^{n}\sum_{l=1}^{2} s_k(x_{i,l,k} - \mu_i)(x_{j,h,k} - \mu_j) +$$

$$\frac{1}{s'}\sum_{k=1}^{n}\sum_{l=1}^{2} s_k(\mu_i(1 - \frac{s}{s'}) - \mu_{i,a} + \mu_{i,d})(\mu_j(1 - \frac{s}{s'}) - \mu_{j,a} + \mu_{j,d})$$

$$= \frac{1}{s'}\sum_{k=1}^{n}\sum_{l=1}^{2} s_k(x_{i,l,k} - \mu_i)(x_{j,h,k} - \mu_j) +$$

$$2\frac{s}{s'}(\mu_i(1 - \frac{s}{s'}) - \mu_{i,a} + \mu_{i,d})(\mu_j(1 - \frac{s}{s'}) - \mu_{j,a} + \mu_{j,d}). \tag{68}$$

From (67) and (68), we obtain

$$\sigma'_{ij,1} = \sigma'_{ij,11} + \sigma'_{ij,12}$$

$$= \sigma_{ij} + 6\frac{s}{s'}\left(\mu_i(1 - \frac{s}{s'}) - \mu_{i,a} + \mu_{i,d}\right)\left(\mu_j(1 - \frac{s}{s'}) - \mu_{j,a} + \mu_{j,d}\right). \tag{69}$$

Note that $\sigma'_{ij,1}$ can be computed in $O(1)$ time. The components $\sigma'_{ij,21}$ and $\sigma'_{ij,22}$ can be computed in $O(n_a)$ time, while $O(n_d)$ time is needed to compute $\sigma'_{ij,31}$ and $\sigma'_{ij,32}$. Thus, $\vec{\mu}'$ and

$$\sigma'_{ij} = \frac{1}{6}(\sigma'_{ij,11} + \sigma'_{ij,12} + \sigma'_{ij,21} + \sigma'_{ij,22} + \sigma'_{ij,31} + \sigma'_{ij,32})$$

$$= \frac{1}{6}(\sigma_{ij} + \sigma'_{ij,21} + \sigma'_{ij,22} + \sigma'_{ij,31} + \sigma_{ij,32}) +$$

$$\frac{s}{s'}\left(\mu_i(1 - \frac{s}{s'}) - \mu_{i,a} + \mu_{i,d}\right)\left(\mu_j(1 - \frac{s}{s'}) - \mu_{j,a} + \mu_{j,d}\right) \tag{70}$$

can be computed in $O(n_a + n_d)$ time.

Deleting points

Let the new convex hull be obtained by deleting n_d tetrahedra from and added n_a tetrahedra to the old convex hull. Consequently, the same formulas and time complexity, as by adding points, follow.

7. References

Chan, T. F., Golub, G. H. & LeVeque, R. J. (1979). Updating formulae and a pairwise algorithm for computing sample variances, *Technical Report STAN-CS-79-773*, Department of Computer Science, Stanford University.

Cheng, S.-W. & Y. Wang, Z. W. (2008). Provable dimension detection using principal component analysis, *Int. J. Comput. Geometry Appl.* 18: 415–440.

Dimitrov, D., Holst, M., Knauer, C. & Kriegel, K. (2009). Closed-form solutions for continuous PCA and bounding box algorithms, *A. Ranchordas et al. (Eds.): VISIGRAPP 2008, CCIS, Springer* 24: 26–40.

Dimitrov, D., Holst, M., Knauer, C. & Kriegel, K. (2011). Efficient dynamical computation of principal components, *Proceedings of International Conference on Computer Graphics Theory and Applications - GRAPP*, pp. 85–93.

Dimitrov, D., Knauer, C., Kriegel, K. & Rote, G. (2009). Bounds on the quality of the PCA bounding boxes, *Computational Geometry* 42: 772–789.

Duda, R., Hart, P. & Stork, D. (2001). *Pattern classification*, John Wiley & Sons, Inc., 2nd ed.

Gottschalk, S., Lin, M. C. & Manocha, D. (1996). OBBTree: A hierarchical structure for rapid interference detection, *Computer Graphics* 30: 171–180.

Jolliffe, I. (2002). *Principal Component Analysis*, Springer-Verlag, New York, 2nd ed.

Knuth, D. E. (1998). *The art of computer programming, volume 2: seminumerical algorithms*, Addison-Wesley, Boston, 3rd ed.

Parlett, B. N. (1998). *The symmetric eigenvalue problem*, Society of Industrial and Applied Mathematics (SIAM), Philadelphia, PA.

Pébay, P. P. (2008). Formulas for robust, one-pass parallel computation of covariances and arbitrary-order statistical moments, *Technical Report SAND2008-6212*, Sandia National Laboratories.

Press, W. H., Teukolsky, S. A., Veterling, W. T. & Flannery, B. P. (1995). *Numerical recipes in C: the art of scientific computing*, Cambridge University Press, New York, USA, 2nd ed.

Vranić, D. V., Saupe, D. & Richter, J. (2001). Tools for 3D-object retrieval: Karhunen-Loeve transform and spherical harmonics, *IEEE 2001 Workshop Multimedia Signal Processing*, pp. 293–298.

Welford, B. P. (1962). Note on a method for calculating corrected sums of squares and products, *Technometrics* 4: 419–420.

West, D. H. D. (1979). Updating mean and variance estimates: an improved method, *Communications of the ACM* 22: 532–535.

14

On-Line Monitoring of Batch Process with Multiway PCA/ICA

Xiang Gao

Yantai Nanshan University

P. R. China

1. Introduction

Batch processes play an important role in the production and processing of low-volume, high-value products such as specialty polymers, pharmaceuticals and biochemicals. Generally, a batch process is a finite-duration process that involves charging of the batch vessel with specified recipe of materials; processing them under controlled conditions according to specified trajectories of process variables, and discharging the final product from the vessel.

Batch processes generally exhibit variations in the specified trajectories, errors in the charging of the recipe of materials, and disturbances arising from variations in impurities. If the problem not being detected and remedied on time, at least the quality of one batch or subsequent batches productions is poor under abnormal conditions during these batch operations. Prior to completion of the batch or before the production of subsequent batches, batch processes need effective strategy of real-time, on-line monitoring to be detected and diagnosed the faults and hidden troubles earlier and identified the causes of the problems for safety and quality.

Based on multivariable statistical analysis, several chemometric techniques have been proposed for online monitoring and fault detection in batch processes. Nomikos and MacGregor (1994, 1995) firstly developed a powerful approach known as multiway principal component analysis (MPCA) by extending the application of principal component analysis (PCA) to three-dimensional batch processes. By again projecting the information contained in the process-variable trajectories onto low-dimensional latent-variable space that summarizes both the variables and their time trajectories, the main idea of their approach is to compress the normal batch data and extract information from massive batch data. A batch process can be monitored by comparing with its time progression of the projections in the reduced space with those of normal batch data after having set up normal batch behaviour. Several studies have investigated the applications of MPCA (Chen & Wang, 2010; Jung-hui & Hsin-hung, Chen, 2006; Kosanovich et al., 1996; Kourti, 2003; Westerhuis et al., 1999).

Many of the variables monitored in one process are not independent in some cases, may be combination of independent variables not being measured directly. Independent component analysis (ICA) can extract the underlying factors or components from non-Gaussian multivariate statistical data in the process, and define a generative model for massive observed data, where the variables are assumed to be linear or nonlinear mixtures of unknown latent variables called as independent components (ICs) (Lee et al., 2004; Ikeda and Toyama, 2000). Unlike capturing the variance of the data and extracting uncorrelated latent variable from correlated data on PCA algorithm, ICA seeks to extract separated ICs that constitute the variables. Furthermore, without orthogonality constraint, ICA is different from PCA whose direction vectors should be orthogonal. Yoo et al. (2004) extended ICA to batch process on proposing on-line batch monitoring using multiway independent component analysis (MICA), and regarded that ICA may reveal more information in non-Gaussian data than PCA.

Although the approach proposed by Nomikos and MacGregor (1994, 1995) is based on the strong assumption that all the batches in process should be equal duration and synchronized, every operational period of the batches is almost different from others actually because of batch-to-batch variations in impurities, initial charges of the recipe component, and heat removal capability from seasonal change, therefore operators have to adjust the operational time to get the desired product quality. There are several methods to deal with the different durations for the algorithm MPCA. However, neither stretching all the data length to the maximum by simply attaching the last measurements nor cutting down all 'redundant trajectories' to the minimum directly could construct the process model perfectly. Kourti et al. (1996) used a sort of indicator variable which is followed by other variables to stretch or compress them applied on industrial batch polymerization process. Kassidas et al. (1998) presented an effective dynamic time warping (DTW) technique to synchronize trajectories, which is flexible to transform the trajectories optimally modelling and monitoring with the concept of MPCA. DTW appropriately translates, expands and contracts the process measurements to generate equal duration, based on the principle of optimally of dynamic programming to compute the distance between two trajectories while time aligning the two trajectories (Labiner et al., 1978). Chen and Liu (2000) put forward an approach to transform all the variables in a batch into a series of orthonormal coefficients with a technique of orthonormal function approximation (OFA), and then use those coefficients for MPCA and multiway partial least square (MPLS) modelling and monitoring (Chen and Liu, 2000, 2001). One group of the extracted coefficients can be thought as abbreviation of its source trajectory, and subsequent relevant information of the projection from PCA can reveal the variation information of process well.

About the measures of online monitoring MPCA, Nomikos and MacGregor (1995) presented three solutions: filling the future observation with mean trajectories from the reference database; attaching the current deviation as the prediction values of incomplete process; and partial model projection that the known data of appeared trajectories are projected onto the corresponding partial loading matrix. The former two schemes are introduced to estimate the future group of data by just filling hypothesis information simply, without consideration of possible subsequent variations; and on the latter scheme only part information of MPCA model is used with the appeared trajectories projection onto the corresponding part of loading matrix of MPCA to analyze the variation of local segments. Therefore the indices of monitoring may be inaccurate on the above three solutions. To eliminate the errors of monitoring, Gao and Bai (2007) developed an innovative measure to estimate the future data of one new batch by calculation of the Generalized Correlation Coefficients (GCC) between the new batch trajectory and historical trajectories, to fill the subsequent unknown portion of the new batch trajectory with the corresponding part of the history one with maximum GCC.

Recently, for online monitoring of batch process, some papers were involved in GCC prediction after DTW synchronization with MPCA/MICA (Bai et al., 2009a, 2009b; Gao et al., 2008b), other works were concerned with GCC prediction after OFA synchronization with MPCA/MICA (Bian 2008; Bian et al., 2009; Gao et al., 2008a). These examples proved that both DTW and OFA are integrated with GCC prediction perfectly with MPCA/MICA.

In this chapter, a set of online batch process monitoring approaches are discussed. On real industrial batch process, the process data is not always followed Gaussian distribution, Compared with MPCA, MICA may reveal more hidden variation than MPCA though its complexity of computation; the methods of synchronization DTW and OFA, are applied in compound monitoring approaches respectively; four solutions for missing data of future value, are applied in an example comparatively.

The chapter is organized as follows. Section 2 gives introduction of the principle of DTW and relevant method of synchronization. In section 3, the principle of OFA is also introduced in advance and narration of how the extracted coefficients from the trajectories are used for model and monitoring. Then the traditional three solutions of Nomikos and MacGregor (1995) and GCC estimation are discussed in Section 4. An industrial polyvinyl chloride (PVC) polymerization process is employed to illustrate the integrative approaches in Section 5. Finally, a conclusion is presented in Section 6.

2. Dynamic time warping

Dynamic Time warping (DTW) is a flexible, deterministic pattern matching method for comparing two dynamic patterns that may not perfectly aligned and are characterized by similar, but locally transformed, compressed and expanded, so that similar features within (Kassidas et al.,1998) the two patterns are matched. The problem can be discussed from two general trajectories, R and T.

2.1 Symmetric and asymmetric DTW algorithm

Let R and T express the multivariate trajectories of two batches, whose matrices of dimension $t \times N$ and $r \times N$, separately, where t and r are the number of observations and N is the number of measured variables. In most case, t and r are not always equal, so that the two batches are not synchronized because they have not common length. Even if $t=r$, their trajectories may not be synchronized because of their different local characteristics. If one applies the monitoring scheme of MPCA (Nomikos and MacGregor, 1994), or the scheme of MICA (Yoo et al., 2004), by simply add or delete some measured points artificially, unnecessary variation will be included in statistical model and the subsequent statistical tests will not detect the faulty batches sensitively.

On the principle of dynamic programming to minimize a distance between two trajectories, DTW warps the two trajectories so that similar events are matched and a minimum distance between them is obtained, because DTW will shift, compress or expand some feature vectors to achieve minimum distance (Nadler and Smith, 1993).

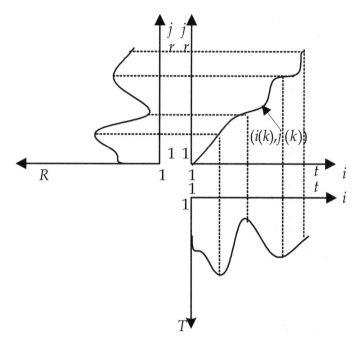

Fig. 1. Sketch map of nonlinear time alignment for two univariate trajectories R and T with DTW

Let i and j denote the time index of the T and R trajectories, respectively. DTW will find optimal route in sequence F^* of K points on a $t \times r$ grid.

$$F^* = \{c(1), c(2), \cdots c(K)\}, \max(t, r) \le K \le t + r \tag{1}$$

where

$$c(k) = [i(k), j(k)] \tag{2}$$

and each point $c(k)$ is an ordered pair indicating a position in the grid. Two univariate trajectories T and R in Figure 1 show the main idea of DTW.

Most of DTW algorithms can be classified either as symmetric or as asymmetric. Although on the former scheme, both of the time index i of T and the time index j of R are mapped onto a common time index k, shown as Eqs.1, 2, the result of synchronization is not ideal, because the time length of synchronized trajectories often exceeds referenced trajectories. On the other hand, the latter maps the time index of T on the time index of R or vice-versa, to expand or compress more one trajectory towards the other. Compared with Eqs.1, 2, the sequence becomes as follow:

$$F^* = \{c(1), c(2), \cdots, c(j), \cdots c(r)\} \tag{3}$$

and

$$c(j) = (i(j), j) \tag{4}$$

This implies that the path will go through each vector of R, but it may skip some vectors of T.

2.2 Endpoints, local and global constraints

In order to find the best path through the grid of $t \times r$ grid, three rules of the DTW algorithm should be specified.

(1) Endpoint constraints: $c(1)=(1,1)$, $c(K)=(t, r)$.

(2) Local constraints: the predecessor of each (i, j) point of F^* except $(1,1)$ is only one from $(i-1, j)$, $(i-1, j-1)$ or $(i, j-1)$, which is shown in Fig.2.

(3)Global constraints: the searching area is $\pm M(M \geq |t - r|)$ widening strip area around the diagonal of the $t \times r$ grid, which is shown in Fig.3.

The endpoint constraints illustrate that the initial and final points in both trajectories are located with certainty. The local continuity constrains consider the characteristics of time indices to avoid excessive compression or expansion of the two time scales (Myers et al. 1980).

On the requirement of monotonous and non-negative path, the local constrains also prevent excessive compression or expansion from the several latest neighbors (Itakura, 1975). The global constraints prevent large deviation from the linear path.

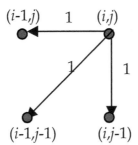

Fig. 2. Local continuity constraint with no constraint on slope

2.3 Minimum accumulated distance of the optimal path

As mentioned above, for the best path through a grid of vector-to-vector distances searched by DTW algorithm, some total distance measured between the two trajectories should be minimized. The calculation of the optimal normalized total distance is impractical, a feasible substitute is minimum accumulated distance, $D_A(i, j)$ from point $(1,1)$ to point (i, j)(Kassidas et al., 1998). The suitable one is:

$$D_A(i, j) = d(i, j) + \min[D_A(i-1, j), D_A(i-1, j-1), D_A(i, j-1)], D_A(1,1) = d(1,1) \tag{5}$$

where

$$d(i, j) = [T(i, :) - R(j, :)] * W * [T(i, :) - R(j, :)]^T \tag{6}$$

$d(i, j)$ is the weighted local distance between the i vector of the T trajectory and the j vector of the R trajectory, therein W is a positive definite weight matrix that reflects the relative importance of each measured variables.

2.4 Synchronization based on combination of symmetric and asymmetric DTW

2.4.1 The advantage and disadvantage of symmetric and asymmetric DTW

As mentioned above, DTW works with pairs of patterns. Therefore, the problem of whether symmetric or asymmetric is suitable for synchronization.

Let $B_i(b_i \times N)$, $i=1,2,...,I$ be a training set of good quality batches for MPCA/MICA models, where b_i is the number of observations and N is number of measured variables, and one defined reference batch trajectories B_{REF}, the objective is to synchronize each B_i with B_{REF} $(b_{REF} \times N)$.

Symmetric DTW algorithms include all points in the original trajectories, but expanded trajectories of various lengths, because the length is determined by DTW. After synchronization, each B_i will be individually synchronized with B_{REF}, but not with each other unfortunately.

Although asymmetric may eliminate some points, they will produce synchronized trajectories of equal length, because each time axis of B_i will be mapped with the one of B_{REF} so that they all are synchronized with reference trajectories B_{REF} and synchronized with each other.

Unavoidably, the asymmetric algorithms have to skip some points in the optimal path, so the characteristics of some segments may be left out after synchronization to construct incomplete MPCA/MICA model from 'trimmed' trajectories to cause miss/false alarm.

2.4.2 The circumstance of combination of symmetric and asymmetric DTW

The essence of DTW is to match the pairs of two trajectories on synchronization. At first, on symmetric DTW algorithm, the optimal path is reconstructed following above 3 constraints and Eq.5,6. Aligning points of B_i with B_{REF} on asymmetric synchronization, some statuses would appear:

(a) Some point of B_i may be copied multiply, because it matches several points of B_{REF};

(b) Some point of B_i may be matched with the point in various time index of B_{REF}, which means it will be transferred after synchronization;

(c) More than one point of B_i may be averaged to a point that will be aligned with the particular point of B_{REF}, because they are aligned with only one point of B_{REF} in symmetric DTW algorithm. Although some local feature of points may be smoothed, it is proved that ensure that all B_i after asymmetric operation have the same duration b_{REF}.

2.4.3 An improvement of DTW algorithm for more measurements

In some processes, the measurement may be relative too large to be satisfied with the need of memory of many calculated minimum accumulate distance $D_A(i, j)$. Gao et al. (2001) presented a solution to overcome the problem 'out of memory'. Their idea is that $D_A(i, j)$

should not be worked out until the final result $D_A(t, r)$ to accumulate a large number of the medium result. The programming can be composed with local dynamic programming in strip of adjacent time intervals, following is the improved algorithm under the three constraints and eq.5, 6, which is shown in Fig.3.

1) When $i=1$, compute $D_A(i, :)$ from $D_A(1, 1)$, let $I_P=1$, $J_P=1$;

2) Then $i \leftarrow (i+1)$, compute $D_A(i, :)$ with the aid of the result of $D_A(i-1, :)$;

3) The local optimal path could be searched between the columns $(i-1, :)$ and $(i, :)$. The start point of the path is (I_P, J_P) and the relay end point is (I_E, J_E), where $I_E=I_P+1$, J_E is ascertained on the following comparison:

$$J_E = \arg\min_{F'}[D_A(I_E, J_P), D_A(I_E, J_P + 1), \cdots D_A(I_E, q)$$
$$q = \min\{r, fix[I_E * (r-1) / (t-1) + M]\} \tag{7}$$

where fix is the function that keeps only the integer fraction of the result of computation.

4) Delete the column of $D_A(i-1, :)$, then set $I_P \leftarrow I_E$, $J_P \leftarrow J_E$;

5) Repeat step 2 to step 4 till $i=t$ (t is one end point of pair);

6) If (I_P, J_P) is (t, r), searching stops; otherwise if the (I_P, J_P) is (t, p) $(p < r)$, the rest path is from the point (t, p) to the final point (t, r).

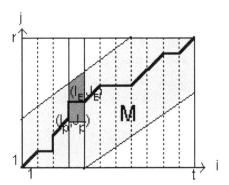

Fig. 3. The local optimization between two columns in the improved DTW

2.5 Procedure of synchronization of batch trajectories

The iterative procedure proposed for the synchronization of unequal batch trajectories (Kassidas et al., 1998) is a practical approach for industrial process, which is now being presented.

First of all, each variable from each batch should be scaled as preparation. Let B_i, $i=1,...,I$ be the result of scaled batch trajectories from I good quality raw batches, the scaling method is to find the average range of each variable in raw batches by averaging the range form each batch, then to divide each variable in all batches with its average range, and store average ranges for monitoring. Then synchronization begins.

Step 0: Select one of the scaled trajectories B_k as the referenced trajectories B_{REF} on the technic requirement. Set weight matrix W equal to the identity matrix. Then execute the following steps for a specified maximum number of iterations.

Step 1: Apply the DTW method between B_i, $i=1,...,I$, and B_{REF}. Let $\tilde{B}_i, i=1,...I$ be the synchronized trajectories whose common durations is same as the one of B_{REF}.

Step 2: Compute the average trajectory $\overline{B}$ from average values of all B_i.

Step 3: For each variable, compute the sum of squared deviations from $\overline{B}$, whose inverse will be the newer weight of the particular variable for the next iteration.

$$W(j,j)=\left[\sum_{i=1}^{I}\sum_{k=1}^{b_{REF}}[\tilde{B}_i(k,j)-\overline{B}(k,j)]^2\right]^{-1} \tag{8}$$

As a diagonal matrix, W should be normalized so that the sum of the weight is equal to the number of variables, that is, W could be replaced as:

$$W \leftarrow W\left\{N \Big/ \left[\sum_{j=1}^{N}W(j,j)\right]\right\} \tag{9}$$

Step 4: In most case, the times of iterations are not greater than 3, so keep the same referenced trajectory: $B_{REF}=B_k$. If the more iterations are needed, set the reference equal to the average trajectory: $B_{REF}=\overline{B}$.

2.6 Offline implementation of DTW for batch monitoring

Now, a available complete trajectory of one new batch $B_{RAW, NEW}$ ($b_{NEW}\times N$) needs to be monitored using MPCA/MICA. It has to be synchronized before the monitoring scheme is applied because most probably the new batch trajectory $B_{RAW, NEW}$ hardly accord with the referenced trajectory B_{REF}.

When being scaled, each variable in the new batch $B_{RAW, NEW}$ is divided with the average range from referenced trajectory to get the resulting scaled new trajectory, B_{NEW}. B_{NEW} is synchronized with referenced trajectory B_{REF} using W from Eq.8, 9 in the synchronization procedure to get the result $\tilde{B}_{NEW}$ ($b_{NEW}\times N$) which can be used in MPCA/MICA model.

3. Orthonormal function approximation

Under the condition of synchronous batch processes, the data from batch process are supposed to take the form of three-way array: $j=1,2...J$ variables are measured at $k=1,2,...K$ time intervals throughout $i=1,2,...I$ batch runs. The most effective unfolding the three-way data on monitoring is to put its slices $(I\times)$ side by side to the right, starting with the one corresponding to the first interval, then to generate a large two–dimensional matrix $(I\times JK)$ (Nomikos and MacGregor 1994, 1995; Wold et al., 1987). The variable in the two-dimensional matrix is treated as a new variable for building PCA model. Nevertheless, the batch processes are asynchronous in some cases so that two–dimensional matrix $(I\times JK)$ can not be formed. Unlike translation, expansion and contraction of process measurements to generate equal duration in DTW, orthonormal function is employed to eliminate the problem resulted from the different operating time to turn the implicit system information into several key parameters which cover the necessary part of the operating conditions for each variable in each batch (Chen and Liu, 2000; Neogi and Schlags, 1998).

3.1 Orthonormal function

On the concept of Orthonormal Function Approximation (OFA), the process measurements of each variable in each batch run can be mapped onto the same number of orthonormal coefficients to represent the key information. As an univariate trajectory, the profile of each variable in each batch run can be represented as a function $F(t)$, which can be approximated in terms of an orthonormal set $\{\varphi_n\}$ of continuous function:

$$F(t) \cong F_n(C,t) = \sum_{n=0}^{N-1} \alpha_n \varphi_n(t) \qquad (10)$$

where the coefficients, $C = \{\alpha_n\}$, $\alpha_n = \int F(t)\varphi_n(t)dt$ are the projection of $F(t)$ onto each basis function. Therefore, the coefficients C of the orthogonal function is representative of the measured variable $F(t)$ of one batch run. Not being calculated from a set of K measurements, the coefficient a_n can be derived practically with orthonormal decomposition of $F(t)$:

$$
\begin{aligned}
&E_0(t_k) = F(t_k) \\
&\alpha_n = [\Phi_n^T \Phi_n]^{-1} \Phi_n E_n \\
&E_{n+1}(t_k) = E_n(t_k) - \alpha_n \varphi_n(t_k) \\
&k = 1, 2, \ldots, K_i \\
&n = 0, 1, \ldots, N-1; i = 1, 2, \ldots, I
\end{aligned}
\qquad (11)
$$

where $E_n = [E_n(t_1)\ E_n(t_2)\ldots E_n(t_{ki})]^T$ and $\Phi_n = [\varphi_n(t_1)\ \varphi_n(t_2)\ldots\varphi_n(t_{ki})]^T$. The Legendre polynomial basis function is regard as an effective function to be used due to the finite time interval for each batch run (Chen and Liu, 2000):

$$
\begin{aligned}
&\varphi_n(t) = \sqrt{\frac{2n+1}{2}} P_n(t) \\
&P_n(t) = \frac{1}{2^n n!} \frac{d^n}{dt^n}[(t^2 - 1)^n]
\end{aligned}
\qquad (12)
$$

where $t \in [-1,1]$. When $n=0$, the constant coefficient a_0 is for $\varphi_0(t) = P_0(t)/\sqrt{2}$ and $P_0(t)=1$. Before applying the orthonormal function approximation, the variables of the system with different units needs to be pretreated in order to be put on an equal basis. However, mean centering of the measurement data is not necessary because the constant coefficient a_0 is for φ_0 orthonormal basis function. Mean centering will affect the constant coefficient for φ_0 corresponding to zero. The ratio convergence test for mathematical series is applied to determine the approximation error associated with the reduction in the number of the basis spaces (Moore and Anthony, 1989). The measure of approximation effectiveness can be obtained as:

$$G(N) = \frac{\|F_N\|^2 - \|F_{N-1}\|^2}{\|F_N\|^2} = \frac{\alpha_{N-1}^2}{\sum_{n=0}^{N-1} \alpha_n^2} \tag{13}$$

where $\|F_N\|^2$ is the square of the Euclidean function norm of approximation $F_N(C, t)$. When a consistent minimum $G_{ij}(N)$ is reached, the required optimal number of terms N_{ij} can be chosen for the measurement variable j at batch i (Moore and Anthony, 1989). Therefore, most of the behavior of the original $F(t)$ is extracted from the coefficients C. Nevertheless, the maximum number of terms of the approximated function for each variable in all batch runs is taken to obtain enough more terms whose expansion $F_N(t)$ extracts the main behavior of $F(t)$.

$$N_j = \max_i \{N_{ij}\} \tag{14}$$

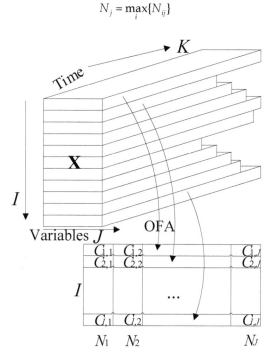

Fig. 4. The three-way array **X** in each batch run of different duration maps into a coefficient matrix Θ

Therefore, the problem originated from the different operational time in each batch run is eliminated with the orthonormal approximation method when the same number of coefficients is used for the same measured variable. In this way, the key parameters contain the necessary part of the operating condition for each variable in each batch run. Like the multiway method, the coefficients are reorganized into time-ordered block and the blocks can be put in order with multiway matrices $\Theta(I \times \sum_{j=1}^{J} N_j)$:

$$\Theta = \begin{bmatrix} C_{1,1} & C_{1,2} & \cdots & C_{1,J} \\ C_{2,1} & C_{2,2} & \cdots & C_{2,J} \\ & & \vdots & \\ C_{I,1} & C_{I,2} & \cdots & C_{I,J} \end{bmatrix} \tag{15}$$

where $C_{i,j} = [\alpha_{i,0}, \alpha_{i,1} \ldots \alpha_{i,N_j-1}]$, represents the coefficient vector of the approximation function for the measurement variable j at batch i, and N_j is the needed number of terms for variable j.

3.2 Offline implementation of OFA for batch monitoring

When one new batch is completed, after being applied orthonormal function transformation, all the variables of the batch along the time trajectory become a row vector composed of a series of coefficients $\Theta_{new} = [C_{new,1}, C_{new,2}, \ldots, C_{new,J}]$ that can be projected onto Θ to implement PCA/ICA algorithm.

4. Online monitoring schemes

4.1 Traditional online monitoring schemes

It is assumed that the future measurements are in perfect accordance with their mean trajectories as calculated from reference database, the first approach is to fill the unknown part of x_{new} with zeros. In other words, batch is supposed to operate normally for the rest of its duration with no deviations in its mean trajectories. On the analysis of Nomikos and MacGregor (1995), the advantage of this approach is a good graphical representation of the batch operation in the t plots and the quick detection of an abnormality in the SPE plot, whereas the drawback of this approach is that the t scores are reluctant, especially at the beginning of the batch run, to detect an abnormal operation.

On the hypothesis that the future deviations form the mean trajectories will retain for the rest of the batch duration at their current values at the time interval k, the second approach is to fill the unknown part of x_{new} with current scaled values under the assumption that the same errors will persist for the rest of the batch run. Although the SPE chart is not relative sensitive than one in the first approach, the t scores pick up an abnormality more quickly (Nomikos and MacGregor, 1995). Nomikos and MacGregor (1995) had to suggest that the future deviations will decay linearly or exponentially from their current values to the end of the batch run, to share the advantages and disadvantages of the first two approaches.

The unknown future observations can be regarded as missing data from a batch in MPCA on the third approach. To be consistent with the already measured values up to current time k, and with the correlation structure of the observation variables in the database as defined by the p-loading matrices of MPCA model, one can use the sub model of principal components of the reference database without excessive consideration of the unknown future values. MPCA projects the already known measurements $(x_{new,k}(kJ \times 1))$ into the reduced space and calculates the t scores at each time interval as:

$$t_{R,k} = (P_k^T P_k)^{-1} P_k^T x_{new,k} \tag{16}$$

where $P_k(kJ \times R)$ is a matrix whose all elements in each columns of p-loading vectors $(\mathbf{p}_r)$ from all the principal component are from start to the current time interval k. The matrix $(P_k^T P_k)^{-1}$ is well conditioned even for the early times, and approaches the identity matrix as k approaches the final time interval K because of the orthogonality property of the loading vectors $\mathbf{p}_r$ (Nomikos and MacGregor, 1995). The advantage of this method is that at least 10% known measurements of new batch trajectory are enough for computation and perfect t scores near to the actual final values. However, Nomikos and MacGregor (1995) also indicated that little information will result in quite large and unexplainable t scores at the early stage of the new batch run. Similarly, the third approach can be applied to MICA model that the deterministic part of independent component vector, $\hat{s}_{d,k}(d \times Jk)$, can be calculated as:

$$\hat{s}_{d,k} = W_d x_{new,k} \tag{17}$$

where $W_d(Jk \times 1)$ is the deterministic part of W_s, a separating matrix in ICA algorithm.

It is uncertain that which one of above mentioned schemes is most suitable for batch process. Nomikos and MacGregor (1995) stated that each scheme is fit for respective condition: the third for non frequent discontinuities, the second for persistent disturbances and the first for non persistent disturbances. They also suggested combining these schemes when online monitoring.

4.2 Online monitoring with filling similar subsequent trajectory

Generally, as measurements of correlation degree between two vectors, Correlation Coefficients (CC) are numerical values which stand for the similarity in some sense. However, because each multivariable trajectory can be expressed as one matrix whose columns are variables with time going on, the relationship of corresponding two matrices of two multivariable trajectories can not be distinctly denoted with CC in the form of a numerical value but a matrix that one can not examine the similarity between the matrices by comparing the CC value. A sort of Generalized Correlation Coefficients measuring method was presented to the solution of the mentioned problem by computation of the traces of covariances, because as the sums of the eigenvalues of the matrices, their traces expresses the features of corresponding matrices in some ways (Gao and Bai., 2007). Suppose that a monitoring trajectory $\mathbf{V}$ ($k \times m$), where k is the current time interval, and m is the number of variables, another trajectory $\mathbf{Y}$ ($k \times m$) from history model database is chosen to match with $\mathbf{V}$ ($k \times m$), their GCC can be defined as:

$$\rho(\mathbf{V}, \mathbf{Y}) = \frac{tr[\text{cov}(\mathbf{V}, \mathbf{Y})]}{\sqrt{tr[\text{cov}(\mathbf{V})]tr[\text{cov}(\mathbf{Y})]}} \tag{18}$$

where tr is the function of trace, $\rho(\mathbf{V}, \mathbf{Y})$ is the GCC. In eq.18, the definitions of cov(V), cov(Y), cov(V, Y) are:

$$\text{cov}(\mathbf{V}) = \frac{[\mathbf{V} - \text{E}(\mathbf{V})]^T [\mathbf{V} - \text{E}(\mathbf{V})]}{n-1} \tag{19}$$

$$\text{cov}(\mathbf{Y}) = \frac{[\mathbf{Y} - E(\mathbf{Y})]^T [\mathbf{Y} - E(\mathbf{Y})]}{n - 1} \tag{20}$$

$$\text{cov}(\mathbf{V}, \mathbf{Y}) = E\{[\mathbf{V} - E(\mathbf{V})]^T [\mathbf{Y} - E(\mathbf{Y})]\} \tag{21}$$

When two trajectories align with each other from start, the range of GCC is (0, 1], they are more similar as the value of their GCC near to 1. Caution must be paid when two trajectories are asynchronous so that the two matrices which have different dimensions have to be dealt with in eq. 22.

4.3 The procedure of online monitoring of asynchronous batch

The first step is to deal with the lack of data of online batch. The trouble of online monitoring of asynchronous batch is to choose the scheme properly. As above mentioned, traditional schemes are relative easy to be implemented whereas GCC approach need more computation time than others. The ongoing new batch $\mathbf{V}$ ($k \times m$) needs to compare with many normal batches and abnormal batches included in history model database Ω contained more matrices for prepared in many cases. Due to different dimensions of matrices between the new batch run and history batch run $\mathbf{N}_i(K_n \times m) \in \Omega$, $i=1,2,...,h$, h is the number of stored history batches in Ω, the pseudo covariance is introduced to be calculated instead of Eq. 21 (Gao et al., 2008b).

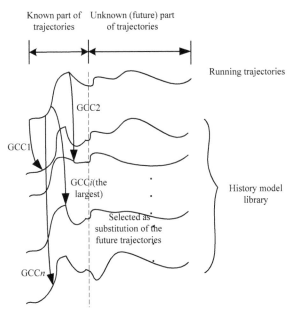

Fig. 5. A sketch of GCC matching to decide the substitute of future measurements inhistory model library

$$psd\,\text{cov}(\mathbf{N}_i, \mathbf{V}) = [\mathbf{N}_i - E(\mathbf{N}_i)][\mathbf{V} - E(\mathbf{V})]^T / \max(K_n, k) \tag{22}$$

Then one of trajectories, $N_i(K_n \times m)$, that have the largest GCC with V $(k \times m)$ is chosen. If $k<K_n$, extend V $(k \times m)$ by copying from $k+1$ to K_n part of $N_i(K_n \times m)$ to follow V $(k \times m)$, otherwise maintain V $(k \times m)$. Although k is far less than K_n sometimes, the result of Eq.22 reveals the homologous relationship like covariance between the two matrices. Hence, the insufficiency of data of online batch run can be solved by filling the assumptive values in different ways.

The second step is pre-treatment of data. Before synchronization, all the measurements of new batch should be scaled.

The third step is synchronization; one can choose DTW or OFA to deal with the asynchronous running trajectory. After that, the new test batch is similar to offline batch so as to be projected onto MPCA/MICA model.

5. Case study

5.1 Brief introduction of technics of PVC polymerization process

As a thermoplastic resin, when its vinyl chloride molecules are associated, the production of PVC is forming chains of macromolecules, whose process is called polymerization. The vinyl chloride (VC) monomer, dipped in aqueous suspension, is polymerized in a rector shown as Fig.6.

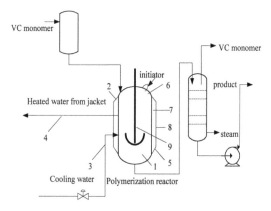

Fig. 6. Flow diagram of PVC polymerization progress

The polymerization process reaction changes violently because the container in the rector goes through water phase, liquid VC phase and solid PVC phase on different stage of reaction. At the start of reaction, water, VC, suspension of stabilizers and initiator are on request loaded into the reactor through respective inlets, and then they are stirred adequately to create a kind of milky solution, suspension of VC droplets.

It is noticed that several indices should be monitored and controlled on each stages of the reaction, especially temperatures. Nine important variables of all the batches depicted on Table 1, are shown in Fig.7 from one batch. At the beginning of the reaction, the hot water is pumped into the jacket of reactor to heat the reactor content to the set temperature (57°C). The indirect heating does not continue until the sufficient reaction heat has been generated from the reaction. PVC in the solution will precipitate quickly to form solid phase PVC granules inside almost each VC monomer droplets on the polymerization, because it is not soluble in water, but little dissolved in the VC.

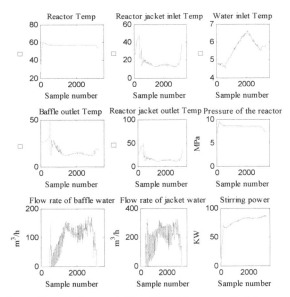

Fig. 7. Typical batch profiles of nine variable of PVC form one batch

Variable NO.	Sensor NO.	Variable name	Unit
1	TIC-P101	Temperature of the reactor	°C
2	TIC-P102	Temperature of the reactor jacket inlet	°C
3	TI-P107	Temperature of the water inlet	°C
4	TI-P108	Temperature of the baffle outlet	°C
5	TI-P109	Temperature of the reactor jacket outlet	°C
6	PIC-P102	Pressure of the reactor	MPa
7	FIC-P101	Flow rate of baffle water	m³/h
8	FIC-P102	Flow rate of jacket water	m³/h
9	JI-P101	Stirring power	KW

Table 1. Polymerization reactor variables

Due to the exothermic reaction, the temperature of the reactor will rise gradually so that the redundant reaction heat should be removed at once to keep constant temperate. In order to cool down the reactor, a flow of cooling water is pumped into the jacket surrounding the reactor. The condenser on the top the reactor also concentrates VC monomer from vapor to liquid. If temperature of reactor is lower than the set point temperature, the hot water is commanded to be injected in the jacket again, which is the automatic control of process by the parameters of the important variables. At the end of the polymerization, there is a little monomer of remained gaseous VC. With the VC being absorbed from the byproduct of exhaust gas, the polymerization does not continue until the action of terminator.

5.2 The essential of the batches of training set and test set

Although the PVC process last just several hours ($3h\sim8h$), the sampling frequency is comparatively higher because it is necessary to online monitor time-variant batch process. The sampling interval is 5 seconds, so that all the measurements of any one batch is on the scope of (2000, 6000) due to the adjustment of the duration for different requirements of products. After more observation of the production, most of the durations of batches are around 3200 measurements and the distribution of the batches does not follow normal distribution.

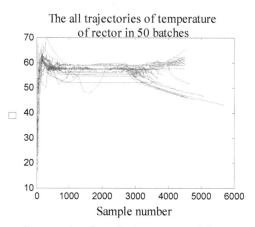

The all trajectories of temperature of rector in 50 batches

Fig. 8. The profile of asynchronous batches of temperature of the reactor of PVC

Batch	Situation	Details
# 1	abnormal	deviation from mean trajectories
# 2	abnormal	fluctuation of temperature of reactor, in the metaphase of reaction
# 3	abnormal	lower stirring power in the early stage of reaction, bring down various flow rates and various temperatures
# 4	abnormal	lower temperate of the water inlet in the early stage of reaction, cause the fluctuation of flow rate of baffle water and lower pressure of the reactor
others	normal	

Table 2. The situations of 10 tested batches of PVC

The data of 50 normal batches are selected from the examples of the practical process as a training dataset. There are other 50 batches sent to the history library Ω, so the number of the batches in Ω is 100.

From Fig.8 we can observe clearly the asynchronous chosen batches from temperature of the reactor (variable 1). There are 10 batches (#1~#10) taken as test data from the batch process in the plant. Some problems of these batches are listed in Table 2 in the polymerization of batch process, one tries to discriminate the abnormal of them with two statistics of SPE and T^2 of MPCA, or SPE and I^2 of MICA, and then find whose variables were affected.

5.3 The offline monitoring of batches without intelligent synchronization

For those asynchronous batches modeling and monitoring, without intelligent synchronization of DTW or OFA, the rough method of synchrozation, to prune so-called redundant data over the specified terminal or to extend the short trajectories with the last values, is experimented. All the durations of reference batches and test batches should be 3200 measurements.

Then the reference data set is arranged as a three-way $\underline{X}$ ($I \times J \times K$), where I corresponds to 50 batches, J corresponds to 9 process variables, and K corresponds to 3200 th time intervals. With the reference batch data $\underline{X}$, the MPCA and MICA models are constructed initially. Offline analysis of ten test batches is executed to show if this kind of rough construction of data for MPCA or MICA is appropriate or not. After batch-wise unfolding, 8 principal components of the MPCA model are determined by the cross-validation method (Nomikos and MacGregor, 1994), which explain 82.61% of the variability in the data. 8ICs are selected for the MICA for 77.54% variation of the whole data. Fig.9 shows the results of SPE based on MPCA and MICA under 99% control limit. It is clear that neither of MPCA nor MICA does well on the incorrect asynchronous multivariate statistic model: MPCA misses the detection of the batch #2, and MICA reports false alarm batches #4,#5, and misses #1,#2.

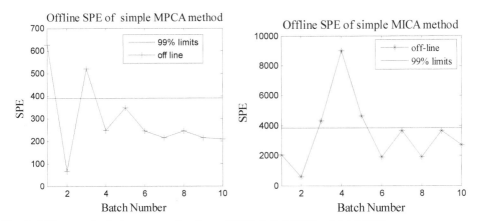

Fig. 9. Offline analysis for ten test batches of PVC, left: MPCA, right: MICA

5.4 Online monitoring of PVC batch process

5.4.1 Online monitoring of PVC with DTW-MPCA and DTW-MICA

On synchronization of DTW operation, all durations of the batches should be 3200. The weight matrix $\mathbf{W}$= [1.1527, 1.8648, 0.2390, 1.4778, 0.1742, 0.2118, 0.8186, 0.2760, 0.4592, 3.3258] from Eq.8, 9 for twice iterations. The MPCA model is built and its retained principal number is 8 to show 88.44% the variation of the batch process, whereas MICA retains 3 IC to explain the 93.85% of variation of data. All three solutions of of Nomikos and MacGregor (1995) and GCC are simulated compared with the offline analysis to find which one is the most appropriate in the batch process.

Fig.10 shows several online monitoring SPE indices of the 10 test batches compared with offline in MPCA and MICA, respectively. It can be shown that the MPCA results of first solution always misses faults in abnormal batches because of its smoothing the variation, the MICA result also misses the alarm of #2 and #3; while the results of second and third soltions are too large to alarm by mistake.

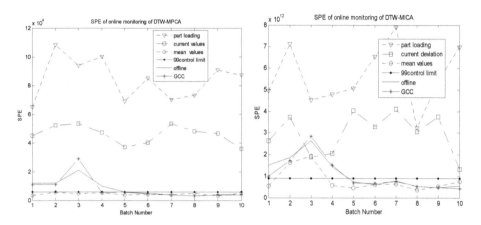

Fig. 10. SPE indices of online monitoring solutions, GCC and three solutions, left: DTW-MPCA; right: DTW-MICA

Comparatively, SPE of GCC prediction has adequate information of variations to identify the abnormal, only its MPCA results miss the abnormal of #4, the MICA results perform well.

5.4.2 Online monitoring of PVC with OFA-MPCA and OFA-MICA

After OFA synchronization, the information of original trajectories are extracted. Each variable of each batch run can be transformed into two coefficients, therefore in stead of irregular time length of three-dimensional data block, the two-dimensional coefficients matrix Θ (50×18) inherits the main features from the primative three-dimensional data block. Based on the new data of coefficients, the MPCA and MICA are experimented respectively. The online monitoring time point is set to 800 th measurement. MPCA algorithm holds 12 PCs to explain the 89.52% variation of the data, whereas MICA reserved 3ICs to illustrate the 51.92% variability in the data. The first two solutions of Nomikos and MacGregor (1995) and GCC are experimented in contrast with the offline analysis to find the best one in the batch process. It is noticed that the third solution does not fit for the coefficients matrix because the loading matrix is not from the coefficients, but from primative variables.

From various on-line monitoring solutions and offline analysis, Q-statistics-the SPE indices of 10 test batches are drawn in Fig.11, with MPCA and MICA, respectively. Similarly, the first solution of Nomikos and MacGregor (1995) erases many fine characters of the process so that it cannot detect the problem of many batches correctly, and the values of results of second online monitoring method are too large to be drawn in Fig.11, and always make false alarm to these batches, so it has to list them in Table 3. GCC performs well that it followed offline with a little difference. OFA-MPCA approach misses the abnormal of #1, #2, #4, whereas OFA-MICA detects four abnormal all.

No.	1	2	3	4	5
MPCA	232.89	101.48	84.305	206.15	153.59
MICA	738.69	844.37	254.93	437.87	301.89
No.	6	7	8	9	10
MPCA	81.499	143.74	81.499	143.74	127.83
MICA	160.76	322.75	160.76	322.75	401.52

Table 3. The SPE of online monitoring in MPCA and MICA of the second solution

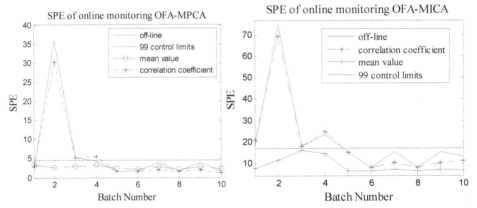

Fig. 11. SPE indices of online monitoring solutions, GCC and the mean values of the first solution, left: OFA-MPCA; right: OFA-MICA

The D-statistics of PVC, T^2 of OFA-MPCA and I^2 of OFA-MICA are drawn in Fig.12 as well. GCC performs well in the D-statistics in the same way, either T^2 or I^2, which are both close to the counterparts of offline. The first traditional solution can not predict any little variation after the time of detection, and the second one always has too larger error to be drawn in Fig. 12 that the results of the second solution has to be enumerated in Table 4.

No.	1	2	3	4	5
T^2	158.23	155.09	143.63	113.78	102.80
I^2	252.67	926.58	263.94	114.91	104.79
No.	6	7	8	9	10
T^2	53.445	129.22	53.445	129.22	131.87
I^2	41.569	70.977	41.569	70.977	110.13

Table 4. The T^2 in MPCA and I^2 in MICA of the second solution online monitoring

From Fig.12, it can be seen that OFA-MICA misses alarm #1 and #4, but OFA-MPCA has more errors: missed #1 and #3, and has a false alarm about #5, #7, #9 and #10.

Consquently, it is proved that the effect of OFA-MICA is better than ones of OFA-MPCA on both of Q-statistics and D-statistics in Fig.11 and Fig.12.

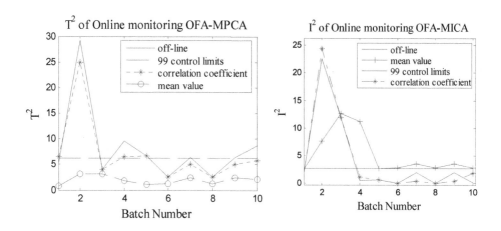

Fig. 12. T^2 or I^2 indices of online monitoring solutions, GCC and the mean values of the first solution, left: OFA-MPCA; right: OFA-MICA

5.5 Contribution plot of SPE and I^2 in OFA-MICA

The contribution plots can be used to dignose the event from non-conforming batches so as to assign a cause of abnormal by indication of which variables are predominatly responsible for the deviations (Jackson and Mudholkar, 1979). For instance, based on the approach of OFA-MICA, when the 800 th measurements of a diseased batch #3, the online SPE and I^2 contribution plots of 9 process variables are shown in Fig. 13 and Fig.14. It is obvious that the ratio of GCC (upper right) looks like the one of offline (upper left) which is different from the others (lower) distinctly. From Fig.13, The comparative larger ones of SPE is temperature of the baffle outlet (variable 4), flow rate of jacket water (variable 8) and stirring power (variable 9). Meanwhile we can find that the notable contribution of I^2 in Fig.14 are temperature of the reactor jacket inlet (variable 2), baffle outlet (variable 4) and flow rate of jacket water (variable 8). Therefore, contrasted with the report from plant in Table 2, the root cause is lower stirring power (the most conspicuous one in bar plot of Fig.13), which decreased other variables such as variable 4 and variable 8 consequently. It is inferred that lower stirring power decreased the rate of the reaction and generated less heat and needed smaller quantity of cooling water.

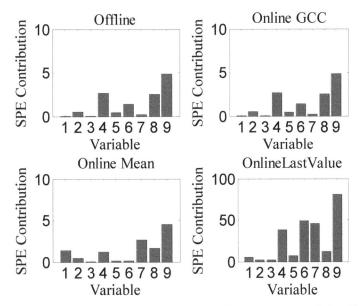

Fig. 13. SPE Contribution plots of 4 monitoring methods in PVC. Upper left: offline; Upper right: online GCC; Lower left: online mean trajectories of first solution; Lower right: online current values of second solution

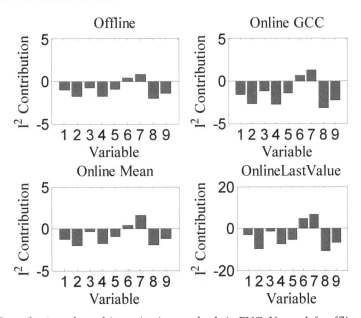

Fig. 14. I^2 Contribution plots of 4 monitoring methods in PVC. Upper left: offline; Upper right: online GCC; Lower left: online mean trajectories of first solution; Lower right: online current values of second solution

6. Conclusion

This chapter introduces online monitoring approaches of batch process to detect fine abnormal at early stage. MICA reveals more nature that occurs abnormal than MPCA. By DTW/OFA, two kinds of synchronization method, more accurate multivariate statistical models are constructed and new batch run is manipulated as much for correct monitoring. GCC method speculates the unknown data of future for MPCA/MICA well when batch process is online. However, in spite of its accuracy, the computation of MICA is more complicated than one of MPCA. It is not suggested to use the methods of synchronization if it is not serious asynchronous among the batch processes, because any method of synchronization consumes a large amount time and memory. Similarly, than other three traditional solutions, GCC needs more time of computation to compare with each other, and huge history model database. None of methods is predominant on the online monitoring of batch processes. The future work may combine the integrative approaches with SDG (Signed Direct Graph) to detect the root cause of the faults (Vedam & Venkatasubramanian, 1999).

7. Acknowledgment

The author wishes to acknowledge the assistance of Miss Lina Bai and Mr. Fuqiang Bian, who have done some work of the simulation of the chapter.

8. References

A., Kassidas; J. F., MacGregor; P. A., Taylor (1998). Synchronization of Batch Trajectories Using Dynamic Time Warping. *AIChE Journal*, Vol.44, No.4, pp. 864-875.

Chang Kyoo, Yoo; Jong-Min, Lee; Peter A., Vanrolleghem; In-Beum, Lee (2004) On-line Monitoring of Batch Processes Using Multiway Independent Analysis. *Chemometrics and Intelligent Laboratory Systems*, Vol.71, Issue 2, pp. 151-163.

Debashis, Neogi & CoryE., Schlags (1998) Multivariate Statistical Analysis of an Emulsion Batch Process. *Industrial & Engineering Chemistry Research*. Vol.37, Issue 10, pp. 3971-3979.

Fuqiang, Bian (2008). OFA Synchronize Method and Its Improvement for Batch Process. *Journal of Beijing Union University (Nature Sciences)*, Vol. 22, No.4, pp. 48-53.

Fuqiang, Bian; Xiang, Gao; Ming Zhe, Yuan (2009). Monitoring based on improved OFA-MPCA. *Journal of the Graduate School of the Chinese Academy of Sciences*, Vol.26, No.2, pp. 209-214.

Hiranmayee, Vedam & Venkat, Venkatasubramanian (1999) PCA-SDG based process monitoring and fault diagnosis. *Control Engineering Practice*, Vol.7, No.8, pp. 903-917.

Italura, F. (1975) Mimimum Prediction Residential Principle Applied to Speech Recognition. *IEEE Trans. on Acoustics, Speech and Signal Processing*, Vol.ASSP-23, No.1, pp. 67-72.

J. A., Westerhuis; T., Kourti, J. F., MacGregor (1999). Comparing Alternative Approaches for Multivariate Statistical Analysis of Batch Process Data. *J. Chemometrics*, Vol.13, Issue 3-4, pp. 397-413.

Junghui, Chen & Hsin-hung, Chen (2006). On-line Batch Process Monitoring Using MHMT-based MPCA. *Chemical Engineering Science*, Vol.61, Issue 10, pp. 3223-3239.

Junghui, Chen & Jialin Liu (2000). Post Analysis on Different Operating Time Processes Using Orthonormal Function Approximation and Multiway Principal Component Analysis. *J. Process Control*, Vol.10, pp. 411-418.

Junghui, Chen & Jialin Liu (2001). Multivariate Calibration Models Based on Functional Space and Partial Least Square for Batch Processes", *4th IFAC workshop on On-line fault detection and supervision in the chemical process industries*, pp. 161-166, ISBN 0080436803, ISBN-13: 9780080436807, Jejudo, Korea, June 7-9, 2001.

Junghui, Chen & Wei-Yann, Wang. (2010). Performance monitoring of MPCA-based Control for Multivariable Batch Control Processes. *Journal of the Taiwan Institute of Chemical Engineers*, Vol.41, Issue 4, pp. 465-474.

J., Lee; C., Yoo; I., Lee (2004). Statistical Process Monitoring with Independent Component Analysis. *Journal of Process control*, Vol.14, Issue 5, pp. 467-485.

J. Edward, Jackson & Govind S., Mudholkar (1979). Control Procedures for Residuals Associated with Principal Component Analyssis. *Technometrics*, Vol.21, No.3, pp. 341-349.

K. A., Kosanovich; K. S., Dahl; M. J., Piovoso (1996). Improved Process Understanding Using Multiway Principal Component Analysis. *Ind. Eng. Chem. Res.*, Vol.35, No.1, pp. 138-146.

L. R., Rabiner; A. E., Rosenberg; S. E., Levinson(1978). Considerations in Dynamic Time Warping Algorithms for Discrete Word Recognition. *IEEE Trans. on Acoustics, Speech and Signal Processing*, Vol.26, Issue 6, pp. 575-582.

Lina, Bai; Xiang, Gao; Jian Jiang, Cui (2009a). Online MPCA Monitoring Approach Based on Generalized Correlation Coefficients. *Control Engineering of China*, Vol.16, No.1, pp.113-116.

Lina, Bai; Xiang, Gao; Ming Zhe, Yuan; Jing Xing, Cao (2009b). Simulation of MICA Based on a New Type of Data Prediction. *Computer Simulation*, Vol.26, No.2, pp.134-138.

Morton Nadler & Eric P., Smith (1993) *Pattern Recognition Engineering*, ISBN-10: 0471622931, ISBN-13: 978-0471622932, Wiley Press, New York, USA.

Myers, C.; L. R., Rabiner; A. E. Rosenberg (1980) Performance Tradeoffs in Dynamic Time Warping Algorithms for Isolated Word Recognition. *IEEE Trans. on Acoustics, Speech and Signal Processing*, Vol.ASSP-28, No.6, pp. 623-635.

Patrick K., Moore & Rayford G., Anthory (1989). The Continuous-lumping Method for Vapor-liquid Equilibrium Calculations. *AIChE Journal*, Vol. 35, Issue 7, pp. 1115-1124.

Paul, Nomikos & John F., MacGregor (1994). Monitoring Batch Processes Using Multi-way Principal Component Analysis, *AIChE Journal*, Vol.40, No.8, pp.1361-1375.

Paul, Nomikos & John F., MacGregor (1995). Multivariate SPC Charts for Monitoring Batch Processes, *Technometrics*, Vol.37, No.1, pp. 41-59.

S., Ikeda & K., Toyama (2000). Independent Component Analysis for Noisy Data—MEG Data Analysis. *Neural Network*, Vol.13, Issue 10, pp.1063-1074.

Wold, S.; Geladi, P.; Esbensen, K.; Ohman, J.(1987) Multi-Way Principal Components and PLS Analysis. *Journal of Chemometrics*, Vol.1, Issue 1, pp. 41-56.

Xiang, Gao; Fuqiang, Bian; Jing Xing, Cao (2008a). PVC Polymerization MICA Monitoring and Fault Diagnosis Based on OFA. *Proceeding of the 27th Chinese Control Conference*, Volume 6, pp. 107-111, ISBN 978-7-8112-4390-1, Kunming, China, July, 16-18, 2008.

Xiang, Gao; Gang, Wang; Yuan, Li; Ji-hu, Ma (2001) Multivariate Statistical Process Monitoring Based on Synchronization of Batch Trajectories Using DTW. *4th IFAC workshop on On-line fault detection and supervision in the chemical process industries*, pp. 383-387, ISBN 0080436803, ISBN-13: 9780080436807, Jejudo, Korea, June 7-9, 2001.

Xiang, Gao & Lina, Bai (2007). Approach of Prediction and Recovery of Multivariate Trajectories Based on Generalized Correlation Coefficients. *Proceeding of 2007 Chinese Control and Decision Conference*, pp. 120-122, ISBN 978-7-81102-396-1 Wuxi, China, July 3-7, 2007.

Xiang, Gao; Lina, Bai; Jian Jiang, Cui (2008b). The DTW Synchronized MPCA On-line Monitoring and Fault Detection Predicted with GCC. *Proceeding of 2008 Chinese Control and Decision Conference*, pp. 551-555, ISBN 978-1-4244-1733-9, Yantai, China, July, 2-4, 2008.

Permissions

The contributors of this book come from diverse backgrounds, making this book a truly international effort. This book will bring forth new frontiers with its revolutionizing research information and detailed analysis of the nascent developments around the world.

We would like to thank Parinya Sanguansat, for lending his expertise to make the book truly unique. He has played a crucial role in the development of this book. Without his invaluable contribution this book wouldn't have been possible. He has made vital efforts to compile up to date information on the varied aspects of this subject to make this book a valuable addition to the collection of many professionals and students.

This book was conceptualized with the vision of imparting up-to-date information and advanced data in this field. To ensure the same, a matchless editorial board was set up. Every individual on the board went through rigorous rounds of assessment to prove their worth. After which they invested a large part of their time researching and compiling the most relevant data for our readers. Conferences and sessions were held from time to time between the editorial board and the contributing authors to present the data in the most comprehensible form. The editorial team has worked tirelessly to provide valuable and valid information to help people across the globe.

Every chapter published in this book has been scrutinized by our experts. Their significance has been extensively debated. The topics covered herein carry significant findings which will fuel the growth of the discipline. They may even be implemented as practical applications or may be referred to as a beginning point for another development. Chapters in this book were first published by InTech; hereby published with permission under the Creative Commons Attribution License or equivalent.

The editorial board has been involved in producing this book since its inception. They have spent rigorous hours researching and exploring the diverse topics which have resulted in the successful publishing of this book. They have passed on their knowledge of decades through this book. To expedite this challenging task, the publisher supported the team at every step. A small team of assistant editors was also appointed to further simplify the editing procedure and attain best results for the readers.

Our editorial team has been hand-picked from every corner of the world. Their multi-ethnicity adds dynamic inputs to the discussions which result in innovative outcomes. These outcomes are then further discussed with the researchers and contributors who give their valuable feedback and opinion regarding the same. The feedback is then collaborated with the researches and they are edited in a comprehensive manner to aid the understanding of the subject.

Apart from the editorial board, the designing team has also invested a significant amount of their time in understanding the subject and creating the most relevant covers. They scrutinized every image to scout for the most suitable representation of the subject and create an appropriate cover for the book.

The publishing team has been involved in this book since its early stages. They were actively engaged in every process, be it collecting the data, connecting with the contributors or procuring relevant information. The team has been an ardent support to the editorial, designing and production team. Their endless efforts to recruit the best for this project, has resulted in the accomplishment of this book. They are a veteran in the field of academics and their pool of knowledge is as vast as their experience in printing. Their expertise and guidance has proved useful at every step. Their uncompromising quality standards have made this book an exceptional effort. Their encouragement from time to time has been an inspiration for everyone.

The publisher and the editorial board hope that this book will prove to be a valuable piece of knowledge for researchers, students, practitioners and scholars across the globe.

List of Contributors

Parinya Sanguansat
Faculty of Engineering and Technology, Panyapiwat Institute of Management, Thailand

Maria Monfreda
Italian Customs Agency Central Directorate for Chemical Analysis and Development of Laboratories, Rome, Italy

Cuauhtémoc Araujo-Andrade
Unidad Académica de Física, Universidad Autónoma de Zacatecas, México

Claudio Frausto-Reyes
Centro de Investigaciones en Óptica, A.C. Unidad Aguascalientes, México

Edgar L. Esparza-Ibarra, Rumen Ivanov-Tsonchev
Unidad Académica de Física, Universidad Autónoma de Zacatecas, México

Esteban Gerbino, Pablo Mobili, Elizabeth Tymczyszyn and Andrea Gómez-Zavaglia
Centro de Investigación y Desarrollo en Criotecnología de Alimentos (CIDCA), Argentina

Ramana Vinjamuri and Wei Wang
Department of Physical Medicine and Rehabilitation, USA
Center for Neural Basis of Cognition University of Pittsburgh, Pittsburgh, PA, USA

Mingui Sun
Department of Neurological Surgery, USA

Zhi-Hong Mao
Department of Electrical and Computer Engineering, USA

Yoshikazu Washizawa
The University of Electro-Communications, Japan

Xian-Hua Han and Yen-Wei Chen
Ritsumeikan University, Japan

Masahiro Kuroda, Yuichi Mori and Michio Sakakihara
Okayama University of Science, Japan

Masaya Iizuka
Okayama University, Japan

Mauridhi Hery Purnomo, Diah P. Wulandari and I. Ketut Eddy Purnama
Electrical Engineering Department – Industrial Engineering Faculty, Institut Teknologi Sepuluh Nopember, Surabaya, Indonesia

Arif Muntasa
Informatics Engineering Department – Engineering Faculty, Universitas Trunojoyo Madura, Indonesia

Yaya Keho
Ecole Nationale Supérieure de Statistique et d'Economie Appliquée (ENSEA), Abidjan Côte d'Ivoire

Shiow-Jyu Lin
Department of Computer Science and Information Engineering, National Taiwan Normal University, Taiwan
Department of Electronic Engineering, National Ilan University, Taiwan

Kun-Hung Lin and Wen-Jyi Hwang
Department of Computer Science and Information Engineering, National Taiwan Normal University

Omar Arif
National University of Sciences and Technology, Pakistan

Patricio A. Vela
Georgia Institute of Technology, USA

Charles Guyon, Thierry Bouwmans and El-hadi Zahzah
Lab. MIA - Univ. La Rochelle, France

Darko Dimitrov
Freie Universität Berlin, Germany

Xiang Gao
Yantai Nanshan University, P. R. China

Printed in the USA
CPSIA information can be obtained
at www.ICGtesting.com
JSHW011502221024
72173JS00005B/1175